JOHN FREDERICK WILLIAM HERSCHEL

A

TREATISE ON ASTRONOMY

Elibron Classics
www.elibron.com

Elibron Classics series.

ISBN 1-4021-6266-9 (paperback)
ISBN 1-4021-2183-0 (hardcover)

This Elibron Classics Replica Edition is an unabridged facsimile of the edition published in 1833 by Longman, Rees, Orme, Brown, Green, & Longmans; and John Taylor, London.

THE

CABINET CYCLOPÆDIA.

London:
Printed by A. Spottiswoode,
New-Street-Square.

THE

CABINET CYCLOPÆDIA.

CONDUCTED BY THE

REV. DIONYSIUS LARDNER, LL.D. F.R.S. L. & E.

M.R.I.A. F.R.A.S. F.L.S. F.Z.S. Hon. F.C.P.S. &c. &c.

ASSISTED BY

EMINENT LITERARY AND SCIENTIFIC MEN.

Natural Philosophy.

ASTRONOMY.

BY

SIR JOHN F. W. HERSCHEL, KN^T. GUELP.

F.R.S.L. & E. M.R.I.A. F.R.A.S. F.G.S. M.C.U.P.S.

CORRESPONDENT OF THE ROYAL ACADEMY OF SCIENCES OF PARIS, AND OTHER FOREIGN SCIENTIFIC INSTITUTIONS.

A NEW EDITION.

LONDON:

PRINTED FOR

LONGMAN, ORME, BROWN, GREEN, & LONGMANS,

PATERNOSTER-ROW.

AND JOHN TAYLOR,

UPPER GOWER STREET.

1838.

"ET QUONIAM EADEM NATURA CUPIDITATEM INGENUIT HOMINIBUS VERI INVENIENDI, QUOD FACILLIME APPARET, CUM VACUI CURIS, ETIAM QUID IN CŒLO FIAT, SCIRE AVEMUS: HIS INITIIS INDUCTI OMNIA VERA DILIGIMUS; ID EST, FIDELIA, SIMPLICIA, CONSTANTIA; TUM VANA, FALSA, FALLENDIA ODIMUS."

CICERO, DE FIN. BON. ET MAL. ii. 14.

AND FORASMUCH AS NATURE ITSELF HAS IMPLANTED IN MAN A CRAVING AFTER THE DISCOVERY OF TRUTH, (WHICH APPEARS MOST CLEARLY FROM THIS, THAT, WHEN UNOPPRESSED BY CARES, WE DELIGHT TO KNOW EVEN WHAT IS GOING ON IN THE HEAVENS,) — LED BY THIS INSTINCT, WE LEARN TO LOVE ALL TRUTH FOR ITS OWN SAKE; THAT IS TO SAY, WHATEVER IS FAITHFUL, SIMPLE, AND CONSISTENT; WHILE WE HOLD IN ABHORRENCE WHATEVER IS EMPTY, DECEPTIVE, OR UNTRUE.

A TREATISE ON ASTRONOMY.

By

SIR JOHN F. W. HERSCHEL, KNT GUELP.

F.R.S. L&E. - M.R.I.A. - F.R.A.S. - F.G.S. - M.C.U.P.S.

CORRESPONDENT OF THE ROYAL ACADEMY OF SCIENCES OF PARIS,
AND OTHER FOREIGN SCIENTIFIC INSTITUTIONS.

New Edition.

H. Corbould, delt. E. Finden, sculpt.

London:
PUBLISHED BY LONGMAN, REES, ORME, BROWN, GREEN & LONGMAN, PATERNOSTER ROW,
AND JOHN TAYLOR, UPPER GOWER STREET.
1833.

CONTENTS.

CHAP. V.

OF THE SUN'S MOTION.

CHAP. VI.

CHAP. VII.

CHAP. VIII.

OF THE SOLAR SYSTEM.

CHAP. IX.

OF THE SATELLITES.

CHAP. X.

OF COMETS.

CHAP. XI.

OF PERTURBATIONS.

CHAP. XII.

OF SIDEREAL ASTRONOMY.

CHAP. XIII.

A

TREATISE

ON

ASTRONOMY.

INTRODUCTION.

(1.) In entering upon any scientific pursuit, one of the student's first endeavours ought to be, to prepare his mind for the reception of truth, by dismissing, or at least loosening his hold on, all such crude and hastily adopted notions respecting the objects and relations he is about to examine as may tend to embarrass or mislead him; and to strengthen himself, by something of an effort and a resolve, for the unprejudiced admission of any conclusion which shall appear to be supported by careful observation and logical argument, even should it prove of a nature adverse to notions he may have previously formed for himself, or taken up, without examination, on the credit of others. Such an effort is, in fact, a commencement of that intellectual discipline which forms one of the most important ends of all science. It is the first movement of approach towards that state of mental purity which alone can fit us for a full and steady perception of moral beauty as well as physical adaptation. It is the "euphrasy and rue" with which we must "purge our sight" before we can receive and contemplate as they are the lineaments of truth and nature.

(2.) There is no science which, more than astronomy, stands in need of such a preparation, or draws more largely on that intellectual liberality which is ready to adopt whatever is demonstrated, or concede whatever is rendered highly probable, however new and uncommon the points of view may be in which objects the most familiar may thereby become placed. Almost all its conclusions stand in open and striking contradiction with those of superficial and vulgar observation, and with what appears to every one, until he has understood and weighed the proofs to the contrary, the most positive evidence of his senses. Thus, the earth on which he stands, and which has served for ages as the unshaken foundation of the firmest structures, either of art or nature, is divested by the astronomer of its attribute of fixity, and conceived by him as turning swiftly on its centre, and at the same time moving onwards through space with great rapidity. The sun and the moon, which appear to untaught eyes round bodies of no very considerable size, become enlarged in his imagination into vast globes,—the one approaching in magnitude to the earth itself, the other immensely surpassing it. The planets, which appear only as stars somewhat brighter than the rest, are to him spacious, elaborate, and habitable worlds; several of them vastly greater and far more curiously furnished than the earth he inhabits, as there are also others less so; and the stars themselves, properly so called, which to ordinary apprehension present only lucid sparks or brilliant atoms, are to him suns of various and transcendent glory — effulgent centres of life and light to myriads of unseen worlds: so that when, after dilating his thoughts to comprehend the grandeur of those ideas his calculations have called up, and exhausting his imagination and the powers of his language to devise similes and metaphors illustrative of the immensity of the scale on which his universe is constructed, he shrinks back to his native sphere; he finds it, in comparison, a mere point; so lost — even in the minute system to which it belongs — as to be invisible and unsuspected from some of its principal and remoter members.

(3.) There is hardly any thing which sets in a stronger light the inherent power of truth over the mind of man, when opposed by no motives of interest or passion, than the perfect readiness with which all these conclusions are assented to as soon as their evidence is clearly apprehended, and the tenacious hold they acquire over our belief when once admitted. In the conduct, therefore of this volume, we shall take it for granted that our reader is more desirous to learn the system which it is its object to teach as it now stands, than to raise or revive objections against it; and that, in short, he comes to the task with a willing mind; an assumption which will not only save ourselves the trouble of piling argument on argument to convince the sceptical, but will greatly facilitate his actual progress, inasmuch as he will find it at once easier and more satisfactory to pursue from the outset a straight and definite path, than to be constantly stepping aside, involving himself in perplexities and circuits, which, after all, can only terminate in finding himself compelled to adopt our road.

(4.) The method, therefore, we propose to follow is neither strictly the analytic nor the synthetic, but rather such a combination of both, with a leaning to the latter, as may best suit with a *didactic* composition. Our object is not to convince or refute opponents, nor to enquire, under the semblance of an assumed ignorance, for principles of which we are all the time in full possession,— but simply to *teach* what we know. The moderate limit of a single volume, and the necessity of being on every point, within that limit, rather diffuse and copious in explanation, as well as the eminently matured and ascertained character of the science itself, render this course both practicable and eligible. Practicable, because there is now no danger of any revolution in astronomy, like those which are daily changing the features of the less advanced sciences, supervening, to destroy all our hypotheses, and throw our statements into confusion. Eligible, because the space to be bestowed, either in combating refuted systems, or in leading the reader forward by slow and measured steps from the known to the unknown, may be more advantageously devoted to such explanatory illustrations as will impress on him a familiar and, as it were, a practical sense of the sequence of phenomena, and the manner in which they are produced. We shall not, then, reject the analytic course where it leads more easily and directly to our objects, or in any way fetter ourselves by a rigid adherence to method. Writing only to be understood, and to communicate as much information in as little space as possible, consistently with its *distinct* and *effectual* communication, we can afford to make no sacrifice to system, to form, or to affectation.

(5.) We shall take for granted, from the outset, the Copernican system of the world; relying on the easy, obvious, and natural explanation it affords of all the phenomena as they come to be described, to impress the student with a sense of its truth, without either the formality of demonstration or the superfluous tedium of eulogy, calling to mind that important remark of Bacon: — "Theoriarum vires, arcta et quasi se mutuo sustinente partium adaptatione, quâ, quasi in orbem cohærent, firmantur *;" nor failing, however, to point out to the reader, as occasion offers, the contrast which its superior simplicity offers to the complication of other hypotheses.

(6.) The preliminary knowledge which it is desirable that the student should possess, in order for the more advantageous perusal of the following pages, consists in the familiar practice of decimal and sexagesimal arithmetic; some moderate acquaintance with geometry and trigonometry, both plane and spherical; the elementary principles of mechanics; and enough of optics to understand the construction and use of the telescope, and some other of the simpler instruments. For the

* The confirmation of theories relies on the compact adaptation of their parts, by which, like those of an arch or dome, they mutually sustain each other, and form a coherent whole.

acquisition of these we may refer him to those other parts of this Cyclopædia which profess to treat of the several subjects in question. Of course, the more of such knowledge he brings to the perusal, the easier will be his progress, and the more complete the information gained; but we shall endeavour in every case, as far as it can be done without a sacrifice of clearness, and of that useful brevity which consists in the absence of prolixity and episode, to render what we have to say as independent of other books as possible.

(7.) After all, we must distinctly caution such of our readers as may commence and terminate their astronomical studies with the present work (though of such,—at least in the latter predicament,—we trust the number will be few), that its utmost pretension is to place them on the threshold of this particular wing of the temple of Science, or rather on an eminence exterior to it, whence they may obtain something like a general notion of its structure; or, at most, to give those who may wish to enter, a ground-plan of its accesses, and put them in possession of the pass-word. Admission to its sanctuary, and to the privileges and feelings of a votary, is only to be gained by one means,—*a sound and sufficient knowledge of mathematics, the great instrument of all exact enquiry, without which no man can ever make such advances in this or any other of the higher departments of science, as can entitle him to form an independent opinion on any subject of discussion within their range.* It is not without an effort that those who possess this knowledge can communicate on such subjects with those who do not, and adapt their language and their illustrations to the necessities of such an intercourse. Propositions which to the one are almost identical, are theorems of import and difficulty to the other; nor is their evidence presented in the same way to the mind of each. In teaching such propositions, under such circumstances, the appeal has to be made, not to the pure and abstract reason, but to the sense of analogy,—to practice and experience:

principles and modes of action have to be established, not by direct argument from acknowledged axioms, but by bringing forward and dwelling on simple and familiar instances in which the same principles and the same or similar modes of action take place; thus erecting, as it were, in each particular case, a separate induction, and constructing at each step a little body of science to meet its exigencies. The difference is that of pioneering a road through an untraversed country and advancing at ease along a broad and beaten highway; that is to say, if we are determined to make ourselves distinctly understood, and will appeal to reason at all. As for the method of *assertion,* or a direct demand on the *faith* of the student (though in some complex cases indispensable, where illustrative explanation would defeat its own end by becoming tedious and burdensome to both parties), it is one which we shall neither adopt ourselves nor would recommend to others.

(8.) On the other hand, although it is something new to abandon the road of mathematical demonstration in the treatment of subjects susceptible of it, and teach any considerable branch of science entirely or chiefly by the way of illustration and familiar parallels, it is yet not impossible that those who are already well acquainted with our subject, and whose knowledge has been acquired by that confessedly higher and better practice which is incompatible with the avowed objects of the present work, may yet find their account in its perusal,—for this reason, that it is always of advantage to present any given body of knowledge to the mind in as great a variety of different lights as possible. It is a property of illustrations of this kind to strike no two minds in the same manner, or with the same force; because no two minds are stored with the same images, or have acquired their notions of them by similar habits. Accordingly, it may very well happen, that a proposition, even to one best acquainted with it, may be placed not merely in a new and uncommon, but in a more impressive and satisfactory light by such a course—some obscurity may be dissipated, some inward misgiving cleared up, or even some link supplied which may lead to the perception of connections and deductions altogether unknown before. And the probability of this is increased when, as in the present instance, the illustrations chosen have not been studiously selected from books, but are such as have presented themselves freely to the author's mind as being most in harmony with his own views; by which, of course, he means to lay no claim to originality in all or any of them beyond what they may really possess.

(9.) Besides, there are cases in the application of mechanical principles with which the mathematical student is but too familiar, where, when the data are before him, and the numerical and geometrical relations of his problems all clear to his conception,—when his forces are estimated and his lines measured, — nay, when even he has followed up the application of his technical processes, and fairly arrived at his conclusion,— there is still something wanting in his mind — not in the evidence, for he has examined each link, and finds the chain complete — not in the principles, for those he well knows are too firmly established to be shaken — but precisely in the *mode of action.* He has followed out a train of reasoning by logical and technical rules, but the signs he has employed are not pictures of nature, or have lost their original meaning as such to his mind: he has not seen, as it were, the process of nature passing under his eye in an instant of time, and presented as a whole to his imagination. A familiar parallel, or an illustration drawn from some artificial or natural process, of which he has that direct and individual impression which gives it a reality and associates it with a name, will, in almost every such case, supply in a moment this deficient feature, will convert all his symbols into real pictures, and infuse an animated meaning into what was before a lifeless succession of words and signs. We cannot, indeed, always promise ourselves to attain this degree of vividness in our illustrations, nor are the points to be elucidated themselves always capable of being so *para-*

phrased (if we may use the expression) by any single instance adducible in the ordinary course of experience; but the object will at least be kept in view; and, as we are very conscious of having, in making such attempts gained for ourselves much clearer views of several of the more concealed effects of planetary perturbation than we had acquired by their mathematical investigation in detail, we may reasonably hope that the endeavour will not always be unattended with a similar success in others.

(10.) From what has been said, it will be evident that our aim is not to offer to the public a technical treatise, in which the student of practical or theoretical astronomy shall find consigned the minute description of methods of observation, or the formulæ he requires prepared to his hand, or their demonstrations drawn out in detail. In all these the present work will be found meagre, and quite inadequate to his wants. Its aim is entirely different; being to present in each case the mere ultimate *rationale* of facts, arguments, and processes; and, in all cases of mathematical application, avoiding whatever would tend to encumber its pages with algebraic or geometrical symbols, to place under his inspection that central thread of common sense on which the pearls of analytical research are invariably strung; but which, by the attention the latter claim for themselves, is often concealed from the eye of the gazer, and not always disposed in the straightest and most convenient form to follow by those who string them. This is no fault of those who have conducted the enquiries to which we allude. The contention of mind for which they call is enormous; and it may, perhaps, be owing to their experience of *how little* can be accomplished in carrying such processes on to their conclusion, by mere ordinary *clearness of head;* and how necessary it often is to pay more attention to the purely mathematical conditions which ensure success,—the hooks-and-eyes of their equations and series,—than to those which enchain causes with their effects, and both with the human reason,— that we must attribute something of that indistinctness of view which is often complained of as a grievance by the earnest student, and still more commonly ascribed ironically to the native cloudiness of an atmosphere too sublime for vulgar comprehension. We think we shall render good service to both classes of readers, by dissipating, so far as our power lies, that accidental obscurity, and by showing ordinary untutored comprehension clearly what it *can*, and what it *cannot*, hope to attain.

CHAPTER I.

GENERAL NOTIONS. — FORM AND MAGNITUDE OF THE EARTH.— HORIZON AND ITS DIP.— THE ATMOSPHERE. — REFRACTION.— TWILIGHT. — APPEARANCES RESULTING FROM DIURNAL MOTION.—PARALLAX.— FIRST STEP TOWARDS FORMING AN IDEA OF THE DISTANCE OF THE STARS. — DEFINITIONS.

(11.) THE magnitudes, distances, arrangement, and motions of the great bodies which make up the visible universe, their constitution and physical condition, so far as they can be known to us, with their mutual influences and actions on each other, so far as they can be traced by the effects produced, and established by legitimate reasoning, form the assemblage of objects to which the attention of the astronomer is directed. The term astronomy * itself, which denotes the *law* or rule of the *astra* (by which the ancients understood not only the stars properly so called, but the sun, the moon, and all the visible constituents of the heavens), sufficiently indicates this ; and, although the term astrology, which denotes the *reason, theory,* or *interpretation* of the stars †, has become degraded in its application, and confined to

* Αστηρ, *a star* ; νομος, *a law* ; or νεμειν, to tend, as a shepherd his flock ; so that αστρονομος means "shepherd of the stars." The two etymologies are, however, coincident.

† Λογος, *reason*, or *a word*, the vehicle of reason ; the interpreter of thought.

superstitious and delusive attempts to divine future events by their dependence on pretended planetary influences, the same meaning originally attached itself to that epithet.

(12.) But, besides the stars and other celestial bodies, the earth itself, regarded as an individual body, is one principal object of the astronomer's consideration, and, indeed, the chief of all. It derives its importance, in a practical as well as theoretical sense, not only from its proximity, and its relation to us as animated beings, who draw from it the supply of all our wants, but as the station from which we see all the rest, and as the only one among them to which we can, in the first instance, refer for any determinate marks and measures by which to recognize their changes of situation, or with which to compare their distances.

(13.) To the reader who now for the first time takes up a book on astronomy, it will no doubt seem strange to class the earth with the heavenly bodies, and to assume any community of nature among things apparently so different. For what, in fact, can be more apparently different than the vast and seemingly immeasurable extent of the earth, and the stars, which appear but as points, and seem to have no size at all? The earth is dark and opaque, while the celestial bodies are brilliant. We perceive in it no motion, while in them we observe a continual change of place, as we view them at different hours of the day or night, or at different seasons of the year. The ancients, accordingly, one or two of the more enlightened of them only excepted, admitted no such community of nature; and, by thus placing the heavenly bodies and their movements without the pale of analogy and experience, effectually intercepted the progress of all reasoning from what passes here below, to what is going on in the regions where they exist and move. Under such conventions, astronomy, as a science of cause and effect, could not exist, but must be limited to a mere registry of appearances, unconnected with any attempt to account for them on reasonable principles. To get rid of this prejudice, therefore, is the first step towards acquiring a knowledge of what is really the case; and the student has made his first effort towards the acquisition of sound knowledge, when he has learnt to familiarize himself with the idea that the earth, after all, *may be* nothing but a great star. How correct such an idea may be, and with what limitations and modifications it is to be admitted, we shall see presently.

(14.) It is evident, that, to form any just notions of the arrangement, in space, of a number of objects which we cannot approach and examine, but of which all the information we can gain is by sitting still and watching their evolutions, it must be very important for us to know, in the first instance, whether what we call sitting still is *really* such: whether the station from which we view them, with ourselves, and all objects which immediately surround us, be not itself in motion, unperceived by us; and if so, of what nature that motion is. The apparent places of a number of objects, and their apparent arrangement with respect to each other, will of course be materially dependent on the situation of the spectator among them; and if this situation be liable to change, unknown to the spectator himself, an appearance of change in the respective situations of the objects will arise, without the reality. If, then, such be actually the case, it will follow that *all* the movements we think we perceive among the stars will not be real movements, but that some part, at least, of whatever changes of relative place we perceive among them must be merely apparent, the results of the shifting of our own point of view; and that, if we would ever arrive at a knowledge of their real motions, it can only be by first investigating our own, and making due allowance for its effects. Thus, the question whether the earth is in motion or at rest, and if in motion, what that motion is, is no idle enquiry, but one on which depends our only chance of arriving at true conclusions respecting the constitution of the universe.

(15.) Nor let it be thought strange that we should speak of a motion existing in the earth, unperceived by its inhabitants: we must remember that it is of the earth *as a whole,* with all that it holds within its substance, or sustains on its surface, that we are speaking; of a motion common to the solid mass beneath, to the ocean which flows around it, the air that rests upon it, and the clouds which float above it in the air. Such a motion, which should displace no terrestrial object from its relative place among others, interfere with no natural processes, and produce no sensations of shocks or jerks, might, it is very evident, subsist undetected by us. There is no peculiar sensation which advertises us that we are *in motion.* We perceive *jerks,* or *shocks,* it is true, because these are sudden *changes* of motion, produced, as the laws of mechanics teach us, by sudden and powerful forces acting during short times; and these forces, applied to our bodies, are what we *feel.* When, for example, we are carried along in a carriage with the blinds down, or with our eyes closed (to keep us from seeing external objects), we perceive a tremor arising from inequalities in the road, over which the carriage is successively lifted and let fall, but we have no sense of *progress.* As the road is smoother, our sense of motion is diminished, though our rate of travelling is accelerated. Those who have travelled on the celebrated railroad between Manchester and Liverpool testify that but for the noise of the train, and the rapidity with which external objects seem to dart by them, the *sensation* is almost that of perfect rest.

(16.) But it is on shipboard, where a great system is maintained in motion, and where we are surrounded with a multitude of objects which participate with ourselves and each other in the common progress of the whole mass, that we feel most satisfactorily the identity of sensation between a state of motion and one of rest. In the cabin of a large and heavy vessel, going smoothly before the wind in still water, or drawn along a canal, not the smallest indication acquaints us with the way it is making. We read, sit, walk, and perform every customary action as if we were on land. If we throw a ball into the air, it falls back into our hand; or, if we drop it, it lights at our feet. Insects buzz around us as in the free air; and smoke ascends in the same manner as it would do in an apartment on shore. If, indeed, we come on deck, the case is, in some respects, different; the air, not being carried along with us, drifts away smoke and other light bodies—such as feathers abandoned to it—apparently, in the opposite direction to that of the ship's progress; but, in reality, *they* remain at rest, and we leave them behind in the air. Still, the illusion, so far as massive objects and our own movements are concerned, remains complete; and when we look at the shore, we then perceive the effect of our own motion transferred, in a contrary direction, to external objects—*external,* that is, *to the system of which we form a part.*

"Provehimur portu, terræque urbesque recedunt."

(17.) Not only do external objects at rest appear in motion generally, with respect to ourselves when we are in motion among them, but they appear to move one among the other—they shift their *relative* apparent places. Let any one travelling rapidly along a high road fix his eye steadily on any object, but at the same time not entirely withdraw his attention from the general landscape,—he will see, or think he sees, the whole landscape thrown into *rotation,* and moving round that object as a centre; all objects between it and himself appearing to move *backwards,* or the contrary way to his own motion; and all beyond it, forwards, or in the direction in which he moves: but let him withdraw his eye from that object, and fix it on another,—a nearer one, for instance,—immediately the appearance of rotation shifts also, and the apparent centre about which this illusive circulation is performed is transferred to the new object, which, for the moment, appears to rest. This apparent change of situation of objects with respect to one another, arising from a motion of the spectator, is called a *parallactic* motion; and it is, therefore, evident that, before we can ascertain whether external objects are really in motion or not, or what their motions are, we must subduct, or allow for, any such *parallactic* motion which may exist.

(18.) In order, however, to conceive the earth as in motion, we must form to ourselves a conception of its shape and size. Now, an object cannot have shape and size, unless it is *limited* on all sides by some definite outline, so as to admit of our imagining it, at least, disconnected from other bodies, and existing insulated in space. The first rude notion we form of the earth is that of a flat surface, of indefinite extent in all directions from the spot where we stand, *above* which are *the air* and *sky;* below, to an indefinite profundity, solid matter. This is a prejudice to be got rid of, like that of the earth's immobility;—but it is one much easier to rid ourselves of, inasmuch as it originates only in our own mental inactivity, in not questioning ourselves *where* we will place a limit to a thing we have been accustomed from infancy to regard as immensely large; and does not, like that, originate in the testimony of our senses unduly interpreted. On the contrary, the direct testimony of our senses lies the other way. When we see the sun set in the evening in the west, and rise again in the east, as we cannot doubt that is the *same* sun we see after a temporary absence, we must do violence to all our notions of solid matter, to suppose it to have made its way *through* the substance of the earth. It must, therefore, have gone *under* it, and that not by a mere subterraneous *channel;* for if we notice the points where it sets and rises for many successive days, or for a whole year, we shall find them constantly shifting, round a very large extent of the horizon; and, besides, the moon and stars also set and rise again in *all* points of the visible horizon. The conclusion is plain: the earth cannot extend indefinitely in depth downwards, nor indefinitely in surface laterally; it must have not only bounds in a horizontal direction, but also an *under side* round which the sun, moon, and stars can pass; and that side must, at least, be so far like what we see, that it must have a sky and sunshine, and a day when it is night to us, and *vice versâ;* where, in short,

> —"redit à nobis Aurora, diemque reducit.
> Nosque ubi primus equis oriens afflavit anhelis,
> Illic sera rubens accendit lumina Vesper." *Georg.*

(19.) As soon as we have familiarized ourselves with the conception of an earth without *foundations* or fixed supports—existing insulated in space from contact of every thing external, it becomes easy to imagine it in motion—or, rather, difficult to imagine it otherwise; for, since there is nothing to *retain* it in one place, should any causes of motion exist, or any *forces* act upon it, it must obey their impulse. Let us next see what obvious circumstances there are to help us to a knowledge of the *shape* of the earth.

(20.) Let us first examine what we can actually *see* of its shape. Now, it is not on land (unless, indeed, on uncommonly level and extensive plains) that we can see any thing of the *general* figure of the earth;—the hills, trees, and other objects which roughen its surface, and break and elevate the line of the horizon, though obviously bearing a most minute proportion to the *whole* earth, are yet too considerable, with respect to ourselves and to that small portion of it which we can see at a single view, to allow of our forming any judgment of the form of the whole, from that of a part so disfigured. But with the surface of the sea, or any vastly extended level plain, the case is otherwise. If we sail out of sight of land, whether we stand on the deck of the ship or climb the mast, we see the surface of the sea—not losing itself in distance and mist, but terminated by a sharp, clear, well defined line, or *offing* as it is called, which runs all round us in a circle, having our station for its centre. That this line is really a circle, we conclude, first, from the perfect apparent similarity of all its parts; and, secondly, from the fact of all its parts appearing at the same distance from us, and that, evidently a moderate one; and, thirdly, from this, that its apparent *diameter*, measured with an instrument called the *dip sector*, is the same (except under some singular atmospheric circumstances, which produce a temporary distortion of the outline), in whatever direction the measure is taken,—properties which belong only to the circle among geometrical figures. If we ascend a high eminence on a plain (for instance, one of the Egyptian pyramids), the same holds good.

(21.) Masts of ships, however, and the edifices erected by man, are trifling eminences compared to what nature itself affords; Ætna, Teneriffe, Mowna Roa, are eminences from which no contemptible *aliquot* part of the whole earth's surface can be seen; but from these again—in those few and rare occasions when the transparency of the air will permit the real boundary of the horizon, the true sea-line, to be seen—the very same appearances are witnessed, but with this remarkable addition, viz. that the angular *diameter* of the visible area, as measured by the dip sector, is materially *less* than at a lower level; or, in other words, that the *apparent size* of the earth has sensibly diminished as we have receded from its surface, while yet the *absolute quantity* of it seen at once has been increased.

(22.) The same appearances are observed universally, in every part of the earth's surface visited by man. Now, the figure of a body which, however seen, appears always *circular*, can be no other than a sphere or globe.

(23.) A diagram will elucidate this. Suppose the earth to be represented by the sphere L H N Q, whose centre is C, and let A, G, M be stations at different elevations above various points of its surface, represented by *a*, *g*, *m* respectively. From each of them (as from M) let a line be drawn, as M N *n*, a tangent to the surface at N, then will this line represent the visual ray along which the spectator at M will see the visible horizon; and as this tangent sweeps round M, and comes successively into the positions M O *o*, M P *p*,

M Q *q*, the point of contact N will mark out on the surface the circle N O P Q. The area of this circle is the

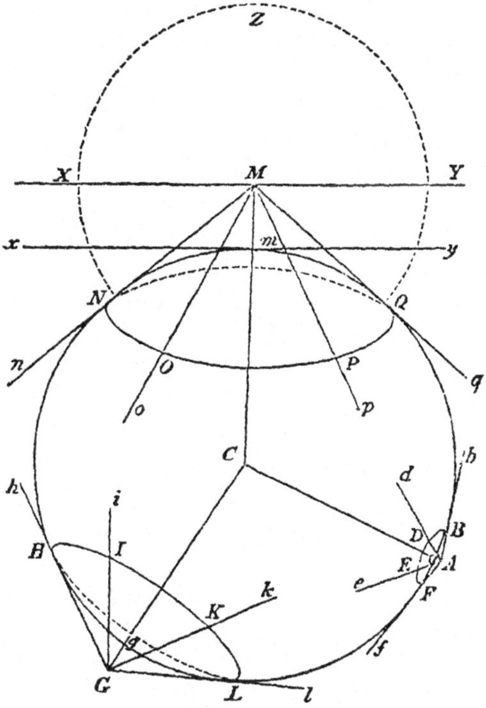

portion of the earth's surface visible to a spectator at M, and the angle N M Q included between the two extreme visual rays is the measure of its apparent angular diameter. Leaving, at present, out of consideration the effect of refraction in the air below M, of which more hereafter, and which always tends, in some degree, to *increase* that angle, or render it more *obtuse*, this is the angle measured by the dip sector. Now, it is evident, 1st, that as the point M is more elevated above *m*, the point immediately below it on the sphere, the visible area, *i.e.* the spherical segment or slice N O P Q, increases; 2dly, that the distance of the visible *horizon** or boundary of our view from the eye, viz. the line M N, increases; and, 3dly, that the angle N M Q becomes

* Ὁρίζω, *to terminate.*

less obtuse, or, in other words, the apparent angular diameter of the earth diminishes, being nowhere so great as 180°, or two right angles, but falling short of it by some sensible quantity, and that more and more the higher we ascend. The figure exhibits three states or stages of elevation, with the horizon, &c. corresponding to each, a glance at which will explain our meaning; or, limiting ourselves to the larger and more distinct, MNOPQ, let the reader imagine *n*NM, MQ*q* to be the two legs of a ruler jointed at M, and kept extended by the globe N*m*Q between them. It is clear, that as the joint M is urged home towards the surface, the legs will open, and the ruler will become more nearly *straight*, but will not attain *perfect* straightness till M is brought fairly up to contact with the surface at *m*, in which case its whole length will become a *tangent* to the sphere at *m*, as is the line *x y*.

(24.) This explains what is meant by the *dip of the horizon*. M *m*, which is perpendicular to the general surface of the sphere at *m*, is also the direction in which a *plumb-line** would hang; for it is an observed fact, that in all situations, in every part of the earth, the direction of a plumb-line is exactly perpendicular to the surface of still water; and, moreover, that it is also exactly perpendicular to a line or surface truly adjusted by a *spirit-level.** Suppose, then, that at our station M we were to adjust a line (a wooden ruler for instance) by a spirit-level, with perfect exactness; then, if we suppose the direction of this line indefinitely prolonged both ways, as X M Y, the line so drawn will be at right angles to M *m*, and therefore parallel to *x m y*, the tangent to the sphere at *m*. A spectator placed at M will therefore see not only all the vault of the sky *above* this line, as X Z Y, but also that portion or zone of it which lies between X N and Y Q; in other words, his sky will be more than a hemisphere by the zone Y Q X N. It is the angular breadth of this redundant zone — the angle Y M Q, by which the *visible* horizon appears de-

* See this instrument described in Chap. II.

pressed below the direction of a spirit-level—that is called the *dip of the horizon.* It is a correction of constant use in nautical astronomy.

(25.) From the foregoing explanations it appears, 1st, That the general figure of the earth (so far as it can be gathered from this kind of observation) is that of a sphere or globe. In this we also include that of the sea, which, wherever it extends, covers and fills in those inequalities and local irregularities which exist on land, but which can of course only be regarded as trifling deviations from the general outline of the whole mass, as we consider an orange not the less round for the roughnesses on its rind. 2dly, That the appearance of a *visible* horizon, or sea offing, is a consequence of the curvature of the surface, and does not arise from the inability of the eye to follow objects to a greater distance, or from atmospheric indistinctness. It will be worth while to pursue the general notion thus acquired into some of its consequences, by which its consistency with observations of a different kind, and on a larger scale, will be put to the test, and a clear conception be formed of the manner in which the parts of the earth are related to each other, and held together as a whole.

(26.) In the first place, then, every one who has passed a little while at the sea side is aware that objects may be seen perfectly well beyond the *offing* or visible horizon—but not the *whole* of them. We only see their upper parts. Their bases where they rest on, or rise out of the water, are hid from view by the spherical surface of the sea, which protrudes between them and ourselves. Suppose a ship, for instance, to sail directly away from our station;—at first, when the distance of the ship is small, a spectator, S, situated at some certain height above the sea, sees the whole of the ship, even to the *water line* where it rests on the sea, as at A. As it recedes it diminishes, it is true, in apparent size, but still the *whole* is seen down to the water line, till it reaches the *visible* horizon at B. But as soon as it has passed this distance, not only does the visible portion still continue to diminish in apparent *size*, but the hull begins to disappear bodily, as if sunk below the surface.

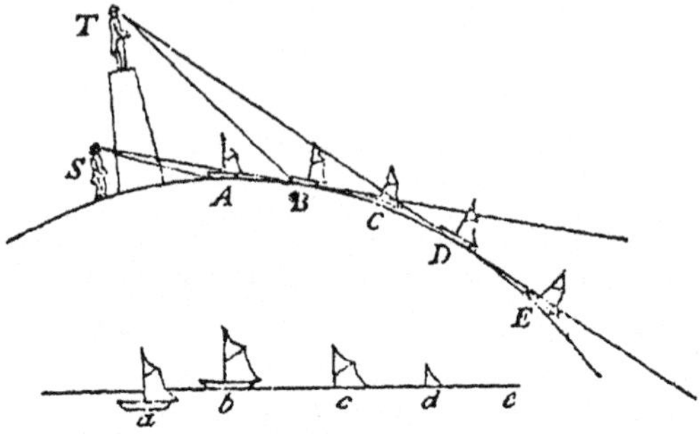

When it has reached a certain distance, as at C, its hull has entirely vanished, but the masts and sails remain, presenting the appearance *c*. But if, in this state of things, the spectator quickly ascends to a higher station, T, whose visible horizon is at D, the hull comes again in sight; and when he descends again he loses it. The ship still receding, the lower sails seem to sink below the water, as at *d*, and at length the whole disappears: while yet the distinctness with which the last portion of the sail *d* is seen is such as to satisfy us that were it not for the interposed segment of the sea, ABCDE, the distance TE is not so great as to have prevented an equally perfect view of the whole.

(27.) In this manner, therefore, if we could measure the heights and exact distance of two stations which could barely be discerned from each other over the edge of the horizon, we could ascertain the actual size of the earth itself: and, in fact, were it not for the effect of refraction, by which we are enabled to see in some small degree *round* the interposed segment (as will be hereafter explained), this would be a tolerably good method of ascertaining it. Suppose A and B to be two eminences, whose perpendicular heights A*a* and B*b* (which, for simplicity, we will suppose to be exactly equal) are known, as well as their exact horizontal interval *a* D *b*, by measurement; then it is clear that D, the visible horizon of both, will lie just half-way between them, and if we suppose *a* D *b* to be the sphere of the earth, and C its centre in the figure C D *b* B, we

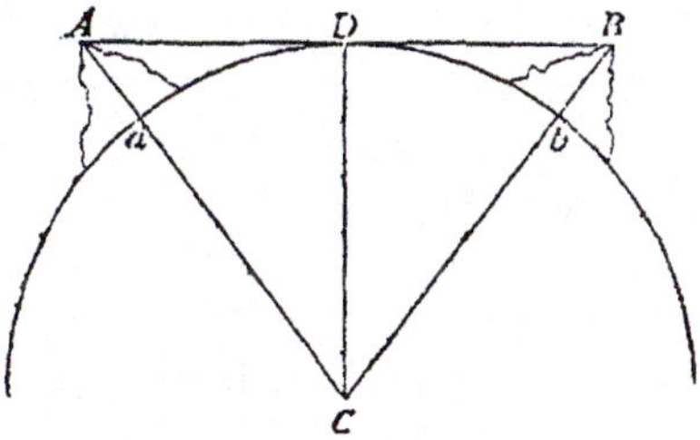

know D *b*, the length of the arch of the circle between D and *b*,—viz. half the measured interval, and *b* B, the excess of its secant above its radius—which is the height of B,—data which, by the solution of an easy geometrical problem, enable us to find the length of the radius D C. If, as is really the case, we suppose both the heights and distance of the stations inconsiderable in comparison with the size of the earth, the solution alluded to is contained in the following proposition:—

The earth's diameter bears the same proportion to the distance of the visible horizon from the eye as that distance does to the height of the eye above the sea level.

When the stations are unequal in height the problem is a little more complicated.

(28.) Although, as we have observed, the effect of refraction prevents this from being an exact method of ascertaining the dimensions of the earth, yet it will suffice to afford such an approximation to it as shall be of use in the present stage of the reader's knowledge, and help him to many just conceptions, on which account we shall exemplify its application in numbers. Now, it appears by observation, that two points, each ten feet above the surface, cease to be visible from each other over still water, and in average atmospheric circumstances, at a distance of about 8 miles. But 10 feet is the 528th part of a mile, so that half their distance, or 4 miles, is to the height of each as 4×528 or $2112:1$, and therefore in the same proportion to 4 miles is the length of the earth's diameter. It must, therefore, be equal to $4 \times 2112 = 8448$, or, in round numbers, about 8000 miles, which is not very far from the truth.

(29.) Such is the first rough result of an attempt to ascertain the earth's magnitude; and it will not be amiss if we take advantage of it to compare it with objects we have been accustomed to consider as of vast size, so as to interpose a few steps between it and our ordinary ideas of dimension. We have before likened the inequalities on the earth's surface, arising from mountains, valleys, buildings, &c. to the roughnesses on the rind of an orange, compared with its general mass. The comparison is quite free from exaggeration. The highest mountain known does not exceed five miles in perpendicular elevation: this is only one 1600th part of the earth's diameter; consequently, on a globe of sixteen inches in diameter, such a mountain would be represented by a protuberance of no more than one hundredth part of an inch, which is about the thickness of ordinary drawing-paper. Now as there is no entire continent, or even any very extensive tract of land, known, whose general elevation above the sea is any thing like half this quantity, it follows, that if we would construct a correct model of our earth, with its seas, continents, and mountains, on a globe sixteen inches in diameter, the whole of the land, with the exception of a few prominent points and ridges, must be comprised on it within the thickness of thin writing-paper; and the highest hills would be represented by the smallest visible grains of sand.

(30.) The deepest mine existing does not penetrate half a mile below the surface: a scratch, or pin-hole, duly representing it, on the surface of such a globe as our model, would be imperceptible without a magnifier.

(31.) The greatest depth of sea, probably, does not much exceed the greatest elevation of the continents; and would, of course, be represented by an excavation, in about the same proportion, into the substance of the globe: so that the ocean comes to be conceived as a mere film of liquid, such as, on our model, would be left by a brush dipped in colour and drawn over those parts intended to represent the sea: only, in so conceiving it, we must bear in mind that the resemblance extends no farther than to proportion in point of quantity. The mechanical laws which would regulate the distribution and movements of such a film, and its adhesion to the surface, are altogether different from those which govern the phenomena of the sea.

(32.) Lastly, the greatest extent of the earth's surface which has ever been seen at once by man, was that exposed to the view of MM. Biot and Gay-Lussac, in their celebrated aeronautic expedition to the enormous height of 25,000 feet, or rather less than five miles. To estimate the proportion of the area visible from this elevation to the whole earth's surface, we must have recourse to the geometry of the sphere, which informs us that the convex surface of a spherical segment is to the whole surface of the sphere to which it belongs as the versed sine, or thickness of the segment, is to the diameter of the sphere; and further, that this thickness, in the case we are considering, is almost exactly equal to the perpendicular elevation of the point of sight above the surface. The proportion, therefore, of the visible area, in this case, to the whole earth's surface, is that of five miles to 8000, or 1 to 1600. The portion visible from Ætna, the Peak of Teneriffe, or Mowna Roa, is about one 4000th.

(33.) When we ascend to any very considerable elevation above the surface of the earth, either in a balloon, or on mountains, we are made aware, by many uneasy sensations, of an insufficient supply of *air*. The barometer, an instrument which informs us of the weight of air incumbent on a given horizontal surface, confirms this impression, and affords a direct measure of the rate of diminution of the quantity of air which a given space includes as we recede from the surface. From its indications we learn, that when we have ascended to the height of 1000 feet, we have left below us about one thirtieth of the whole mass of the atmosphere: — that at 10,600 feet of perpendicular elevation (which is rather less than that of the summit of Ætna*) we have ascended through about one third; and at 18,000 feet (which is nearly that of Cotopaxi) through one half the material, or, at least, the ponderable, body of air incumbent on the earth's surface. From the progression of these numbers, as well as, *à priori*, from the nature of the air itself, which is *compressible*, i. e. capable of being condensed, or crowded into a smaller space in proportion to the incumbent pressure, it is easy to see that, although by rising still higher we should continually get above more and more of the air, and so relieve ourselves more and more from the pressure with which it weighs upon us, yet the amount of this additional relief, or the *ponderable quantity* of air surmounted, would be by no means in proportion to the additional height ascended, but in a constantly decreasing ratio. An easy calculation, however, founded on our experimental knowledge of the properties of air, and the mechanical laws which regulate its dilatation and compression, is sufficient to show that, at an altitude above the surface of the earth not exceeding the hundredth part of its diameter, the tenuity, or rarefaction, of the air must be so excessive, that not only animal life could not subsist, or combustion be maintained in it, but that the most delicate means we possess of ascertaining the existence of *any air at all* would fail to afford the slightest perceptible indications of its presence.

(34.) Laying out of consideration, therefore, at present, all nice questions as to the probable existence of a definite limit to the atmosphere, beyond which there is, absolutely and rigorously speaking, *no* air, it is clear, that, for all practical purposes, we may speak of those regions which are more distant above the earth's surface than the hundredth part of its diameter as void of air, and of course of clouds (which are nothing but

* The height of Ætna above the Mediterranean (as it results from a barometrical measurement of my own, made in July, 1824, under very favourable circumstances) is 10,872 English feet. — *Author.*

visible vapours, diffused and *floating* in the air, sustained by it, and rendering it *turbid* as mud does water). It seems probable, from many indications, that the greatest height at which visible clouds *ever exist* does not exceed ten miles; at which height the density of the air is about an eighth part of what it is at the level of the sea.

(35.) We are thus led to regard the atmosphere of air, with the clouds it supports, as constituting a coating of equable or nearly equable thickness, enveloping our globe on all sides; or rather as an aërial ocean, of which the surface of the sea and land constitutes the bed, and whose inferior portions or strata, within a few miles of the earth, contain by far the greater part of the whole mass, the density diminishing with extreme rapidity as we recede upwards, till, within a very moderate distance (such as would be represented by the sixth of an inch on the model we have before spoken of, and which is not more in proportion to the globe on which it rests, than the downy skin of a peach in comparison with the fruit within it), all sensible trace of the existence of air disappears.

(36.) Arguments, however, are not wanting to render it, if not absolutely certain, at least in the highest degree probable, that the surface of the aërial, like that of the aqueous ocean, has a real and definite limit, as above hinted at; beyond which there is positively *no* air, and above which a fresh quantity of air, could it be added from without, or carried aloft from below, instead of dilating itself indefinitely upwards, would, after a certain very enormous but still finite enlargement of volume, sink and merge, as water poured into the sea, and distribute itself among the mass beneath. With the truth of this conclusion, however, astronomy has little concern; all the effects of the atmosphere in modifying astronomical phenomena being the same, whether it be supposed of definite extent or not.

(37.) Moreover, whichever idea we adopt, it is equally certain that, within those limits in which it possesses any appreciable density, its constitution is the same over all points of the earth's surface; that is to say, on the great scale, and leaving out of consideration temporary and local causes of derangement, such as winds, and great fluctuations, of the nature of waves, which prevail in it to an immense extent: in other words, that the law of diminution of the air's density as we recede upwards *from the level of the sea* is the same in every column into which we may conceive it divided, or from whatever point of the surface we may set out. It may therefore be considered as consisting of successively superposed strata or layers, each of the form of a spherical shell, concentric with the general surface of the sea and land, and each of which is *rarer*, or specifically lighter, than that immediately beneath it; and *denser*, or specifically heavier, than that immediately above it. This kind of distribution of its ponderable mass is necessitated by the laws of the equilibrium of fluids, whose results barometric observations demonstrate to be in perfect accordance with experience.

It must be observed, however, that with this distribution of its strata the inequalities of mountains and valleys have no concern: these exercise no more influence in modifying their general spherical figure than the inequalities at the bottom of the sea interfere with the general sphericity of its surface.

(38.) It is the power which air possesses, in common with all transparent media, of *refracting* the rays of light, or bending them out of their straight course, which renders a knowledge of the constitution of the atmosphere important to the astronomer. Owing to this property, objects seen obliquely through it appear otherwise situated than they would to the same spectator, had the atmosphere no existence: it thus produces a false impression respecting their places, which must be rectified by ascertaining the amount and direction of the displacement so apparently produced on each, before we can come at a knowledge of the true directions in which they are situated from us at any assigned moment.

(39.) Suppose a spectator placed at A, any point

of the earth's surface KA*k*; and let L*l*, M*m*, N*n*, represent the successive strata or layers, of decreasing density, into which we may conceive the atmosphere to be divided, and which are spherical surfaces concentric with K*k*, the earth's surface. Let S represent a star, or other heavenly body, beyond the utmost limit of the atmosphere; then, if the air were away, the spectator would see it in the direction of the straight line AS. But, in reality, when the ray of light SA reaches the atmosphere, suppose at *d*, it will, by the laws of optics, begin to bend *downwards*, and take a more inclined direction, as *d c*. This bending will at first be imperceptible, owing to the

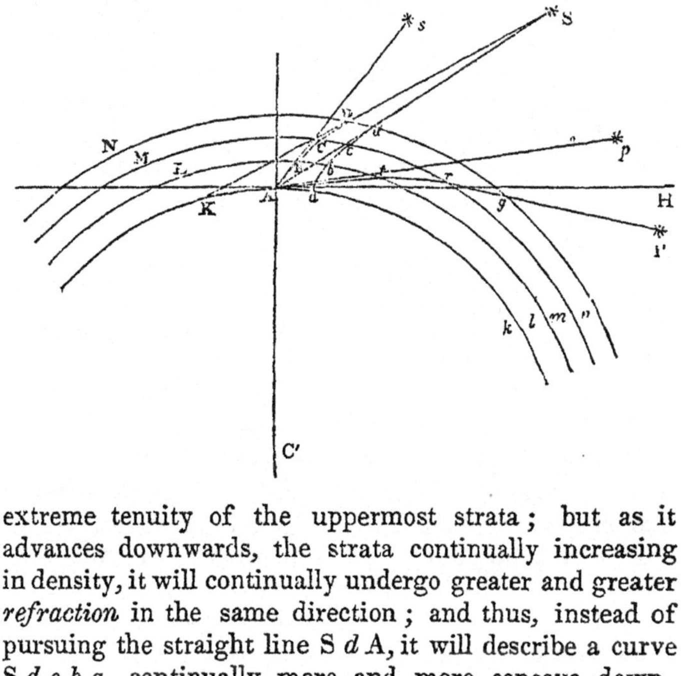

extreme tenuity of the uppermost strata; but as it advances downwards, the strata continually increasing in density, it will continually undergo greater and greater *refraction* in the same direction; and thus, instead of pursuing the straight line S *d* A, it will describe a curve S *d c b a*, continually more and more concave downwards, and will reach the earth, not at A, but at a certain point *a*, nearer to S. *This* ray, consequently, will not reach the spectator's eye. The ray by which he will see the star is, therefore, not S *d*A, but another ray which, had there been no atmosphere would have struck the earth at K, a point *behind* the spectator; but which, being bent by the air into the curve S D C B A, actually strikes on A. Now, it is a law of optics, that an object is seen in the direction which the visual ray has at the instant of *arriving at the eye,* without regard to what may have been otherwise its course between the object and the eye. Hence the star S will be seen, not in the direction A S, but in that of A *s*, a *tangent* to the curve S D C B A, at A. But because the curve described by the refracted ray is concave downwards, the tangent A*s* will lie *above* A S, the unrefracted ray: consequently the object S will appear more elevated above the horizon A H, when seen through the refracting atmosphere, than it would appear were there no such atmosphere. Since, however, the disposition of the strata is the same in all directions around A, the visual ray will not be made to deviate *laterally,* but will remain constantly in the same vertical plane, S A C′, passing through the eye, the object, and the earth's centre.

(40.) The effect of the air's refraction, then, is to *raise* all the heavenly bodies higher above the horizon in appearance than they are in reality. Any such body, situated actually *in* the true horizon, will appear *above* it, or will have some certain apparent *altitude* (as it is called). Nay, even some of those actually below the horizon, and which would therefore be invisible but for the effect of refraction, are, by that effect, raised above it and brought into sight. Thus, the sun, when situated at P below the true horizon, A H, of the spectator, becomes visible to him, as if it stood at *p*, by the refracted ray P *q r t* A, to which A *p* is a tangent.

(41.) The exact estimation of the amount of atmospheric refraction, or the strict determination of the angle S A*s*, by which a celestial object at any assigned altitude, H A S, is raised in appearance above its true place, is, unfortunately, a very difficult subject of physical enquiry, and one on which geometers (from whom alone we can look for any information on the subject) are not yet entirely agreed. The difficulty arises from this, that the *density* of any stratum of air (on which its refracting power depends) is affected not *merely* by the superincumbent pressure, but also by its *temperature* or degree of heat. Now, although we know that as we recede from the earth's surface the temperature of the air is constantly diminishing, yet the *law*, or amount of this diminution at different heights, is not yet fully ascertained. Moreover, the refracting power of air is perceptibly affected by its *moisture;* and this, too, is not the same in every part of an aërial column; neither are we acquainted with the laws of its distribution. The consequence of our ignorance on these points is to introduce a corresponding degree of uncertainty into the determination of the amount of refraction, which affects, to a certain appreciable extent, our knowledge of several of the most important *data* of astronomy. The uncertainty thus induced is, however, confined within such very narrow limits as to be no cause of embarrassment, except in the most delicate enquiries, and to call for no further allusion in a treatise like the present.

(42.) A "Table of Refractions," as it is called, or a statement of the amount of apparent displacement arising from this cause, at all altitudes, or in every situation of a heavenly body, from the horizon to the *zenith* *, or point of the sky vertically above the spectator, and, under all the circumstances in which astronomical observations are usually performed which may influence the result, is one of the most important and indispensable of all astronomical tables, since it is only by the use of such a table we are enabled to get rid of an illusion which must otherwise pervert all our notions respecting the celestial motions. Such have been, accordingly, constructed with great care, and are to be found in every collection of astronomical tables.† Our design, in the present treatise, will not admit of the introduction of tables; and we must, therefore, content ourselves here, and in si-

* From an Arabic word of this signification.

† Vide "Requisite Tables to be used with the Nautical Almanac." See also Nautical Almanac for 1833, Dr. Pearson's Astronomical Tables, and Mr. Baily's Astronomical Tables and Formulæ.

milar cases, with referring the reader to works especially destined to furnish these useful aids to calculation. It is, however, desirable that he should bear in mind the following general notions of its amount, and law of variation.

(43.) 1st. In the *zenith* there is no refraction; a celestial object, situated vertically over head, is seen in its true direction, as if there were no atmosphere.

2dly. In descending from the *zenith* to the horizon, the refraction continually increases; objects near the horizon appearing more elevated by it above their true directions than those at a high altitude.

3dly. The *rate* of its increase is nearly in proportion to the tangent of the apparent angular distance of the object from the zenith. But this rule, which is not far from the truth, at moderate *zenith distances*, ceases to give correct results in the vicinity of the horizon, where the law becomes much more complicated in its expression.

4thly. The average amount of refraction, for an object half-way between the zenith and horizon, or at an apparent altitude of 45°, is about 1′ (more exactly 57″), a quantity hardly sensible to the naked eye; but at the visible horizon it amounts to no less a quantity than 33′, which is rather more than the greatest apparent diameter of either the sun or the moon. Hence it follows, that when we see the lower edge of the sun or moon just *apparently* resting on the horizon, its whole disk is in reality below it, and would be entirely out of sight and concealed by the convexity of the earth but for the bending round it, which the rays of light have undergone in their passage through the air, as alluded to in art. 40.

(44.) It follows from this, that one obvious effect of refraction must be to shorten the duration of night and darkness, by actually prolonging the stay of the sun and moon above the horizon. But even after they are set, the influence of the atmosphere still continues to send us a portion of their light; not, indeed, by direct transmission, but by *reflection* upon the vapours, and minute solid particles, which float in it, and, perhaps, also on the actual material atoms of the air itself. To understand how this takes place, we must recollect, that it is not only by the direct light of a luminous object that we see, but that whatever portion of its light which would not otherwise reach our eyes is intercepted in its course, and thrown back, or laterally, upon us, becomes to us a means of illumination. Such reflective obstacles always exist floating in the air. The whole course of a sunbeam penetrating through the chink of a window-shutter into a dark room is *visible* as a bright line in the air; and even if it be stifled, or *let out* through an opposite crevice, the light scattered through the apartment, from this source is sufficient to prevent entire darkness in the room. The luminous lines occasionally seen in the air, in a sky full of partially broken clouds, which the vulgar term "the sun drawing water," are similarly caused. They are sunbeams, through apertures in clouds, partially intercepted and reflected on the dust and vapours of the air below. Thus it is with those solar rays which, after the sun is itself concealed by the convexity of the earth, continue to traverse the higher regions of the atmosphere above our heads, and pass through and out of it, without directly striking on the earth at all. Some portion of them is intercepted and reflected by the floating particles above mentioned, and thrown back, or laterally, so as to reach us, and afford us that secondary illumination, which is twilight. The course of such rays will be immediately understood from the annexed figure, in which A B C D is the earth; A a point on its surface, where the sun S is in the act of setting; its last lower ray S A M just grazing the surface at A, while its superior rays S N, S O, traverse the atmosphere above A without striking the earth, leaving it finally at the points P Q R, after being more or less bent in passing through it, the lower most, the higher less, and that which, like S R O, merely grazes the exterior limit of the atmosphere, not at all. Let us consider several points, A, B, C, D, each more remote than the last from A, and each more deeply involved in the *earth's shadow*, which occupies the whole space from A beneath the line A M. Now, A just receives the sun's last direct ray, and, besides, is illuminated by

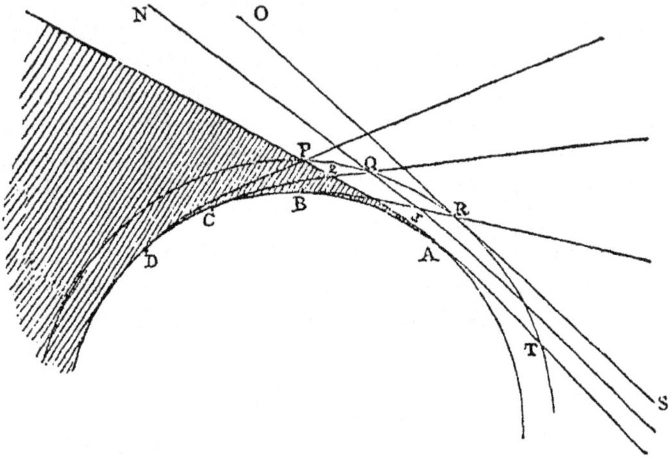

the whole reflective atmosphere P Q R T. It therefore receives twilight from the whole sky. The point B, to which the sun has set, receives no direct solar light, nor any, direct or reflected, from all that part of *its* visible atmosphere which is below A P M; but from the lenticular portion P R *x*, which is traversed by the sun's rays, and which lies above the visible horizon B R of B, it receives a twilight, which is strongest at R, the point immediately below which the sun is, and fades away gradually towards P, as the luminous part of the atmosphere thins off. At C, only the last or thinnest portion, P Q *z* of the lenticular segment, thus illuminated, lies above the horizon, C Q, of that place: here, then, the twilight is feeble, and confined to a small space in and near the horizon, which the sun has quitted, while at D the twilight has ceased altogether.

(45.) When the sun is above the horizon, it illuminates the atmosphere and clouds, and these again disperse and scatter a portion of its light in all directions, so as to send some of its rays to every exposed point, from every point of the sky. The generally diffused light, therefore, which we enjoy in the daytime, is a phenomenon originating in the very same causes as the twilight. Were it not for the reflective and scattering power of the atmosphere, no objects would be visible to us out of direct sunshine; every shadow of a passing cloud would be pitchy darkness; the stars would be visible all day, and every apartment, into which the sun had not direct admission, would be involved in nocturnal obscurity. This scattering action of the atmosphere on the solar light, it should be observed, is greatly increased by the irregularity of temperature caused by the same luminary in its different parts, which, during the daytime, throws it into a constant state of undulation, and, by thus bringing together masses of air of very unequal temperatures, produces partial reflections and refractions at their common boundaries, by which much light is turned aside from the direct course, and diverted to the purposes of general illumination.

(46.) From the explanation we have given, in arts. 39. and 40., of the nature of atmospheric refraction, and the mode in which it is produced in the progress of a ray of light through successive strata, or layers, of the atmosphere, it will be evident, that whenever a ray passes *obliquely* from a higher level to a lower one, or *vice versâ*, its course is not rectilinear, but concave downwards; and of course any object seen by means of such a ray, must appear deviated from its true place, whether that object be, like the celestial bodies, entirely beyond the atmosphere, or, like the summits of mountains, seen from the plains, or other terrestrial stations, at different levels, seen from each other, immersed in it. Every difference of level, accompanied, as it must be, with a difference of density in the aërial strata, must also have, corresponding to it, a certain amount of refraction; less, indeed, than what would be produced by the *whole* atmosphere, but still often of very appreciable, and even considerable, amount. This refraction between terrestrial stations is termed *terrestrial refraction*, to distinguish it from that total effect which is only produced on celestial objects, or such as are beyond the atmosphere, and which is called celestial or astronomical refraction.

(47.) Another effect of refraction is to distort the visible forms and proportions of objects seen near the horizon. The sun, for instance, which, at a considerable altitude, always appears round, assumes, as it approaches the horizon, a flattened or oval outline; its horizontal diameter being visibly greater than that in a vertical direction. When very near the horizon, this flattening is evidently more considerable on the lower side than on the upper; so that the apparent form is neither circular nor elliptic, but a species of oval, which deviates more from a circle below than above. This singular effect, which any one may notice in a fine sunset, arises from the rapid rate at which the refraction increases in approaching the horizon. Were every visible point in the sun's circumference equally raised by refraction, it would still appear circular, though displaced: but the lower portions being *more* raised than the upper, the vertical diameter is thereby shortened, while the two extremities of its horizontal diameter are equally raised, and in parallel directions, so that its apparent length remains the same. The dilated size (generally) of the sun or moon, when seen near the horizon, beyond what they appear to have when high up in the sky, has nothing to do with refraction. It is an illusion of the judgment, arising from the terrestrial objects interposed, or placed in close comparison with them. In that situation we view and judge of them as we do of terrestrial objects—in detail, and with an acquired habit of attention to parts. Aloft we have no associations to guide us, and their insulation in the expanse of sky leads us rather to undervalue than to over-rate their apparent magnitudes. Actual measurement with a proper instrument corrects our error, without, however, dispelling our illusion. By this we learn, that the sun, when just on the horizon, subtends at our eyes almost exactly the same, and the moon a materially *less* angle, than when seen at a great altitude in the sky, owing to the effect of what is called parallax, to be explained presently.

(48.) After what has been said of the small extent of the atmosphere in comparison of the mass of the earth, we shall have little hesitation in admitting those luminaries which people and adorn the sky, and which, while they obviously form no part of the earth, and receive no support from it, are yet not borne along at random like clouds upon the air, nor drifted by the winds, to be external to our atmosphere. As such we have considered them while speaking of their refractions —as existing in the immensity of space beyond, and situated, perhaps, for any thing we can perceive to the contrary, at enormous distances from us and from each other.

(49.) Could a spectator exist unsustained by the earth, or any solid support, he would see around him at one view the whole contents of space—the visible constituents of the universe: and, in the absence of any means of judging of their distances from him, would refer them, in the directions in which they were seen from his station, to the concave surface of an imaginary sphere, having his eye for a centre, and its surface at some vast indeterminate distance. Perhaps he might judge those which appear to him large and bright, to be nearer to him than the smaller and less brilliant; but, independent of other means of judging, he would have no warrant for this opinion, any more than for the idea that all were equidistant from him, and *really* arranged on such a spherical surface. Nevertheless, there would be no impropriety in his referring their places, geometrically speaking, to those points of such a purely imaginary sphere, which their respective visual rays intersect; and there would be much advantage in so doing, as by that means their appearance and relative situation could be accurately measured, recorded, and mapped down. The objects in a landscape are at every variety of distance from the eye, yet we lay them all down in a picture on one plane, and at one distance, in their actual *apparent proportions*, and the likeness is not taxed with incorrectness, though a man in the foreground should be represented larger than a mountain in the distance. So it is to a spectator of the heavenly bodies pictured, *projected*, or mapped down on that imaginary sphere we call the *sky* or *heaven*. Thus, we may easily conceive that the moon, which appears to us as large as the sun, though less bright, *may* owe that apparent equality to its greater proximity, and *may* be really much less; while both the moon and sun may only appear larger and brighter than the stars, on account of the remoteness of the latter.

(50.) A spectator on the earth's surface is prevented, by the great mass on which he stands, from seeing into all that portion of space which is below him, or to see which he must look in any degree downwards. It is true that, if his place of observation be at a great elevation, the dip of the horizon will bring within the scope of vision a little more than a hemisphere, and refraction, wherever he may be situated, will enable him to look, as it were, a little round the corner; but the zone thus added to his visual range can hardly ever, unless in very extraordinary circumstances *, exceed a couple of degrees in breadth, and is always ill seen on account of the vapours near the horizon. Unless, then, by a change of his geographical situation, he should shift his horizon (which is always a plane touching the spherical convexity of the earth at his station); or unless, by some movements proper to the heavenly bodies, they should of themselves come above his horizon; or, lastly, unless, by some rotation of the earth itself on its centre, the point of its surface

* Such as the following, for instance:—The late Mr. Sadler, the celebrated aëronaut, ascended in a balloon from Dublin at about 2 o'clock in the afternoon, and was wafted across the channel. About sunset he approached the English coast, when the balloon descended near the surface of the sea. By this time the sun was set, and the shades of evening began to close in. He threw out nearly all his ballast, and suddenly sprung upwards to a great height, and by so doing witnessed the whole phenomenon of a western sunrise. He subsequently descended in Wales, and witnessed a second sunset on the same evening. I have this anecdote from Dr. Lardner, who was present at his ascent, and read his own account of the voyage.—*Author.*

which he occupies should be carried round, and presented towards a different region of space; he would never obtain a sight of almost one half the objects external to our atmosphere. But if any of these cases be supposed, more, or all, may come into view according to the circumstances.

(51.) A traveller, for example, shifting his locality on our globe, will obtain a view of celestial objects invisible from his original station, in a way which may be not inaptly illustrated by comparing him to a person standing in a park close to a large tree. The massive obstacle presented by its trunk cuts off his view of all those parts of the landscape which it occupies as an object; but by walking round it a complete successive view of the whole panorama may be obtained. Just in the same way, if we set off from any station, as London, and travel southwards, we shall not fail to notice that many celestial objects which are never seen from London come successively into view, as if rising up above the horizon, night after night, from the south, although it is in reality our horizon, which, travelling with us southwards round the sphere, sinks in succession beneath them. The novelty and splendour of fresh

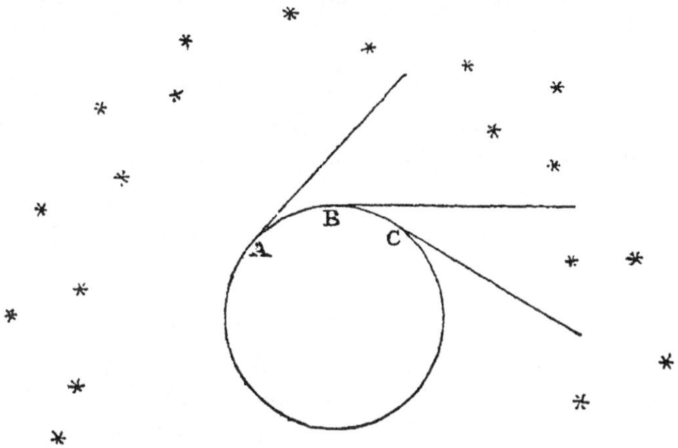

constellations thus gradually brought into view in the clear calm nights of tropical climates, in long voyages to the south, is dwelt upon by all who have enjoyed this spectacle, and never fails to impress itself on the recollection among the most delightful and interesting of the associations connected with extensive travel. A glance at the accompanying figure, exhibiting three successive stations of a traveller, A, B, C, with the horizon corresponding to each, will place this process in clearer evidence than any description.

(52.) Again: suppose the earth itself to have a motion of rotation on its centre. It is evident that a spectator at rest (as it appears to him) on any part of it will, unperceived by himself, be carried round with it: unperceived, we say, because his horizon will constantly contain, and be limited by, the same terrestrial objects. He will have the same landscape constantly before his eyes, in which all the familiar objects in it, which serve him for landmarks and directions, retain, with respect to himself or to each other, the same invariable situations. The perfect smoothness and equality of the motion of so vast a mass, in which every object he sees around him participates alike, will (art. 15.) prevent his entertaining any suspicion of his actual change of place. Yet, with respect to external objects,—that is to say, all celestial ones which do not participate in the supposed rotation of the earth,—his horizon will have been all the while shifting in its relation to them, precisely as in the case of our traveller in the foregoing article. Recurring to the figure of that article, it is evidently the same thing, so far as their visibility is concerned, whether he has been carried by the earth's rotation successively into the situations A, B, C; or whether, the earth remaining at rest, he has transferred himself personally along its surface to those stations. Our spectator in the park will obtain precisely the same view of the landscape, whether he walk round the tree, or whether we suppose it sawed off, and made to turn on an upright pivot, while he stands on a projecting step attached to it, and allows himself to be carried round by its motion. The only difference will be in his view of the tree itself, of which, in the former case, he will see every part, but, in the latter, only that portion of it which remains constantly opposite to him, and immediately under his eye.

(53.) By such a rotation of the earth, then, as we have supposed, the horizon of a stationary spectator will be constantly depressing itself below those objects which lie in that region of space towards which the rotation is carrying him, and elevating itself above those in the opposite quarter; admitting into view the former, and successively hiding the latter. As the horizon of every such spectator, however, appears *to him* motionless, all such changes will be referred by him to a motion in the objects themselves so successively disclosed and concealed. In place of his horizon approaching the stars, therefore, he will judge the stars to approach his horizon; and when it passes over and hides any of them, he will consider them as having sunk below it, or *set*; while those it has just disclosed, and from which it is receding, will seem to be rising above it.

(54.) If we suppose this rotation of the earth to continue in one and the same direction,—that is to say, to be performed round one and the same *axis*, till it has completed an entire revolution, and come back to the position from which it set out when the spectator began his observations,—it is manifest that every thing will then be in precisely the same relative position as at the outset: all the heavenly bodies will appear to occupy the same places in the concave of the sky which they did at that instant, except such as may have actually moved in the interim; and if the rotation still continue, the same phenomena of their successive rising and setting, and return to the same places, will continue to be repeated in the same order, and (if the velocity of rotation be uniform) in equal intervals of time, *ad infinitum*.

(55.) Now, in this we have a lively picture of that grand phenomenon, the most important beyond all comparison which nature presents, the daily rising and setting of the sun and stars, their progress through the vault of the heavens, and their return to the same apparent places at the same hours of the day and night. The accomplishment of this restoration in the regular interval of twenty-four hours is the first instance we encounter of that great law of *periodicity**, which, as we shall see, pervades all astronomy; by which expression we understand the continual reproduction of the same phenomena, in the same order, at equal intervals of time.

(56.) A free rotation of the earth round its centre, if it exist and be performed in consonance with the same mechanical laws which obtain in the motions of masses of matter under our immediate control, and within our ordinary experience, must be such as to satisfy two essential conditions. It must be invariable in its direction *with respect to the sphere itself*, and uniform in its velocity. The rotation must be performed *round an axis* or diameter of the sphere, whose *poles* or extremities, where it meets the surface, correspond always to the same points on the sphere. Modes of rotation of a solid body under the influence of external agency are conceivable, in which the poles of the imaginary line or axis about which it is at any moment revolving shall hold no fixed places on the surface, but shift upon it every moment. Such changes, however, are inconsistent with the idea of a rotation of a body of regular figure about its axis of symmetry, performed in free space, and without resistance or obstruction from any surrounding medium. The complete absence of such obstructions draws with it, of necessity, the strict fulfilment of the two conditions above mentioned.

(57.) Now, these conditions are in perfect accordance with what we observe, and what recorded observation teaches us, in respect of the diurnal motions of the heavenly bodies. We have no reason to believe, from history, that any sensible change has taken place since the earliest ages in the interval of time elapsing between two successive returns of the same star to the same

* Περίοδος, a *going round*, a circulation or revolution.

point of the sky; or, rather, it is demonstrable from astronomical records that no such change *has* taken place. And with respect to the other condition, — *the permanence of the axis of rotation*, — the appearances which any alteration in that respect must produce, would be marked, as we shall presently show, by a corresponding change of a very obvious kind in the apparent motions of the stars; which, again, history decidedly declares them *not* to have undergone.

(58.) But, before we proceed to examine more in detail how the hypothesis of the rotation of the earth about an axis accords with the phenomena which the diurnal motion of the heavenly bodies offers to our notice, it will be proper to describe, with precision, in what that diurnal motion consists, and how far it is participated in by them all; or whether any of them form exceptions, wholly or partially, to the common analogy of the rest. We will, therefore, suppose the reader to station himself, on a clear evening, just after sunset, when the first stars begin to appear, in some open situation whence a good general view of the heavens can be obtained. He will then perceive, above and around him, as it were, a vast concave hemispherical vault, beset with stars of various magnitudes, of which the brightest only will first catch his attention in the twilight; and more and more will appear as the darkness increases, till the whole sky is over-spangled with them. When he has awhile admired the calm magnificence of this glorious spectacle, the theme of so much song, and of so much thought, — a spectacle which no one can view without emotion, and without a longing desire to know something of its nature and purport, — let him fix his attention more particularly on a few of the most brilliant stars, such as he cannot fail to recognize again without mistake after looking away from them for some time, and let him refer their apparent situations to some surrounding objects, as buildings, trees, &c., selecting purposely such as are in different quarters of his horizon. On comparing them *again* with their respective points of reference, after a moderate interval, as the night advances, he will not fail to perceive that they have changed their places, and advanced, as by a general movement, in a westward direction; those towards the eastern quarter appearing to rise or recede from the horizon, while those which lie towards the west will be seen to approach it; and, if watched long enough, will, for the most part, finally sink beneath it, and disappear; while others, in the eastern quarter, will be seen to rise as if out of the earth, and, joining in the general procession, will take their course with the rest towards the opposite quarter.

(59.) If he persists for a considerable time in watching their motions, on the same or on several successive nights, he will perceive that each star appears to describe, as far as its course lies above the horizon, a circle in the sky; that the circles so described are not of the same magnitude for all the stars; and that those described by different stars differ greatly in respect of the parts of them which lie above the horizon. Some, which lie towards the quarter of the horizon which is denominated the SOUTH *, only remain for a short time above it, and disappear, after describing in sight only the small upper segment of their diurnal circle; others, which rise between the south and east, describe larger segments of their circles above the horizon, remain proportionally longer in sight, and set precisely as far to the westward of south as they rose to the eastward; while such as rise exactly in the east remain just twelve hours visible, describe a semicircle, and set exactly in the west. With those, again, which rise between the east and north, the same law obtains; at least, as far as regards the time of their remaining above the horizon, and the proportion of the visible segment of their diurnal circles to their whole circumferences. Both go on increasing; they remain in view more than twelve hours, and their visible diurnal arcs are more than semicircles. But the

* We suppose our observer to be stationed in some northern latitude; somewhere in Europe, for example.

magnitudes of the circles themselves diminish, as we go from the east, northward; the greatest of all the circles being described by those which rise exactly in the east point. Carrying his eye farther northwards, he will notice, at length, stars which, in their diurnal motion, just graze the horizon at its north point, or only dip below it for a moment; while others never reach it at all, but continue always above it, revolving in entire circles round ONE POINT called the POLE, which appears to be the common centre of all their motions, and which alone, in the whole heavens, may be considered immoveable. Not that this point is marked by any star. It is a purely imaginary centre; but there is near it one considerably bright star, called the Pole Star, which is easily recognized by the very small circle it describes; so small, indeed, that, without paying particular attention, and referring its position very nicely to some fixed mark, it may easily be supposed at rest, and be, itself, mistaken for the common centre about which all the others in that region describe their circles; or it may be known by its configuration with a very splendid and remarkable *constellation* or group of stars, called by astronomers the GREAT BEAR.

(60.) He will further observe that the apparent relative situations of all the stars among one another is not changed by their diurnal motion. In whatever parts of their circles they are observed, or at whatever hour of the night, they form with each other the same identical groups or configurations, to which the name of CONSTELLATIONS has been given. It is true, that, in different parts of their course, these groups stand differently with respect to the horizon; and those towards the north, when in the course of their diurnal movement they pass alternately above and below that common centre of motion described in the last article, become actually inverted with respect to the horizon, while, on the other hand, they always turn the same points towards the pole. In short, he will perceive that the whole assemblage of stars visible at once, or in succession, in the heavens, may be regarded as one great constellation, which seems to revolve with a uniform motion, as if it formed one coherent mass; or as if it were attached to the internal surface of a vast hollow sphere, having the earth, or rather the spectator, in its centre, and turning round an axis inclined to his horizon, so as to pass through that fixed point or *pole* already mentioned.

(61.) Lastly, he will notice, if he have patience to outwatch a long winter's night, commencing at the earliest moment when the stars appear, and continuing till morning twilight, that those stars which he observed setting in the west have again risen in the east, while those which were rising when he first began to notice them have completed their course, and are now set; and that thus the hemisphere, or a great part of it, which was then above, is now beneath him, and its place supplied by that which was at first under his feet, which he will thus discover to be no less copiously furnished with stars than the other, and bespangled with groups no less permanent and distinctly recognizable. Thus he will learn that the great constellation we have above spoken of as revolving round the pole is co-extensive with the whole surface of the sphere, being in reality nothing less than a universe of luminaries surrounding the earth on all sides, and brought in succession before his view, and referred (each luminary according to its own visual ray or direction from his eye) to the imaginary spherical surface, of which he himself occupies the centre. (See art. 49.)

(62.) There is, however, one portion or segment of this sphere of which he will not thus obtain a view. As there is a segment towards the north, adjacent to the pole above his horizon, in which the stars *never set*, so there is a corresponding segment, about which the smaller circles of the more southern stars are described, in which they *never rise*. The stars which border upon the extreme circumference of this segment just graze the southern point of his horizon, and show themselves for a few moments above it, precisely as those near the circumference of the northern segment graze his northern horizon, and dip for a moment below it, to re-appear immediately. Every point in a spherical surface has, of course, another diametrically opposite to it; and as the spectator's horizon divides his sphere into two hemispheres—a superior and inferior—there must of necessity exist a depressed pole to the south, corresponding to the elevated one to the north, and a portion surrounding it, perpetually beneath, as there is another surrounding the north pole, perpetually above it.

> " Hic vertex nobis semper sublimis; at illum
> Sub pedibus nox atra videt, manesque profundi."—VIRGIL.
>
> One pole rides high, one, plunged beneath the main,
> Seeks the deep night, and Pluto's dusky reign.

(63.) To get sight of this segment, he must travel southwards. In so doing, a new set of phenomena come forward. In proportion as he advances to the south, some of those constellations which, at his original station, barely grazed the northern horizon, will be observed to sink below it and set; at first remaining hid only for a very short time, but gradually for a longer part of the twenty-four hours. They will continue, however, to circulate about the same point — that is, holding the same invariable position *with respect to them* in the concave of the heavens among the stars; but this point itself will become gradually depressed with respect to the spectator's horizon. The axis, in short, about which the diurnal motion is performed, will appear to have become continually less and less inclined to the horizon; and by the same degrees as the northern pole is depressed the southern will rise, and constellations surrounding it will come into view; at first momentarily, but by degrees for longer and longer times in each diurnal revolution — realizing, in short, what we have already stated in art. 51.

(64.) If he travel continually southwards, he will at length reach a line on the earth's surface, called *the equator*, at any point of which, indifferently, if he take up his station and recommence his observations, he will find that he has both the centres of diurnal motion in his horizon, occupying opposite points, the northern Pole having been depressed, and the southern raised; so that, in this geographical position, the diurnal rotation of the heavens will appear to him to be performed about a horizontal axis, every star describing half its diurnal circle above and half beneath his horizon, remaining alternately visible for twelve hours, and concealed during the same interval. In this situation, *no* part of the heavens is concealed from his *successive* view. In a night of twelve hours (supposing such a continuance of darkness possible at the equator) the whole sphere will have passed in review over him — the whole hemisphere with which he began his night's observation will have been carried down beneath him, and the entire opposite one brought up from below.

(65.) If he pass the equator, and travel still farther southwards, the southern pole of the heavens will become elevated above his horizon, and the northern will sink below it; and the more, the farther he advances southwards; and when arrived at a station as far to the south of the equator as that from which he started was to the north, he will find the whole phenomena of the heavens reversed. The stars which at his original station described their whole diurnal circles above his horizon, and never set, now describe them entirely below it, and never *rise*, but remain constantly invisible to him; and, *vice versâ*, those stars which at his former station he never saw, he will now never cease to see.

(66.) Finally, if, instead of advancing southwards from his first station, he travel northwards, he will observe the northern pole of the heavens to become more elevated above his horizon, and the southern more depressed below it. In consequence, his hemisphere will present a less variety of stars, because a greater proportion of the whole surface of the heavens remains constantly visible or constantly invisible: the circle described by each star, too, becomes more nearly parallel to the horizon; and, in short, every appearance leads to suppose that could he travel far enough to the north, he would at length attain a point *vertically under* the northern pole of the heavens, at which none of the stars would either rise or set, but each would circulate round the horizon in circles parallel to it. Many endeavours have been made to reach this point, which is called the north pole of the earth, but hitherto without success; a barrier of almost insurmountable difficulty being presented by the increasing rigour of the climate: but a very near approach to it has been made: and the phenomena of those regions, though not precisely such as we have described as what must subsist *at* the pole itself, have proved to be in exact correspondence with its near proximity. A similar remark applies to the south pole of the earth, which, however, is more unapproachable, or, at least, has been less nearly approached, than the north.

(67.) The above is an account of the phenomena of the diurnal motion of the stars, as modified by different geographical situations, not grounded on any speculation, but actually observed and recorded by travellers and voyagers. It is, however, in complete accordance with the hypothesis of a rotation of the earth round a fixed axis. In order to show this, however, it will be necessary to premise a few observations on the appearances presented by an assemblage of remote objects, when viewed from different parts of a small and circumscribed station.

(68.) Imagine a landscape, in which a great multitude of objects are placed at every variety of distance from the beholder. If he shift his point of view, though but for a few paces, he will perceive a very great change in the apparent positions of the nearer objects, both with respect to himself and to each other. If he advance northwards, for instance, near objects on his right and left, which were, therefore, to the east and west of his original station, will be left behind him, and appear to have receded southwards; some, which covered each other at first, will appear to separate, and others to approach, and perhaps conceal each other. Remote objects, on the contrary, will exhibit no such great and remarkable changes of relative position. An object to the east of his original station, at a mile or two distance, will

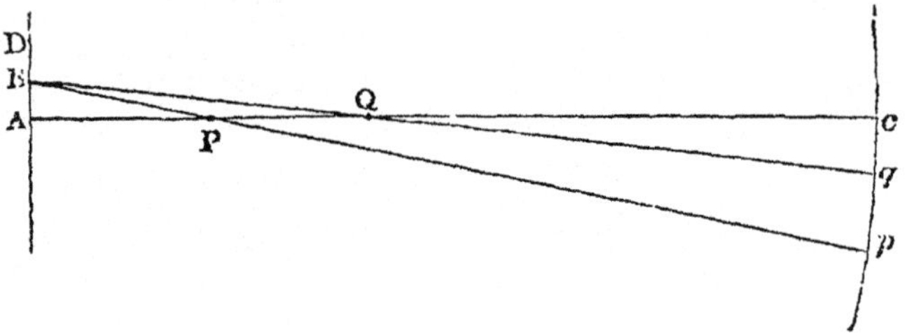

still be referred by him to the east point of his horizon, with hardly any perceptible deviation. The reason of this is, that the position of every object is referred by us to the surface of an imaginary sphere of an indefinite radius, having our eye for its centre; and, as we advance in any direction, A B, carrying this imaginary sphere along with us, the visual rays A P, A Q, by which objects are referred to its surface (at C, for instance), shift their positions with respect to the line in which we move, A B, which serves as an axis or line of reference, and assume new positions, B P *p*, B Q *q*, revolving round their respective objects as centres. Their intersections, therefore, *p*, *q*, with our visual sphere, will appear to recede on its surface, but with different degrees of angular velocity in proportion to their proximity; the same distance of advance A B subtending a greater angle, A P B $=$ *c* P *p*, at the near object P than at the remote one Q.

(69.) This apparent angular motion of an object on our sphere of vision*, arising from a change of our

* The ideal sphere without us, to which we refer the places of objects, and which we carry along with us wherever we go, is no doubt intimately connected by association, if not entirely dependent on that obscure perception of sensation in the retinæ of our eyes, of which, even when closed and unexcited, we cannot entirely divest them. We have a real spherical surface

point of view, is called *parallax*, and it is always expressed by *the angle* B A P *subtended at the object* P by a line joining the two points of view A B under consideration. For it is evident that the difference of angular position of P, with respect to the invariable direction A B D, when viewed from A and from B, is the difference of the two angles D B P and D A P; now, D B P being the exterior angle of the triangle, A B P is equal to the sum of the interior and opposite, D B P = D A P + A P B, whence D B P — D A P = A P B.

(70.) It follows from this, that the amount of parallactic motion arising from any given change of our point of view is, *cæteris paribus*, less, as the distance of an object viewed is greater; and when that distance is extremely great in comparison with the change in our point of view, the parallax becomes insensible; or, in other words, objects do not appear to vary in situation at all. It is on this principle, that in alpine regions visited for the first time we are surprised and confounded at the little progress we appear to make by a considerable change of place. An hour's walk, for instance, produces but a small parallactic change in the relative situations of the vast and distant masses which surround us. Whether we walk round a circle of a hundred yards in diameter, or merely turn ourselves round in its centre, the distant panorama presents almost exactly the same aspect, — we hardly seem to have changed our point of view.

(71.) Whatever notion, in other respects, we may form of the stars, it is quite clear they must be immensely distant. Were it not so, the apparent angular

---

within our eyes, the seat of sensation and vision, corresponding, point for point, to the external sphere. On this the stars, &c. are really mapped down, as we have supposed them in the text to be, on the imaginary concave of the heavens. When the whole surface of the retina is excited by light, habit leads us to associate it with the idea of a real surface existing without us. Thus we become impressed with the notion of *a sky* and *a heaven*, but the concave surface of the retina itself is the true seat of all *visible* angular dimension and angular motion. The substitution of the *retina* for the *heavens* would be awkward and inconvenient in language, but it may always be mentally made. (See Schiller's pretty enigma on the eye in his Turandot.)

interval between any two of them seen over head would be much greater than when seen near the horizon, and the constellations, instead of preserving the same appearances and dimensions during their whole diurnal course, would appear to enlarge as they rise higher in the sky, as we see a small cloud in the horizon swell into a great overshadowing canopy when drifted by the wind across our zenith, or as may be seen in the annexed figure, where *a b*, A B, *a b*, are three different positions

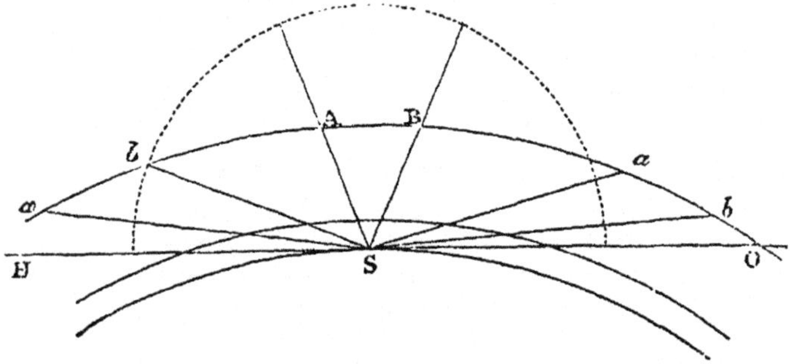

of the same stars, as they would, if near the earth, be seen from a spectator S, under the visual angles *a* S *b*, A S B. No such change of apparent dimension, however, is observed. The nicest measurements of the apparent angular distance of any two stars *inter se*, taken in any parts of their diurnal course, (after allowing for the unequal effects of refraction, or when taken at such times that this cause of distortion shall act equally on both,) manifest *not the slightest* perceptible variation. Not only this, but at whatever point of the earth's surface the measurement is performed, the results are *absolutely identical*. No instruments ever yet invented by man are delicate enough to indicate, by an increase or diminution of the angle subtended, that one point of the earth is nearer to or further from the stars than another.

(72.) The necessary conclusion from this is, that the dimensions of the earth, large as it is, are comparatively *nothing*, absolutely imperceptible, when compared with the interval which separates the stars from the earth. If an observer walk round a circle not more than a few yards in diameter, and from different points in its circumference measure with a sextant, or other more exact instrument adapted for the purpose, the angles PAQ, PBQ, PCQ, subtended at those stations by two well defined points in his visible horizon, PQ,

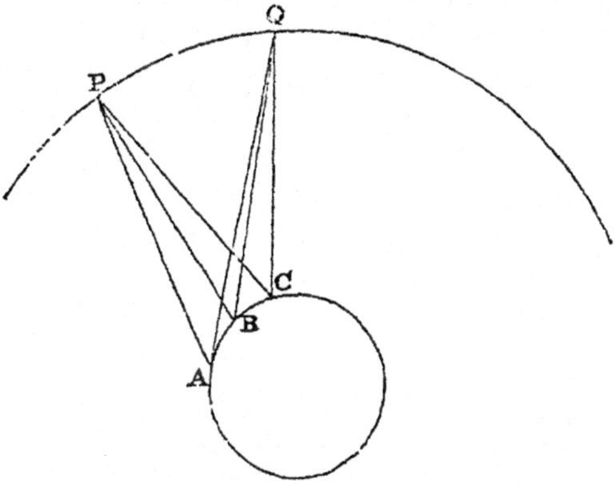

he will at once be advertised, by the difference of the results, of his change of distance from them arising from his change of place, although that difference may be so small as to produce no change in their *general* aspect to his unassisted sight. This is one of the innumerable instances where accurate measurement obtained by instrumental means places us in a totally different situation in respect to matters of fact, and conclusions thence deducible, from what we should hold, were we to rely in all cases on the mere judgment of the eye. To so great a nicety have such observations been carried by the aid of an instrument called a theodolite, that a circle of the diameter above mentioned may thus be rendered *sensible,* may thus be detected to have *a size,* and an ascertainable *place,* by reference to objects distant by fully 100,000 times its own dimensions. Observations, differing, it is true, somewhat in method, but identical in principle, and executed with nearly as much exactness, have been applied to the stars, and with a result such as has been already stated. Hence it follows, incontrovertibly, that the distance of the stars from the earth cannot be *so small* as 100,000 of the earth's diameters. It is, indeed, incomparably greater; for we shall hereafter find it fully demonstrated that the distance just named, immense as it may appear, is yet much under-rated.

(73.) From such a distance, to a spectator with our faculties, and furnished with our instruments, the earth would be imperceptible; and, reciprocally, an object of the earth's size, placed at the distance of the stars, would be equally undiscernible. If, therefore, at the point on which a spectator stands, we draw a plane touching the globe, and prolong it in imagination till it attain the region of the stars, and through the centre of the earth conceive another plane parallel to the former, and co-extensive with it, to pass; these, although separated throughout their whole extent by the same interval, viz. a semidiameter of the earth, will yet, on account of the vast distance at which that interval is seen, be confounded together, and undistinguishable from each other in the region of the stars, when viewed by a spectator on the earth. The zone they there include will be of evanescent breadth to his eye, and will only mark out a great circle in the heavens, which, like the *vanishing point* in perspective to which all parallel *lines* in a picture appear to converge, is, in fact, the *vanishing line* to which all *planes* parallel to the horizon offer a similar appearance of ultimate convergence in the great *panorama* of nature.

(74.) The two planes just described are termed, in astronomy, the *sensible* and *rational* horizon of the observer's station; and the great circle in the heavens which marks their vanishing line, is also spoken of as a circle of the sphere, under the name of the *celestial horizon*, or simply *the* horizon.

From what has been said (art. 72.) of the distance of the stars, it follows, that if we suppose a spectator at the centre of the earth to have his view bounded by the *rational* horizon, in the same manner as that of a corresponding spectator on the surface is by his *sensible* horizon, the two observers will see the same stars in the same relative situations, each beholding that entire hemisphere of the heavens which is above the celestial horizon, corresponding to their common zenith.

(75.) Now, so far as appearances go, it is clearly the same thing whether the heavens, that is, all space, with its contents, revolve round a spectator at rest in the earth's centre, or whether that spectator simply turn round in the opposite direction in his place, and view them in succession. The aspect of the heavens, at every instant, as referred to his horizon (which must be supposed to turn with him), will be the same in both suppositions. And since, as has been shown, appearances are also, so far as the stars are concerned, the same to a spectator on the surface as to one at the centre, it follows that, whether we suppose the heavens to revolve without the earth, or the earth within the heavens, *in the opposite direction*, the diurnal phenomena, to all its inhabitants, will be no way different.

(76.) The Copernican astronomy adopts the latter as the true explanation of these phenomena, avoiding thereby the necessity of otherwise resorting to the cumbrous mechanism of a solid but invisible sphere, to which the stars must be supposed attached, in order that they may be carried round the earth without derangement of their relative situations *inter se*. Such a contrivance would, indeed, suffice to explain the diurnal revolution of the stars, so as to "save appearances;" but the movements of the sun and moon, as well as those of the planets, are incompatible with such a supposition, as will appear when we come to treat of these bodies. On the other hand, that a spherical mass of moderate dimensions, (or, rather, when compared with the surrounding and visible universe, of evanescent magnitude,) held by no tie, and free to move and to revolve, should do so, in conformity with those general laws which, so far as we know, regulate the motions of all material bodies, is so far from being a postulate difficult to be conceded, that the wonder would rather be should the fact prove otherwise. As a postulate, therefore, we shall henceforth regard it; and as, in the progress of our work, analogies offer themselves in its support from what we observe of other celestial bodies, we shall not fail to point them out to the reader's notice. Meanwhile, it will be proper to define a variety of terms which will be continually employed hereafter.

(77.) DEFINITION 1. The *axis* of the earth is that diameter about which it revolves, with a uniform motion, *from west to east;* performing one revolution in the interval which elapses between any star leaving a certain point in the heavens, and returning to the same point again.

(78.) DEF. 2. The *poles* of the earth are the points where its axis meets its surface. The North Pole is that nearest to Europe; the South Pole that most remote from it.

(79.) DEF. 3. The sphere of the heavens, or the sphere of the stars, is an imaginary spherical surface, of infinite radius, and having the centre of the earth, or, which comes to the very same thing, the eye of any spectator on its surface, for its centre. Every point in this sphere may be regarded as the vanishing point of a system of lines parallel to that radius of the sphere which passes through it, seen in perspective from the earth; and any great circle on it, as the vanishing line of a system of planes parallel to its own. This mode of conceiving such points and circles has great advantages in a variety of cases.

(80.) DEF. 4. The *zenith* and *nadir* * are the two points of the sphere of the heavens, vertically over a spectator's head, and vertically under his feet; they are, therefore, the vanishing points of all lines *mathematically* parallel to the direction of a plumb-line at his station. The plumb-line itself is, at every point of the earth, perpendicular to its spherical surface: at no two stations, therefore, can the actual directions of

* From Arabic words. Nadir corresponds evidently to the German *nieder* (down).

two plumb-lines be regarded as mathematically parallel. They converge towards the centre of the earth: but for very small intervals (as in the area of a building — in one and the same town, &c.) the difference from exact parallelism is so small, that it may be practically disregarded. An interval of a mile corresponds to a convergence of plumb-lines amounting to about 1 minute. The zenith and nadir are the *poles* of the celestial horizon; that is to say, points 90° distant from every point in it. The *celestial horizon* itself is the vanishing line of a system of planes parallel to the sensible and rational horizon.

(81.) Def. 5. *Vertical circles* of the sphere are great circles passing through the zenith and nadir, or great circles perpendicular to the horizon. On these are measured the *altitudes* of objects above the horizon — the complements to which are their *zenith distances.*

(82.) Def. 6. The poles of the heavens are the points of the sphere to which the earth's axis is directed; or the vanishing points of all lines parallel thereto.

(83.) Def. 7. The *earth's equator* is a great circle on its surface, equidistant from its poles, dividing it into two hemispheres — a northern and a southern; in the midst of which are situated the respective poles of the earth of those names. The *plane* of the equator is, therefore, a plane perpendicular to the earth's axis, and passing through its centre. The *celestial equator* is a great circle of the heavens, marked out by the indefinite extension of the plane of the terrestrial, and is the vanishing line of all planes parallel to it. This circle is called by astronomers the *equinoctial.*

(84.) Def. 8. The terrestrial *meridian* of a station on the earth's surface is a great circle passing through both the poles and through the place. When its plane is prolonged to the sphere of the heavens, it marks out the *celestial meridian* of a spectator stationed at that place. When we speak of the meridian of a spectator, we intend the celestial meridian, which is a vertical circle passing through the poles of the heavens.

The *plane of the meridian* is the plane of this circle, and its intersection with the sensible horizon of the spectator is called a *meridian line,* and marks the north and south points of his horizon.

(85.) Def. 9. *Azimuth* is the angular distance of a celestial object from the north or south point of the horizon (according as it is the north or south pole which is *elevated*), when the object is referred to the horizon by a vertical circle; or it is the angle comprised between two vertical planes — one passing through the elevated pole, the other through the object. The *altitude* and azimuth of an object being known, therefore, its place in the visible heavens is determined. For their simultaneous measurement, a peculiar instrument has been imagined, called an altitude and azimuth instrument, which will be described in the next chapter.

(86.) Def. 10. The *latitude* of a place on the earth's surface is its angular distance from the equator, measured on its own terrestrial meridian: it is reckoned in degrees, minutes, and seconds, from 0 up to 90°, and northwards or southwards according to the hemisphere the place lies in. Thus, the observatory at Greenwich is situated in 51° 28′ 40″ north latitude. This definition of latitude, it will be observed, is to be considered as only temporary. A more exact knowledge of the physical structure and figure of the earth, and a better acquaintance with the niceties of astronomy, will render some modification of its terms, or a different manner of considering it, necessary.

(87.) Def. 11. Parallels of latitude are small circles on the earth's surface parallel to the equator. Every point in such a circle has the same latitude. Thus, Greenwich is said to be situated *in the parallel of* 51° 28′ 40″.

(88.) Def. 12. The *longitude* of a place on the earth's surface is the inclination of its meridian to that of some fixed station referred to as a point to reckon from. English astronomers and geographers use the observatory at Greenwich for this station; foreigners, the principal observatories of their respective nations. Some geographers have adopted the island of Ferro. Hereafter, when we speak of longitude, we reckon from Greenwich. The longitude of a place is, therefore, measured by the arc of the equator intercepted between the meridian of the place and that of Greenwich; or, which is the same thing, by the spherical angle at the pole included between these meridians.

As *latitude* is reckoned north or south, so *longitude* is usually said to be reckoned west or east. It would add greatly, however, to systematic regularity, and tend much to avoid confusion and ambiguity in computations, were this mode of expression abandoned, and longitudes reckoned invariably *westward* from their origin round the whole circle from 0 to 360°. Thus, the longitude of Paris is, in common parlance, either 2° 20′ 22″ east, or 357° 39′ 38″ west of Greenwich. But, in the sense in which we shall henceforth use and recommend others to use the term, the latter is its proper designation. Longitude is also reckoned in time at the rate of 24 h. for 360°, or 15° per hour. In this system the longitude of Paris is 23h. 50m. 38½s.

(89.) Knowing the longitude and latitude of a place, it may be laid down on an artificial globe; and thus a map of the earth may be constructed. Maps of particular countries are detached portions of this general map, extended into planes; or, rather, they are representations on planes of such portions, executed according to certain conventional systems of rules, called *projections*, the object of which is either to distort as little as possible the outlines of countries from what they are on the globe — or to establish easy means of ascertaining, by inspection or graphical measurement, the latitudes and longitudes of places which occur in them, without referring to the globe or to books — or for other peculiar uses. See Chap. III.

(90.) A globe, or general map of the heavens, as well as charts of particular parts, may also be constructed, and the stars laid down in their proper situations relative to each other, and to the poles of the heavens and the celestial equator. Such a representation, once made, will exhibit a true appearance of the stars as they present themselves in succession to every spectator on the surface, or as they may be conceived to be seen at once by one at the centre of the globe. It is, therefore, independent of all *geographical* localities. There will occur in such a representation neither zenith, nadir, nor horizon—neither east nor west points; and although great circles may be drawn on it from pole to pole, corresponding to terrestrial meridians, they can no longer, in this point of view, be regarded as the celestial meridians of fixed points on the earth's surface, since, in the course of one diurnal revolution, every point in it passes beneath each of them. It is on account of this change of conception, and with a view to establish a complete distinction between the two branches of *Geography* and *Uranography**, that astronomers have adopted different terms (viz. *declination*, and *right ascension*) to represent those arcs in the heavens which correspond to *latitudes* and *longitudes* on the earth. It is for this reason that they term the equator of the heavens the *equinoctial;* that what are meridians on the earth are called *hour circles* in the heavens, and the angles they include between them at the poles are called *hour angles*. All this is convenient and intelligible; and had they been content with this nomenclature, no confusion could ever have arisen. Unluckily, the early astronomers have employed *also* the words latitude and longitude in their uranography, in speaking of arcs of circles not corresponding to those meant by the same words on the earth, but having reference to the motion of the sun and planets among the stars. It is now too late to remedy this confusion, which is ingrafted into every existing work on astronomy: we can only regret, and warn the reader of it, that he may be on his guard when, at a more advanced period of our work, we shall have occasion to

* Γη, the earth; γραφειν, to describe or represent: ουρανος, the heavens.

define and use the terms in their *celestial sense,* at the same time urgently recommending to future writers the adoption of others in their places.

(91.) As terrestrial longitudes reckon from an assumed fixed meridian, or from a determinate point on the equator; so right ascensions in the heavens require some determinate hour circle, or some known point in the equinoctial, as the commencement of their reckoning, or their *zero point.* The hour circle passing through some remarkably bright star might have been chosen; but there would have been no particular advantage in this; and astronomers have adopted, in preference, a point in the *equinoctial,* called the *equinox.* through which they suppose the hour circle to pass, from which all others are reckoned, and which point is itself the zero point of all right ascensions, counted on the equinoctial.

The right ascensions of celestial objects are always reckoned *eastward* from the equinox, and are estimated either in degrees, minutes, and seconds, as in the case of terrestrial longitudes, from 0° to 360°, which completes the circle; or, in time, in hours, minutes, and seconds, from 0h. to 24h. The apparent diurnal motion of the heavens being contrary to the real motion of the earth, this is in conformity with the westward reckoning of longitudes. (Art 87.)

(92.) *Sidereal time* is reckoned by the diurnal motion of the stars, or rather of that point in the equinoctial from which right ascensions are reckoned. This point may be considered as a star, though no star is, in fact, there; and, moreover, the point itself is liable to a certain slow variation,—so slow, however, as not to affect, perceptibly, the interval of any two of its successive returns to the meridian. This interval is called a sidereal day, and is divided into 24 sidereal hours, and these again into minutes and seconds. A clock which marks sidereal time, *i. e.* which goes uniformly at such a rate as always to show 0h. 0m. 0s. when the equinox comes on the meridian, is called sidereal clock, and is an indispensable piece of furniture in every observatory.

(93.) It remains to illustrate these descriptions by reference to a figure. Let C be the centre of the

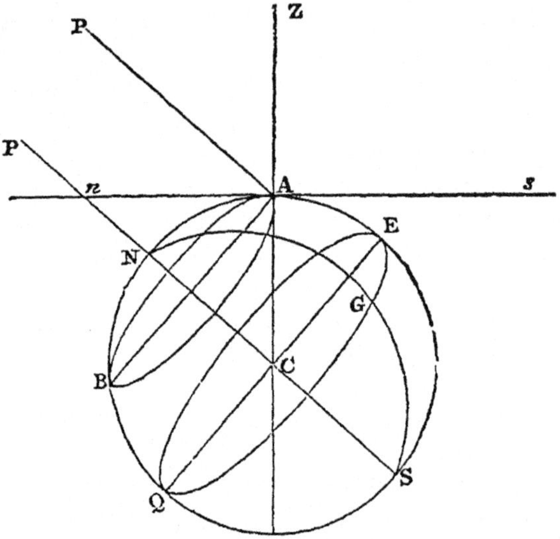

earth, N C S its axis; then are N and S its *poles*; E Q its *equator*; A B the *parallel* of latitude of the station A on its surface; A P parallel to S C N, the direction in which an observer at A will see the *elevated* pole of the heavens; and A Z, the prolongation of the terrestrial radius C A, that of his zenith. N A E S will be his meridian; N G S that of some fixed station, as Greenwich; and G E, or the spherical angle G N E, his longitude, and E A his latitude. Moreover, if *n s* be a plane touching the surface in A, this will be his sensible horizon; *n* A *s* marked on that plane by its intersection with his meridian will be his meridian line, and *n* and *s* the north and south points of his horizon.

(94.) Again, neglecting the size of the earth, or conceiving him stationed at its centre, and referring every thing to his *rational* horizon; let the annexed figure represent the sphere of the *heavens*; C the spectator; Z his zenith; and N his nadir: then will

H A O a great circle of the sphere, whose poles are Z N, be his *celestial horizon;* P *p* the *elevated* and

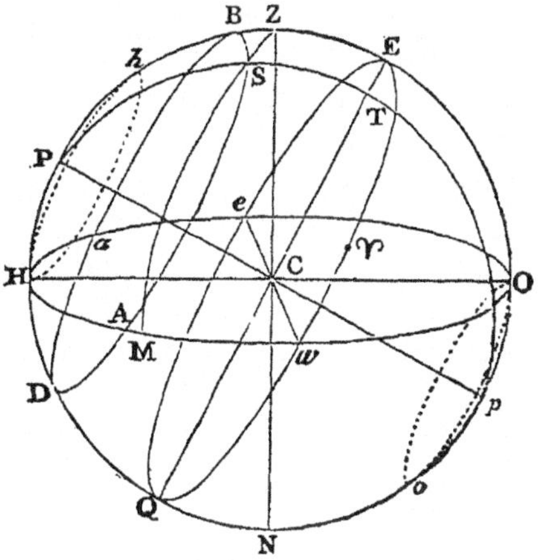

depressed POLES of the heavens; H P the *altitude of the pole,* and H P Z E O his *meridian;* E T Q, a great circle perpendicular to P *p*, will be the *equinoctial;* and if ♈ represent the equinox, ♈ T will be the *right ascension,* T S the *declination,* and P S the *polar distance* of any star or object S, referred to the equinoctial by the *hour circle* P S T *p;* and B S D will be the diurnal circle it will appear to describe about the pole. Again, if we refer it to the horizon by the *vertical circle* Z S A, H A will be its azimuth, A S its altitude, and Z S its zenith distance. H and O are the north and south, and *e w* the east and west points of his horizon, or of the heavens. Moreover, if H *h*, O *o*, be small circles, or *parallels of declination,* touching the horizon in its north and south points, H *h* will be the circle of *perpetual apparition,* between which and the elevated pole the stars never set; O *o* that of *perpetual occultation,* between which and the depressed pole they *never rise.* In all the zone of the heavens between H *h* and O *o*, they rise and set, any one of them, as S, remaining above the horizon, in that part of its diurnal circle represented by A B *a*, and below it throughout all the part represented by A D *a*. It will exercise the reader to construct this figure for several different *elevations of the pole*, and for a variety of positions of the star S in each. The following consequences result from these definitions, and are propositions which the reader will readily bear in mind: —

(95.) The altitude of the elevated pole is equal to the latitude of the spectator's geographical station. For, comparing the figures of arts. 93. and 94., it appears that the angle P A Z, between the pole and zenith, in the one figure, which is the *co-altitude* (complement to 90° of the altitude) of the pole, is equal to the angle N C A in the other; C N and A P being parallels whose *vanishing point* is the pole. Now, N C A is the *co-latitude* of the plane A.

(96.) The same stars, in their diurnal revolution, come to the meridian, *successively*, of every place on the globe once in twenty-four sidereal hours. And, since the diurnal rotation is uniform, the interval, in sidereal time, which elapses between the same star coming upon the meridians of two different places is measured by the difference of longitudes of the places.

(97.) *Vice versâ*—the interval elapsing between two *different stars* coming on the meridian of *one and the same place*, expressed in sidereal time, is the measure of the difference of right ascensions of the stars.

This explains the reason of the double division of the equator and equinoctial into *degrees and hours*.

(98.) The equinoctial intersects the horizon in the east and west points, and the meridian in a point whose altitude is equal to the co-latitude of the place. Thus, at Greenwich, the altitude of the intersection of the equinoctial and meridian is 38° 31′ 20″.

(99.) All the heavenly bodies *culminate* (*i. e.* come to their greatest altitudes) on the meridian; which is, therefore, the best situation to observe them, being least confused by the inequalities and vapours of the atmosphere, as well as least displaced by refraction.

(100.) All celestial objects within the circle of perpetual apparition come twice on the meridian, above the horizon, in every diurnal revolution; once *above* and once *below* the pole. These are called their *upper* and *lower culminations*.

(101.) We shall conclude this chapter by calling the reader's attention to a fact, which, if he now learn it for the first time, will not fail to surprise him, viz. that the stars continue visible through telescopes during the day as well as the night; and that, in proportion to the power of the instrument, not only the largest and brightest of them, but even those of inferior lustre, such as scarcely strike the eye at night as at all conspicuous, are readily found and followed even at noonday,—unless in that part of the sky which is very near the sun,—by those who possess the means of pointing a telescope accurately to the proper places. Indeed, from the bottoms of deep narrow pits, such as a well, or the shaft of a mine, such bright stars as pass the zenith may even be discerned by the naked eye; and we have ourselves heard it stated by a celebrated optician, that the earliest circumstance which drew his attention to astronomy was the regular appearance, at a certain hour, for several successive days, of a considerable star, through the shaft of a chimney.

CHAP. II.

OF THE NATURE OF ASTRONOMICAL INSTRUMENTS AND OBSERVATIONS IN GENERAL. — OF SIDEREAL AND SOLAR TIME. — OF THE MEASUREMENT OF TIME. — CLOCKS, CHRONOMETERS, THE TRANSIT INSTRUMENT. — OF THE MEASUREMENT OF ANGULAR INTERVALS. — APPLICATION OF THE TELESCOPE TO INSTRUMENTS DESTINED TO THAT PURPOSE. — OF THE MURAL CIRCLE. — FIXATION OF POLAR AND HORIZONTAL POINTS. — THE LEVEL. — PLUMB LINE. — ARTIFICIAL HORIZON. — COLLIMATOR. — OF COMPOUND INSTRUMENTS WITH CO-ORDINATE CIRCLES, THE EQUATORIAL. — ALTITUDE AND AZIMUTH INSTRUMENT. — OF THE SEXTANT AND REFLECTING CIRCLE. — PRINCIPLE OF REPETITION.

(102.) OUR first chapter has been devoted to the acquisition chiefly of preliminary notions respecting the globe we inhabit, its relation to the celestial objects which surround it, and the physical circumstances under which all astronomical observations must be made, as well as to provide ourselves with a stock of *technical words* of most frequent and familiar use in the sequel. We might now proceed to a more exact and detailed statement of the facts and theories of astronomy; but, in order to do this with full effect, it will be desirable that the reader be made acquainted with the principal means which astronomers possess, of determining, with the degree of nicety their theories require, the data on which they ground their conclusions; in other words, of ascertaining by measurement the apparent and real magnitudes with which they are conversant. It is only when in possession of this knowledge that he can fully appreciate either the truth of the theories themselves, or the degree of reliance to be placed on any of their conclusions antecedent to trial: since it is only by knowing what amount of error can certainly be perceived and distinctly measured, that he can satisfy himself whether any theory offers so close an approximation, in its numerical results, to actual phenomena, as will justify him in receiving it as a true representation of nature.

(103.) Astronomical instrument-making may be justly regarded as the most refined of the mechanical arts, and that in which the nearest approach to geometrical precision is required, and has been attained. It may be thought an easy thing, by one unacquainted with the niceties required, to turn a circle in metal, to divide its circumference into 360 equal parts, and these again into smaller subdivisions, — to place it accurately on its centre, and to adjust it in a given position; but practically it is found to be one of the most difficult. Nor will this appear extraordinary, when it is considered that, owing to the application of telescopes to the purposes of angular measurement, every imperfection of structure or division becomes magnified by the whole optical power of that instrument; and that thus, not only direct errors of workmanship, arising from unsteadiness of hand or imperfection of tools, but those inaccuracies which originate in far more uncontrollable causes, such as the unequal expansion and contraction of metallic masses, by a change of temperature, and their unavoidable flexure or bending by their own weight, become perceptible and measurable. An angle of one minute occupies, on the circumference of a circle of 10 inches in radius, only about $\frac{1}{350}$th part of an inch, a quantity too small to be *certainly* dealt with without the use of magnifying glasses; yet one minute is a gross quantity in the astronomical measurement of an angle. With the instruments now employed in observatories, a single second, or the 60th part of a minute, is rendered a distinctly visible and appreciable quantity. Now, the arc of a circle, subtended by one second, is less than the 200,000th part of the radius, so that on a circle of 6 feet in diameter it would occupy no greater linear extent than $\frac{1}{5700}$th part of an inch; a quantity requiring a powerful microscope to be *discerned* at all. Let any one figure to himself, therefore, the difficulty of placing on the circumference of a metallic circle of such dimensions (supposing the difficulty of its construction surmounted), 360 marks, dots, or cognizable divisions, which shall be true to their places within such minute limits; to say nothing of the subdivision of the degrees so marked off into minutes, and of these again into seconds. Such a work has probably baffled, and will probably for ever continue to baffle, the utmost stretch of human skill and industry; nor, if executed, could it endure. The ever varying fluctuations of heat and cold have a tendency to produce not merely temporary and transient, but permanent, uncompensated changes of form in all considerable masses of those metals which alone are applicable to such uses; and their own weight, however symmetrically formed, must always be unequally sustained, since it is impossible to apply the sustaining power to *every part* separately: even could this be done, at all events force must be used to move and to fix them; which can never be done without producing temporary and risking permanent change of form. It is true, by dividing them on their centres, and in the identical places they are destined to occupy, and by a thousand ingenious and delicate contrivances, wonders have been accomplished in this department of art, and a degree of perfection has been given, not merely to *chefs d'œuvre*, but to instruments of moderate prices and dimensions, and in ordinary use, which, on due consideration, must appear very surprising. But though we are entitled to look for *wonders* at the hands of scientific artists, we are not to expect *miracles*. The demands of the astronomer will always surpass the power of the artist; and it must, therefore, be constantly the aim of the former to make himself, as far as possible, independent of the imperfections incident to every work the latter can place in his hands. He must, therefore, endeavour so to combine his observations, so to choose his opportunities, and so to familiarize himself with all the causes which may produce instrumental derangement, and with all the peculiarities of structure and material of each instrument he possesses, as not to allow himself to be misled by their errors, but to extract from their indications, as far as possible, all that is *true*, and reject all that is erroneous. It is in this that the art of the practical astronomer consists,—an art of itself of a curious and intricate nature, and of which we can here only notice some of the leading and general features.

(104.) The great aim of the practical astronomer being numerical correctness in the results of instrumental measurement, his constant care and vigilance must be directed to the detection and compensation of errors, either by annihilating, or by taking account of, and allowing for them. Now, if we examine the sources from which errors may arise in any instrumental determination, we shall find them chiefly reducible to three principal heads:—

(105.) 1st, External or incidental causes of error; comprehending such as depend on external, uncontrollable circumstances: such as, fluctuations of weather, which disturb the amount of refraction from its tabulated value, and, being reducible to no fixed law, induce uncertainty to the extent of their own possible magnitude; such as, by varying the temperature of the air, vary also the form and position of the instruments used, by altering relative magnitude and the tension of their parts; and others of the like nature.

(106.) 2dly, *Errors of observation:* such as arise, for example, from inexpertness, defective vision, slowness in seizing the exact *instant* of occurrence of a phenomenon, or precipitancy in anticipating it, &c.; from atmospheric indistinctness; insufficient optical power in the instrument, and the like. Under this head may also be classed all errors arising from momentary instrumental derangement,—slips in clamping, looseness of screws, &c.

(107.) 3dly, The third, and by far the most numerous class of errors to which astronomical measurements are liable, arise from causes which may be deemed instrumental, and which may be subdivided into two principal classes. The *first* comprehends those which arise from an instrument not *being* what it professes to be, which is *error* of *workmanship*. Thus, if a pivot or axis, instead of being, as it ought, exactly cylindrical, be slightly flattened, or elliptical, — if it be not exactly (as it is intended it should) concentric with the circle it carries; — if this circle (so called) be in reality *not* exactly circular, or not in one plane; — if its divisions, intended to be precisely equidistant, should be placed in reality at unequal intervals, — and a hundred other things of the same sort. These are not mere speculative sources of error, but practical annoyances, which every observer has to contend with.

(108.) The *other* subdivision of instrumental errors comprehends such as arise from an instrument not being placed in the *position* it ought to have; and from those of its parts, which are made purposely moveable, not being properly disposed *inter se*. These are *errors of adjustment*. Some are unavoidable, as they arise from a general unsteadiness of the soil or building in which the instruments are placed; which, though too minute to be noticed in any other way, become appreciable in delicate astronomical observations: others, again, are consequences of imperfect workmanship, as where an instrument once well adjusted will not remain so, but keeps deviating and shifting. But the most important of this class of errors arise from the non-existence of natural indications, *other* than those afforded by astronomical observations themselves, whether an instrument has or has not the exact position, with respect to the horizon and its cardinal points, the axis of the earth, or to other principal astronomical lines and circles, which it ought to have to fulfil properly its objects.

(109.) Now, with respect to the first two classes of error, it must be observed, that, in so far as they cannot be reduced to known laws, and thereby become subjects of calculation and due allowance, they actually vitiate, to their full extent, the results of any observations in which they subsist. Being, however, in their nature casual and accidental, their effects necessarily lie sometimes one way, sometimes the other; sometimes diminishing, sometimes tending to increase the results. Hence, by greatly multiplying observations, under varied circumstances, and taking the *mean or average* of their results, this class of errors may be so far *subdued*, by setting them to destroy one another, as no longer sensibly to vitiate any theoretical or practical conclusion. This is the great and indeed only resource against such errors, not merely to the astronomer, but to the investigator of numerical results in every department of physical research.

(110.) With regard to errors of adjustment and workmanship, not only the *possibility*, but the *certainty*, of their existence, in every imaginable form, in all instruments, must be contemplated. Human hands or machines never formed a circle, drew a straight line, or erected a perpendicular, nor ever placed an instrument in *perfect* adjustment, unless accidentally; and then only during an instant of time. This does not prevent, however, that a great approximation to all these desiderata should be attained. But it is the peculiarity of astronomical observation to be the *ultimate means of detection* of all mechanical defects which elude by their minuteness every other mode of detection. What the eye cannot discern, nor the touch perceive, a course of astronomical observations will make distinctly evident. The imperfect products of man's hands are here tested by being brought into comparison with the perfect workmanship of nature; and there is none which will bear the trial. Now, it may seem like arguing in a vicious circle, to deduce theoretical conclusions and laws from observation, and then to turn round upon the instruments with which those observations were made, accuse them of imperfection, and attempt to detect and rectify their errors by means of the very laws and theories which they have helped us to a knowledge of. A little consideration, however, will suffice to show that such a course of proceeding is perfectly legitimate.

(111.) The steps by which we arrive at the laws of natural phenomena, and especially those which depend for their verification on numerical determinations, are necessarily successive. Gross results and palpable laws are arrived at by rude observation with coarse instruments, or without any instruments at all; and these are corrected and refined upon by nicer scrutiny with more delicate means. In the progress of this, subordinate laws are brought into view, which modify both the verbal statement and numerical results of those which first offered themselves to our notice; and when these are traced out, and reduced to certainty, others, again, subordinate to them, make their appearance, and become subjects of further enquiry. Now, it invariably happens (and the reason is evident) that the first glimpse we catch of such subordinate laws — the first form in which they are dimly shadowed out to our minds — is that of *errors*. We perceive a discordance between what we *expect*, and what we *find*. The first occurrence of such a discordance we attribute to accident. It happens again and again; and we begin to suspect our instruments. We then enquire, to what amount of error their determinations can, *by possibility*, be liable. If their *limit of possible error* exceed the observed deviation, we at once condemn the instrument, and set about improving its construction or adjustments. Still the same deviations occur, and, so far from being palliated, are more marked and better defined than before. We are now sure that we are on the traces of a law of nature, and we pursue it till we have reduced it to a definite statement, and verified it by repeated observation, under every variety of circumstances.

(112.) Now, in the course of this enquiry, it will not fail to happen that other discordances will strike us. Taught by experience, we suspect the existence of some natural law, before unknown; we tabulate (*i. e.* draw out in order) the results of our observations; and we perceive, in this synoptic statement of them, distinct indications of a regular progression. Again we improve or vary our instruments, and we now lose sight of this supposed new law of nature altogether, or find it replaced by some other, of a totally different character. Thus we are led to suspect an instrumental cause for what we have noticed. We examine, therefore, the *theory* of our instrument; we suppose defects in its structure, and, by the aid of geometry, we trace their influence in introducing *actual errors* into its indications. These errors have *their laws*, which, so long as we have no knowledge of causes to guide us, may be confounded with laws of nature, and are mixed up with them in their effects. They are not fortuitous, like errors of observation, but, as they arise from sources inherent in the instrument, and unchangeable while it and its adjustments remain unchanged, they are reducible to fixed and ascertainable forms; each particular defect, whether of structure or adjustment, producing its own appropriate *form* of error. When these are thoroughly investigated, we recognize among them one which coincides in its nature and progression with that of our observed discordances. The mystery is at once solved: we have detected, by direct observation, an instrumental defect.

(113.) It is, therefore, a chief requisite for the practical astronomer to make himself completely familiar with the *theory* of his instruments, so as to be able at once to decide what *effect* on his observations any given imperfection of structure or adjustment will produce in any given circumstances under which an observation can be made. Suppose, for example, that the principle of an instrument required that a circle should be exactly concentric with the axis on which it is made to turn. As this is a condition which no workmanship can fulfil, it becomes necessary to enquire what errors will be produced in observations made and registered on the faith of such an instrument, by any assigned deviation in this respect; that is to say, what would be the disagreement between observations made with it and with one absolutely perfect, could such be obtained. Now, a simple theorem in geometry shows that, whatever be the extent of this deviation, it may be annihilated in its effect on the result of observations depending on the graduation of the limb, by the very easy method of reading off the divisions on two diametrically opposite points of the circle, and taking a mean; for the effect of excentricity is always to increase one such reading by just the same quantity by which it diminishes the other. Again, suppose that the proper use of the instrument required that this axis should be exactly parallel to that of the earth. As it never can be *placed* or remain so, it becomes a question, what amount of error will arise in its use from any assigned deviation, whether in a horizontal or vertical plane, from this precise position. Such enquiries constitute the theory of instrumental errors; a theory of the utmost importance to practice, and one of which a complete knowledge will enable an observer, with very moderate instrumental means, to attain a degree of precision which might seem to belong only to the most refined and costly. In the present work, however, we have no further concern with it. The few astronomical instruments we propose to describe in this chapter will be considered as perfect both in construction and adjustment.

(114.) As the above remarks are very essential to a right understanding of the philosophy of our subject and the spirit of astronomical methods, we shall elucidate them by taking a case. Observant persons, before the invention of astronomical instruments, had already concluded the apparent diurnal motions of the stars to be performed in circles about fixed poles in the heavens, as shown in the foregoing chapter. In drawing this conclusion, however, refraction was entirely overlooked, or, if forced on their notice by its great magnitude in the immediate neighbourhood of the horizon, was regarded as a local irregularity, and, as such neglected, or slurred over. As soon, however, as the diurnal paths of the stars were attempted to be traced by instruments, even of the coarsest kind, it became evident that the notion of exact circles described about one and the same pole would not represent the phenomena correctly, but that, owing to some cause or other, the apparent diurnal orbit of every star is distorted from a circular into an oval form, its lower segment being *flatter* than its upper; and the deviation being greater the nearer the star approached the horizon, the effect being the same as if the circle had been squeezed upwards from below, and the lower parts more than the higher. For such an effect, as it was soon found to arise from no casual or instrumental cause, it became necessary to seek a natural one; and refraction readily occurred, to solve the difficulty. In fact, it is a case precisely analogous to what we have already (art. 47.) noticed, of the apparent distortion of the sun near the horizon, only on a larger scale, and traced up to greater altitudes. This new law once established, it became necessary to modify the expression of that anciently received, by inserting in it a *salvo* for the effect of refraction, or by making a distinction between the *apparent* diurnal orbits, as affected by refraction, and the *true* ones cleared of that effect.

(115.) Again: The first impression produced by a view of the diurnal movement of the heavens is, that *all* the heavenly bodies perform this revolution in one common period, viz. *a day*, or 24 hours. But no sooner do we come to examine the matter *instrumentally*, i. e. by noting, by timekeepers, their successive arrivals on the meridian, than we find differences which cannot be accounted for by any error of observation. All the *stars*, it is true, occupy the same interval of time between their successive appulses to the meridian, or to any vertical circle; but *this* is a very different one from that occupied by the sun. It is palpably shorter; being, in fact, only $23^h\ 56'\ 4{\cdot}09''$, instead of 24 hours, such hours as our common clocks mark. Here, then, we have already *two different days*, a *sidereal* and a *solar ;* and if, instead of the sun, we observe the moon, we find a third, much longer than either, — a *lunar* day, whose average duration is $24^h\ 54^m$ of our ordinary time, which last is *solar* time, being of necessity conformable to the *sun's* successive re-appearances, on which all the business of life depends.

(116.) Now, all the stars are found to be unanimous in giving the same exact duration of $23^h\ 56'\ 4''\cdot09$, for the *sidereal day ;* which, therefore, we cannot hesitate to receive as the period in which the earth makes one revolution on its axis. We are, therefore, compelled to look on the sun and moon as exceptions to the general law ; as having a different nature, or at least a different relation to us, from the stars ; and as having motions, real or apparent, of their own, independent of the rotation of the earth on its axis. Thus a great and most important distinction is disclosed to us.

(117.) To establish these facts, almost no apparatus is required. An observer need only station himself to the north of some well-defined vertical object, as the angle of a building, and, placing his eye exactly at a certain fixed point (such as a small hole in a plate of metal nailed to some immoveable support), notice the successive disappearances of any star behind the building, by a watch.* When he observes the sun, he must shade his eye with a dark-coloured or smoked glass, and notice the moments when its western and eastern edges successively come up to the wall, from which, by taking half the interval, he will ascertain (what he cannot directly *observe*) the moment of disappearance of its centre.

(118.) When, in pursuing and establishing this general fact, we are led to attend more nicely to the times of the daily arrival of the sun on the meridian,

* This is an excellent practical method of ascertaining the rate of a clock or watch, being exceedingly accurate if a few precautions are attended to ; the chief of which is, to take care that that part of the edge behind which the star (a bright one, *not a planet*) disappears shall be quite smooth ; as otherwise variable refraction may transfer the point of disappearance from a protuberance to a notch and thus vary the moment of observation unduly : this is easily secured, by nailing up a smooth-edged board.

irregularities (so they first seem) begin to be observed. The intervals between two successive arrivals are not the same at all times of the year. They are sometimes greater, sometimes less, than 24 hours, as shown by the clock; that is to say, the *solar day* is not always of the same length. About the 21st of December, for example, it is half a minute *longer*, and about the same day of September nearly as much *shorter*, than its *average duration*. And thus a distinction is again pressed upon our notice between the *actual* solar day, which is never two days in succession alike; and the *mean solar day* of 24 hours, which is an average of all the solar days throughout the year. Here, then, a new source of enquiry opens upon us. The sun's apparent motion is not only not the same with that of the stars, but it is not (as the latter is) uniform. It is subject to fluctuations, whose laws become matter of investigation. But to pursue these laws, we require nicer means of observation than what we have described, and are obliged to call in to our aid an instrument called the *transit instrument*, especially destined for such observations, and to attend minutely to all the causes of irregularity in the going of clocks and watches which may affect our reckoning of time. Thus we become involved by degrees in more and more delicate instrumental enquiries; and we speedily find that, in proportion as we ascertain the amount and law of one great or leading fluctuation, or inequality, as it is called, of the sun's diurnal motion, we bring into view others continually smaller and smaller, which were before obscured, or mixed up with errors of observation and instrumental imperfections. In short, we may not inaptly compare the *mean* length of the solar day to the mean or average height of water in a harbour, or the general level of the sea unagitated by tide or waves. The great annual fluctuation above noticed may be compared to the daily variations of level produced by the tides, which are nothing but enormous waves extending over the whole ocean, while the smaller subordinate inequalities may be assimilated to waves ordinarily so called, on which, when large, we perceive lesser undulations to ride, and on these, again, minuter ripplings, to the series of whose subordination we can perceive no end.

(119.) With the causes of these irregularities in the solar motion we have no concern at present; their explanation belongs to a more advanced part of our subject: but the distinction between the solar and sidereal days, as it pervades every part of astronomy, requires to be early introduced, and never lost sight of. It is, as already observed, the *mean* or *average* length of the solar day, which is used in the civil reckoning of time. It commences at midnight, but astronomers (at least those of this country), even when they use mean solar time, depart from the civil reckoning, commencing their day at noon, and reckoning the hours from 0 round to 24. Thus, 11 o'clock in the forenoon of the second of January, in the civil reckoning of time, corresponds to January 1 day 23 hours in the astronomical reckoning; and one o'clock in the afternoon of the former, to January 2 days 1 hour of the latter reckoning. This usage has its advantages and disadvantages, but the latter seem to preponderate; and it would be well if, in consequence, it could be broken through, and the civil reckoning substituted.

(120.) Both astronomers and civilians, however, who inhabit different points of the earth's surface, differ from each other in their reckoning of time; as it is obvious they must, if we consider that, when it is noon at one place, it is midnight at a place diametrically opposite; sunrise at another; and sunset, again, at a fourth. Hence arises considerable inconvenience, especially as respects places differing very widely in situation, and which may even in some critical cases involve the mistake of a whole day. To obviate this inconvenience, there has lately been introduced a system of reckoning time by mean solar days and parts of a day counted from a fixed instant, common to all the world, and determined by no *local* circumstance, such as noon or midnight, but by the motion of the sun among the stars. Time, so reckoned, is called equinoctial time; and is numerically the same, at the same instant, in every part of the globe. Its origin will be explained more fully at a more advanced stage of our work.

(121.) Time is an essential element in astronomical observation, in a twofold point of view:— 1st, As the representative of angular motion. The earth's diurnal motion being uniform, every star describes its diurnal circle uniformly; and the time elapsing between the passage of the stars in succession across the meridian of any observer becomes, therefore, a direct measure of their differences of right ascension. 2dly, As the fundamental element (or, independent variable, to use the language of geometers) in all dynamical theories. The great object of astronomy is the determination of the laws of the celestial motions, and their reference to their proximate or remote causes. Now, the statement of the *law* of any observed motion in a celestial object can be no other than a proposition declaring what has been, is, and will be, the real or apparent situation of that object *at any time*, past, present, or future. To compare such laws, therefore, with observation, we must possess a register of the observed situations of the object in question, and of the *times when* they were observed.

(122.) The measurement of time is performed by clocks, chronometers, clepsydras, and hour-glasses: the two former are alone used in modern astronomy. The hour-glass is a coarse and rude contrivance for measuring, or rather counting out, fixed portions of time, and is entirely disused. The clepsydra, which measured time by the gradual emptying of a large vessel of water through a determinate orifice, is susceptible of considerable exactness, and was the only dependence of astronomers before the invention of clocks and watches. At present it is abandoned, owing to the greater convenience and exactness of the latter instruments. In one case only has the revival of its use been proposed; viz. for the accurate measurement of very small portions of time, by the flowing out of mercury from a small orifice in the bottom of a vessel, kept constantly full to a fixed height. The stream is intercepted at the moment of noting any event, and directed aside into a receiver, into which it continues to run, till the moment of noting any other event, when the intercepting cause is suddenly removed, the stream flows in its original course, and ceases to run into the receiver. The *weight* of mercury received, compared with the weight received in an interval of time observed by the clock, gives the interval between the events observed. This ingenious and simple method of resolving, with all possible precision, a problem which has of late been much agitated, is due to captain Kater.

(123.) The pendulum clock, however, and the balance watch, with those improvements and refinements in its structure which constitute it emphatically a *chronometer* *, are the instruments on which the astronomer depends for his knowledge of the lapse of time. These instruments are now brought to such perfection, that an irregularity in the *rate* of going, to the extent of a single second in twenty-four hours in two consecutive days, is not tolerated in one of good character; so that any interval of time less than twenty-four hours may be certainly ascertained within a few tenths of a second, by their use. In proportion as intervals are longer, the risk of error, as well as the amount of error risked, becomes greater, because the accidental errors of many days may accumulate; and causes producing a slow progressive change in the rate of going may subsist unperceived. It is not safe, therefore, to trust the determination of time to clocks, or watches, for many days in succession, without checking them, and ascertaining their errors by reference to natural events which we know to happen, day after day, at equal intervals. But if this be done, the

* χρονος, time; μετρειν, to measure.

longest intervals may be fixed with the same precision as the shortest; since, in fact, it is then only the times intervening between the first and last moments of such long intervals, and such of those periodically recurring events adopted for our points of reckoning, as occur within twenty-four hours respectively of either, that we measure by artificial means. The whole days are counted out for us by nature; the fractional parts only, at either end, are measured by our clocks. To keep the reckoning of the integer days correct, so that none shall be lost or counted twice, is the object of the calendar. Chronology marks out the order of succession of events, and refers them to their proper years and days; while chronometry, grounding its determinations on the precise observation of such regularly periodical events as can be conveniently and exactly subdivided, enables us to fix the moments in which phenomena occur, with the last degree of precision.

(124.) In the *culmination*, or *transit*, (*i.e.* the passage across the meridian of an observer,) of every star in the heavens, he is furnished with such a regularly periodical natural event as we allude to. Accordingly, it is to the *transits* of the brightest and most conveniently situated fixed stars that astronomers resort to ascertain their exact time, or, which comes to the same thing, to determine the exact amount of error of their clocks.

(125.) The instrument with which the culminations of celestial objects are observed is called a *transit instrument*. It consists of a telescope firmly fastened on a horizontal axis directed to the east and west points of the horizon, or at right angles to the plane of the meridian of the place of observation. The extremities of the axis are formed into cylindrical pivots of exactly equal diameters, which rest in notches formed in metallic supports, bedded (in the case of large instruments) on strong piers of stone, and susceptible of nice adjustment by screws, both in a vertical and horizontal direction. By the former adjustment, the axis can be rendered precisely horizontal, by *level-ling* it with a *level* made to rest on the pivots. By

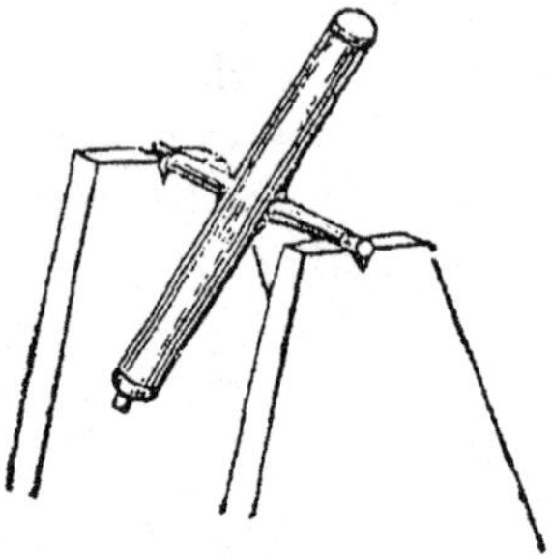

the latter adjustment the axis is brought precisely into the east and west direction, the criterion of which is furnished by the observations themselves made with the instrument, or by a well-defined object, called a *meridian mark*, originally determined by such observations, and then, for convenience of ready reference, permanently established, at a great distance, exactly in a *meridian line* passing through the central point of the whole instrument. It is evident, from this description, that, if the central line of the telescope (that which joins the centres of its object-glass and eye-glass, and which is called in astronomy its *line of collimation*) be once well adjusted at right angles to the axis of the transit, it will never quit the plane of the meridian, when the instrument is turned round on its axis.

(126.) In the focus of the eye-piece, and at right angles to the length of the telescope, is placed a system of one horizontal and five equidistant vertical threads or wires, as represented in the annexed figure, which

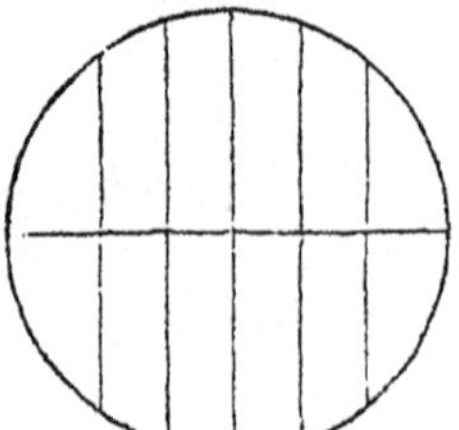

always appear in *the field of view*, when properly illuminated, by day by the light of the sky, by night by that of a lamp introduced by a contrivance not necessary here to explain. The place of this system of wires may be altered by adjusting screws, giving it a lateral (horizontal) motion ; and it is by this means brought to such a position, that the middle one of the vertical wires shall intersect *the line of collimation* of the telescope, where it is arrested and permanently fastened. In this situation it is evident that the middle thread will be a visible representation of that portion of the celestial meridian to which the telescope is pointed; and when a star is seen to cross this wire in the telescope, it is in the act of culminating, or passing the celestial meridian. The instant of this event is noted by the clock or chronometer, which forms an indispensable accompaniment of the transit instrument. For greater precision, the moments of its crossing all the five vertical threads is noted, and a mean taken, which (since the threads are equidistant) would give exactly the same result, were all the observations perfect, and will, of course, tend to subdivide and destroy their errors in an average of the whole.

(127.) For the mode of executing the adjustments, and allowing for the errors unavoidable in the use of this simple and elegant instrument, the reader must consult works especially devoted to this department of practical astronomy.* We shall here only mention one important verification of its correctness, which consists in *reversing* the ends of the axis, or turning it east for west. If this be done, and it continue to give the same results, and intersect the same point on the meridian mark, we may be sure that the line of collimation of the telescope is truly at right angles to the axis, and describes strictly a plane, *i. e.* marks out in the heavens a *great circle*. In good transit observations, an error of two or three tenths of a second of time in the moment

* See Dr. Pearson's Treatise on Practical Astronomy. Also Bianchi Sopra lo Stromento de' Passagi. Ephem. di Milano, 1824.

of a star's culmination is the utmost which need be apprehended, exclusive of the error of the clock: in other words, a clock may be compared with the earth's diurnal motion by a single observation, without risk of greater error. By multiplying observations, of course, a yet greater degree of precision may be obtained.

(128.) The angular intervals measured by means of the transit instrument and clock are arcs of the equinoctial, intercepted between circles of declination passing through the objects observed; and their measurement, in this case, is performed by no artificial graduation of circles, but by the help of the earth's diurnal motion, which carries equal arcs of the equinoctial across the meridian, in equal times, at the rate of 15° per sidereal hour. In all other cases, when we would measure angular intervals, it is necessary to have recourse to circles, or portions of circles, constructed of metal or other firm and durable material, and mechanically subdivided into equal parts, such as degrees, minutes, &c. Let ABCD be such a circle, divided into 360 degrees,

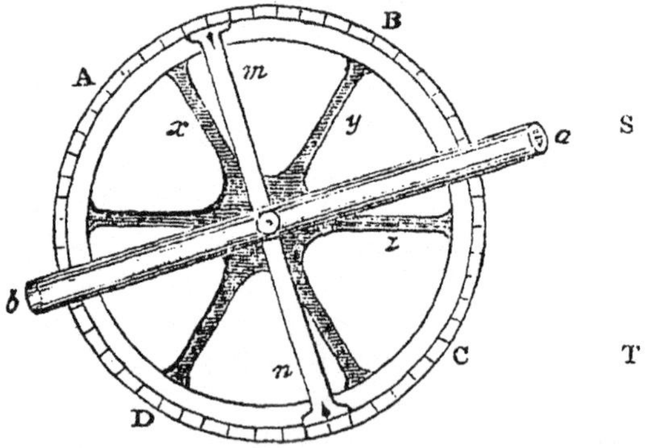

(numbered in order from any point 0° in the circumference, round to the same point again,) and connected with its centre by spokes or rays, *x y z*, firmly united to its circumference or *limb*. At the centre let a circular hole be pierced, in which shall move a pivot exactly fitting it, carrying a tube, whose axis, *a b*, is exactly parallel to the plane of the circle, or perpendicular to the pivot; and also the two arms, *m n*, at right angles to it, and forming one piece with the tube and the axis; so that the motion of the axis on the centre shall carry the tube and arms smoothly round the circle, to be arrested and fixed at any point we please, by a contrivance called *a clamp*. Suppose, now, we would measure the angular interval between two fixed objects, S T. The plane of the circle must first be adjusted so as to pass through them both. This done, let the *axis a b* of the tube be directed to one of them, S, and *clamped*. Then will a mark on the arm *m* point either exactly to some one of the divisions on the limb, or between two of them adjacent. In the former case, the division must be noted, as *the reading* of the arm *m*. In the latter, the fractional part of one whole interval between the consecutive divisions by which the mark on *m surpasses* the last inferior division must be estimated or measured by some mechanical or optical means. (See art. 130.) The division and fractional part thus noted, and reduced into degrees, minutes, and seconds, is to be set down as the *reading of the limb* corresponding to that position of the *tube a b*, where it points to the object S. The same must then be done for the object T; the tube pointed to it, and the *limb "read off."* It is manifest, then, that, if the lesser of these readings be subtracted from the greater, *their difference* will be the angular interval between S and T, as seen from the centre of the circle, at whatever point of the limb the commencement of the graduations on the point 0° be situated.

(129.) The very same result will be obtained, if, instead of making the tube moveable upon the circle, we connect it invariably with the latter, and make both revolve together on an axis concentric with the circle, and forming one piece with it, working in a hollow formed to receive and fit it in some fixed support. Such a combination is represented in section in the annexed sketch. T is the tube or sight, fastened, at *p p*, on the circle A B, whose axis, D, works in the solid metallic centring E, from which originates an arm, F, carrying at its extremity an index, or other proper mark,

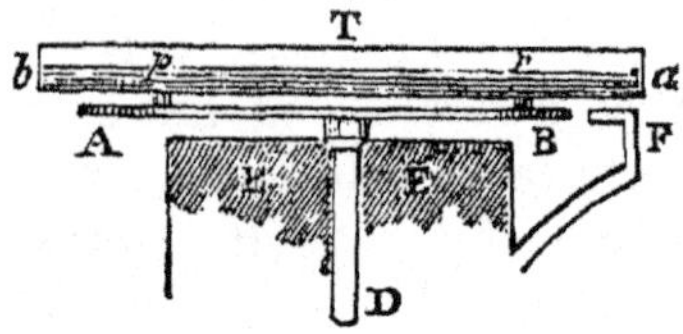

to point out and read off the exact division of the circle at B, the point close to it. It is evident that, as the telescope and circle revolve through any angle, the part of the limb of the latter, which by such revolution is carried past the index F, will measure the angle described. This is the most usual mode of applying divided circles in astronomy.

(130.) The index F may either be a simple pointer, like a clock hand (*fig. a*); or a vernier (*fig. b*); or,

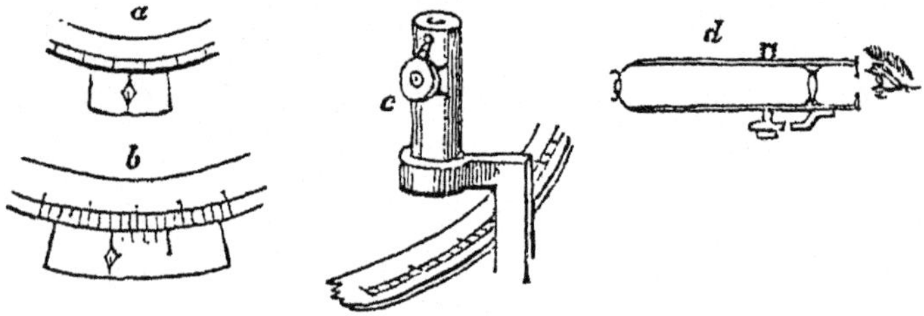

lastly, a compound microscope (*fig. c*), represented in section (in *fig. d*), and furnished with a cross in the common focus of its object and eye-glass, moveable by a fine-threaded screw, by which the intersection of the cross may be brought to exact coincidence with the image of the nearest of the divisions of the circle; and by the turns and parts of a turn of the screw required for this purpose the distance of that division from the original or zero point of the microscope may be estimated. This simple but delicate contrivance gives to the reading off of a circle a degree of accuracy only limited by the power of the microscope, and the perfection with which a screw can be executed, and places the subdivision of angles on the same footing of optical certainty which is introduced into their measurement by the use of the telescope.

(131.) The exactness of the result thus obtained must depend, 1st, on the precision with which the tube *a b* can be pointed to the objects; 2dly, on the accuracy of graduation of the limb; 3dly, on the accuracy with which the subdivision of the intervals between any two consecutive graduations can be accomplished. The mode of accomplishing the latter object with any required exactness has been explained in the last article. With regard to the graduation of the limb, being merely of a mechanical nature, we shall pass it without remark, further than this, that, in the present state of instrument-making, the amount of error from this source of inaccuracy is reduced within very narrow limits indeed. With regard to the first, it must be obvious that, if the sights *a b* be nothing more than what they are represented in the figure (art. 128.) simple crosses or pin-holes at the ends of a hollow tube, or an eye-hole at one end, and a cross at the other, no greater nicety in pointing can be expected than what simple vision with the naked eye can command. But if, in place of these simple but coarse contrivances, the tube itself be converted into *a telescope*, having an object-glass at *b*, and an eye-piece at *a*; and if the motion of the tube on the limb of the circle be arrested when the object is brought just into the centre of the field of view, it is evident that a greater degree of exactness may be attained in the pointing of the tube than by the unassisted eye, in proportion to the magnifying power and distinctness of the telescope used. The last attainable degree of exactness is secured by stretching in the common focus of the object and eye-glasses two delicate fibres, such as fine hairs or spider-lines, intersecting each other at right angles in the centre of the field of view. Their points of intersection afford a permanent mark with which the image of the object can be brought to exact coincidence by a proper degree of caution (aided by mechanical contrivances), in bringing the telescope to its final situation on the limb of the circle, and retaining it there till the "reading off" is finished.

(132.) This application of the telescope may be considered as completely annihilating that part of the error of observation which might otherwise arise from erroneous estimation of the direction in which an object lies from the observer's eye, or from the centre of the instrument. It is, in fact, the grand source of all the precision of modern astronomy, without which all other refinements in instrumental workmanship would be thrown away; the errors capable of being committed in pointing to an object, without such assistance, being far greater than what could arise from any but the very coarsest graduation.* In fact, the telescope thus applied becomes, with respect to angular, what the microscope is with respect to linear dimension. By concentrating attention on its smallest points, and magnifying into palpable intervals the minutest differences, it enables us not only to scrutinize the form and structure of the objects to which it is pointed, but to refer their apparent places, with all but geometrical precision, to the parts of any scale with which we propose to compare them.

(133.) The simplest mode in which the measure-

* The honour of this capital improvement has been successfully vindicated by Derham (Phil. Trans. xxx. 603.) to our young, talented, and unfortunate countryman Gascoigne, from his correspondence with Crabtree and Horrockes, in his (Derham's) possession. The passages cited by Derham from these letters leave no doubt that, so early as 1640, Gascoigne had applied telescopes to his quadrants and sextants, *with threads in the common focus of the glasses;* and had even carried the invention so far as to illuminate the field of view by artificial light, which he found "*very helpful when the moon appeareth not, or it is not otherwise light enough.*" These inventions were freely communicated by him to Crabtree, and through him to his friend Horrockes, the pride and boast of British astronomy; both of whom expressed their unbounded admiration of this and many other of his delicate and admirable improvements in the art of observation. Gascoigne, however, perished, at the age of twenty-three, at the battle of Marston Moor; and the premature and sudden death of Horrockes, at a yet earlier age, will account for the temporary oblivion of the invention. It was revived, or re-invented, in 1667, by Picard and Auzout (Lalande, Astron. 2310.), after which its use became universal. Morin, even earlier than Gascoigne (in 1635), had proposed to substitute the telescope for plain sights; but it is the thread or wire stretched in the focus with which the image of a star can be brought to exact coincidence, which gives the telescope its advantage in practice; and the idea of this does not seem to have occurred to Morin. (See Lalande, *ubi supra.*)

ment of an angular interval can be executed, is what we have just described; but, in strictness, this mode is applicable only to terrestrial angles, such as those occupied on the sensible horizon by the objects which surround our station,—because these only remain stationary during the interval while the telescope is shifted on the limb from one object to the other. But the diurnal motion of the heavens, by destroying this essential condition, renders the direct measurement of angular distance from *object* to *object* by this means impossible. The same objection, however, does not apply if we seek only to determine the interval between the *diurnal circles* described by any two celestial objects. Suppose every star, in its diurnal revolution, were to leave behind it a visible trace in the heavens,—a fine line of light, for instance,—then a telescope once pointed to a star, so as to have its image brought to coincidence with the intersection of the wires, would constantly remain pointed to some portion or other of this line, which would therefore continue to appear in its field as a luminous line, permanently intersecting the same point, till the star came round again. From one such line to another the telescope might be shifted, at leisure, without error; and then the angular interval between the two diurnal circles, *in the plane of the telescope's rotation*, might be measured. Now, though we cannot *see* the path of a star in the heavens, we can *wait* till the star itself crosses the field of view, and seize the moment of its passage to place the intersection of its wires so that the star shall traverse it; by which, when the telescope is well clamped, we equally well secure the position of its diurnal circle as if we continued to *see* it ever so long. The reading off of the limb may then be performed at leisure; and when another star comes round *into the plane* of the circle, we may unclamp the telescope, and a similar observation will enable us to assign the place of *its* diurnal circle on the limb: and the observations may be repeated alternately, every day, as the stars pass, till we are satisfied with their result.

(134.) This is the principle of the mural circle, which is nothing more than such a circle as we have described in art. 129., firmly supported, in the plane of the meridian, on a long and powerful horizontal axis. This axis is let into a massive pier, or wall, of stone (whence the name of the instrument), and so secured by screws as to be capable of adjustment both in a vertical and horizontal direction ; so that, like the axis of the transit, it can be maintained in the exact direction of the east and west points of the horizon, the plane of the circle being consequently truly meridional.

(135.) The meridian, being at right angles to all the diurnal circles described by the stars, its arc intercepted between any two of them will measure the least distance between these circles, and will be equal to the difference of the declinations, as also to the difference of the *meridian altitudes* of the objects—at least when corrected for refraction. These differences, then, are the angular intervals *directly* measured by the mural circle. But from these, supposing the law of refraction known, it is easy to conclude, not their differences only, but the quantities themselves, as we shall now explain.

(136.) The declination of a heavenly body is the complement of its distance from the pole. The pole, being a point in the meridian, might be directly observed on the limb of the circle, if any star stood *exactly* therein; and thence the *polar distances*, and, of course, the declinations of all the rest, might be at once determined. But this not being the case, a bright star as near the pole as can be found is selected, and observed in its *upper* and *lower* culminations ; that is, when it passes the meridian *above* and *below* the pole. Now, as its distance from the pole remains the same, the difference of reading off the circle in the two cases is, of course (when corrected for refraction), equal to twice the polar distance of the star ; the arc intercepted on the limb of the circle being, in this case, equal to the angular diameter of the star's diurnal circle. In the annexed diagram, H P O represents the celestial meridian, P the pole,

B R, A Q, C D the diurnal circles of stars which arrive on the meridian—at B A and C in their upper,

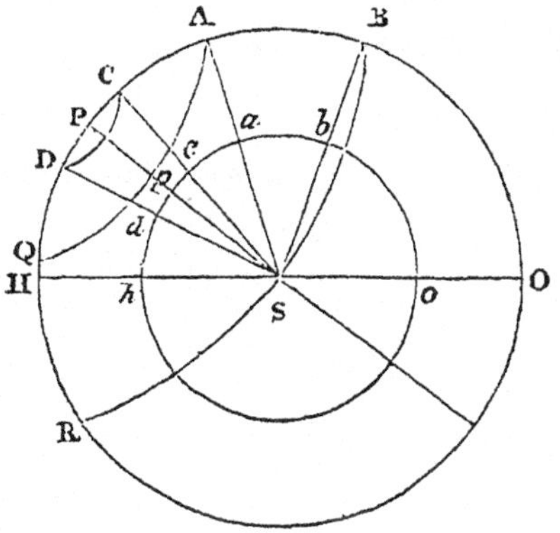

and at R Q D in their lower culminations, of which D happens above the horizon H O. P is the pole; and if we suppose *h p o* to be the mural circle, having S for its centre, *b a c p d* will be the points on its circumference corresponding to B A C P D in the heavens. Now, the arcs *b a*, *b c*, *b d*, and *c d* are given immediately by observation; and since C P = P D, we have also *c p* = *p d*, and each of them $= \frac{1}{2}$ *c d*, consequently the place of *the polar point*, as it is called, upon the limb of the circle becomes known, and the arcs *p b*, *p a*, *p c*, which represent on the circle the *polar distances* required, become also known.

(137.) The situation of the pole star, which is a very brilliant one, is eminently favourable for this purpose, being only about a degree and a half from the pole; it is, therefore, the star usually and almost solely chosen for this important purpose; the more especially because, both its culminations taking place at great and not very different altitudes, the refractions by which they are affected are of small amount, and differ but slightly from each other, so that their correction is easily and safely applied. The brightness of the pole star, too, allows it to be easily observed in the daytime. In consequence of these peculiarities, this star is one of constant resort with astronomers for the adjustment and verification of instruments of almost every description. In the case of the transit, for example, it furnishes a ready means of ascertaining whether the plane of the telescope's motion is coincident with the meridian. For since this latter plane bisects its diurnal circle, the eastern and the western portion of it require equal times for their description. Let, therefore, the moments of its transit above and below the pole be noted; and if they are found to follow at equal intervals of 12 sidereal hours, we may conclude with certainty that the plane of the telescope's motion is meridional, or the position of its horizontal axis exactly east and west. But if it pass from one to the other apparent culmination in unequal intervals of time, it is equally certain that an extra-meridional error must exist, the deviation lying towards that side on which the least interval is occupied. And the axis must be moved *in azimuth* accordingly, till the difference in question disappears on repeating the observations.

(138.) The place of the *polar point* on the limb of the mural circle once determined, becomes an origin, or zero point, from which the polar distances of all objects, referred to other points on the same lines, reckon. It matters not whether the actual commencement 0° of the graduations stand there, or not; since it is only by the *difference* of the readings that the arcs on the limb are determined; and hence a great advantage is obtained in the power of commencing anew a fresh series of observations, in which a different part of the circumference of the circle shall be employed, and different graduations brought into use, by which inequalities of division may be detected and neutralized. This is accomplished practically by detaching the telescope from its old bearings on the circle, and fixing it afresh on a different part of the circumference.

(139.) A point on the limb of the mural circle, not less important than the *polar point*, is the *horizontal point*, which, being once known, becomes in like manner an origin, or zero point, from which altitudes are reckoned. The principle of its determination is ultimately nearly the same with that of the polar point. As no star exists in the celestial horizon, the observer must seek to determine two points on the limb, the one of which shall be precisely as far *below* the horizontal point as the other is above it. For this purpose, a star is observed at its culmination on one night, by pointing the telescope directly to it, and the next, by pointing to *the image of the same star reflected* in the still, unruffled surface of a fluid at perfect rest. Mercury, as the most reflective fluid known, is generally chosen for that use. As the surface of a fluid at rest is necessarily horizontal, and as the angle of reflection, by the laws of optics, is equal to that of incidence, this image will be just as much depressed below the horizon, as the star itself is elevated above it (allowing for the difference of refraction at the moments of observation). The arc intercepted on the limb of the circle between the star and its reflected image thus consecutively observed, when corrected for refraction, is the double altitude of the star, and its point of bisection the horizontal point. The reflecting surface of a fluid so used for the determination of the altitudes of objects is called an *artificial horizon.*

(140.) The mural circle is, in fact, at the same time, a transit instrument; and, if furnished with a proper system of vertical wires in the focus of its telescope, may be used as such. As the axis, however, is only supported at one end, it has not the strength and permanence necessary for the more delicate purposes of a transit; nor can it be verified, as a transit may, by the *reversal* of the two ends of its axis, east for west. Nothing, however, prevents a divided circle being permanently fastened on the axis of a transit instrument, near to one of its extremities, so as to revolve with it, the reading off being performed by a microscope fixed on one of its piers. Such an instrument is called a TRANSIT CIRCLE, or a MERIDIAN CIRCLE, and serves for the simultaneous determination of the right ascensions and polar distances of objects observed with it; the time of transit being noted by the clock, and the circle being read off by the lateral microscope.

(141.) The determination of the horizontal point on the limb of an instrument is of such essential importance in astronomy, that the student should be made acquainted with every means employed for this purpose. These are, the artificial horizon, the plumb-line, the level, and the floating collimator. The artificial horizon has been already explained. The plumb-line is a fine thread or wire, to which is suspended a weight, whose oscillations are impeded and quickly reduced to rest by plunging it in water. The direction ultimately assumed by such a line, admitting *its perfect flexibility,* is that of gravity, or perpendicular to the surface of still water. Its application to the purposes of astronomy is, however, so delicate, and difficult, and liable to error, unless extraordinary precautions are taken in its use, that it is at present almost universally abandoned, for the more convenient and equally exact instrument *the level.*

(142.) The level is nothing more than a glass tube nearly filled with a liquid, (spirit of wine being that

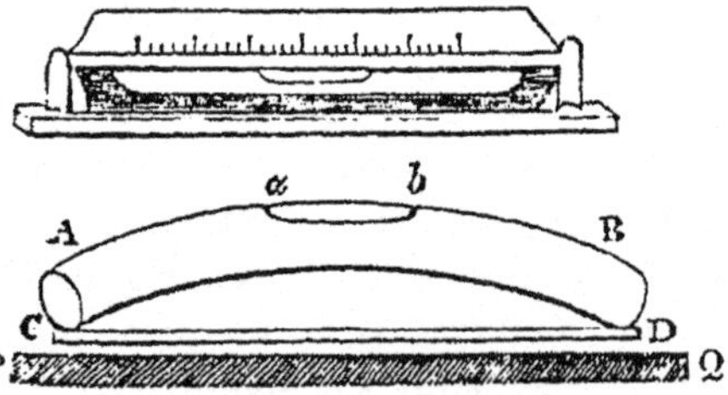

now generally used, on account of its extreme *mobility,* and not being liable to freeze,) the bubble in which, when the tube is placed horizontally, would rest indifferently in any part if *the tube* could be mathematically *straight.* But that being impossible to execute, and every tube having some slight curvature, if the convex side be placed upwards, the bubble will occupy the higher part, as in the figure (where the curvature is purposely exaggerated). Suppose such a tube as A B firmly fastened on a straight bar, C D, and marked at *a b*, two points distant by the length of the bubble; then, if the instrument be so placed that the bubble shall occupy this interval, it is clear that C D can have no other than one definite inclination to the horizon; because, were it ever so little moved one way or other, the bubble would shift its place, and run towards the elevated side. Suppose, now, that we would ascertain whether any given line P Q be horizontal; let the base of the level C D be set upon it, and note the points *a b*, between which the bubble is exactly contained; then turn the level end for end, so that C shall rest on Q, and D on P. If then the bubble continue to occupy the same place between *a* and *b*, it is evident that P Q can be no otherwise than horizontal. If not, the side towards which the bubble runs is highest, and must be lowered. Astronomical levels are furnished with a divided scale, by which the places of the ends of the bubble can be nicely marked; and it is said that they can be executed with such delicacy, as to indicate a single second of angular deviation from exact horizontality.

(143.) The mode in which a level may be applied to find the horizontal point on the limb of a vertical divided circle may be thus explained: Let A B be a telescope firmly fixed to such a circle, D E F, and move-

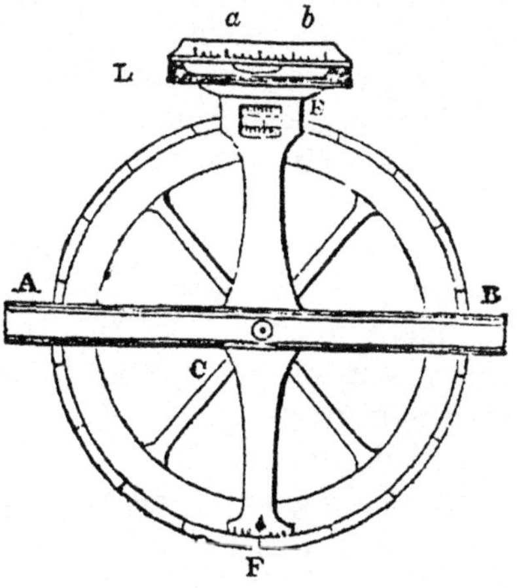

able in one with it on a horizontal axis C, which must be like that of a transit, susceptible of reversal (see art. 127.), and with which the circle is inseparably connected. Direct the telescope on some distant well-defined object S, and bisect it by its horizontal wire, and in this position clamp it fast. Let L be a level fastened at right angles to an arm, LEF, furnished with a microscope, or vernier at F, and, if we please, another at E. Let this arm be fitted by grinding on the axis C, but capable of moving smoothly on it without carrying it round, and also of being clamped fast on it, so as to prevent it from moving until required. While the telescope is kept fixed on the object S, let the level be set so as to bring its bubble to the marks *a b*, and clamp it there. Then will the arm LCF have some certain determinate inclination (no matter what) to the horizon. In this position let the circle be read off at F, and then let the whole apparatus be *reversed* by turning its horizontal axis end for end, *without unclamping the level arm* from the axis. This done, by the motion of the whole instrument (level and all) on its axis, restore the *level* to its horizontal position with the bubble at *a b*. Then we are sure that the telescope has now the same inclination to the horizon *the other way*, that it had when pointed to S, and the reading off at F will not have been changed. Now unclamp the level, and, keeping it nearly horizontal, turn round the circle on the axis, so as to carry back the telescope *through the zenith* to S, and in that position clamp the circle and telescope fast. Then it is evident that an angle equal to twice the zenith distance of S has been moved over by the axis of the telescope from its last position. Lastly, without unclamping the telescope and circle, let the level be once more rectified. Then will the arm LEF once more assume the same definite position with respect to the horizon; and, consequently, if the circle be again read off, the difference between this and the previous reading must measure the arc of its circumference which has passed under the point F, which may be considered as having all the while retained an invariable position. This difference, then, will be the double zenith distance of S, and its half the zenith distance simply, the complement of which is its altitude. Thus the altitude corresponding to a given reading of the limb becomes known, or, in other words, the horizontal point on the limb is ascertained. Circuitous as this process may appear, there is no other mode of employing the level for this purpose which does not in the end come to the same thing. Most commonly, however, the level is used as a mere *fiducial* reference, to preserve a horizontal point once well determined by other means, which is done by adjusting it so as to stand level when the telescope is truly horizontal, and thus leaving it depending on the permanence of its adjustment.

(144.) The last, but probably not the least exact, as it certainly is, in innumerable cases, the most convenient means of ascertaining the *horizontal point*, is that afforded by the floating collimator, a recent invention of captain Kater. This elegant instrument is nothing more than a small telescope furnished with a cross-wire in its focus, and fastened horizontally, or as nearly so as may be, on a flat iron *float*, which is made to swim on mercury, and which, of course, will, when left to itself, assume always one and the same invariable inclination to the horizon. If the cross-wires of the col-

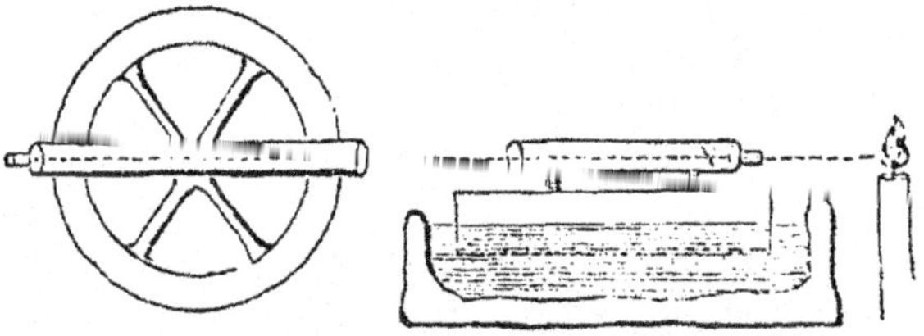

limator be illuminated by a lamp, being in the focus of its object-glass, the rays from them will issue parallel, and will therefore be in a fit state to be brought to a focus by the object-glass of any other telescope, in which they will form an image *as if they came from a celestial object in their direction,* i. e. at an altitude equal to their inclination. Thus the intersection of the cross of the collimator may be observed *as if it were a star,* and that, however near the two telescopes are to each other. By transferring then, the collimator *still floating* on a vessel of mercury from the one side to the other of a circle, we are furnished with two *quasi-celestial* objects, at precisely equal altitudes, on opposite sides of the centre; and if these be observed in succession with the telescope of the circle, bringing its cross to bisect the image of the cross of the collimator (for which end the wires of the latter cross are purposely set

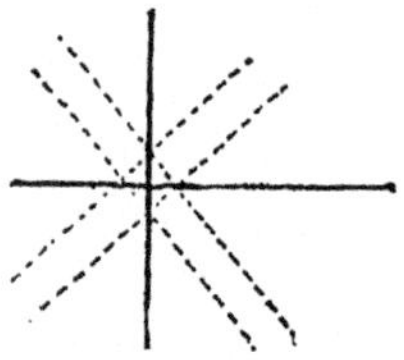

45° inclined to the horizon) the difference of the readings on its limb will be twice the zenith distance of either; whence, as in the last article, the horizontal or zenith point is immediately determined.*

(145.) The transit and mural circle are essentially meridian instruments, being used only to observe the stars at the moment of their meridian passage. Independent of this being the most favourable moment for seeing them, it is that in which their diurnal motion is parallel to the horizon. It is therefore easier at this time than it could be at any other, to place the telescope exactly in their true direction; since their apparent course in the field of view being parallel to the horizontal thread of the system of wires therein, they may, by giving a fine motion to the telescope, be brought to exact coincidence with it, and time may be allowed to examine and correct this coincidence, if not at first accurately hit, which is the case in no other situation. Generally speaking, all angular magnitudes, which it is of importance to ascertain exactly, should, if possible, be observed at their maxima or minima of increase or

* Another, and, in many respects, preferable form of the floating collimator, in which the telescope is *vertical,* and whereby the *zenith* point is directly ascertained, is described in the Phil. Trans. 1828, p. 257., by the same author.

diminution; because at these points they remain not perceptibly changed during a time long enough to complete, and even, in many cases, to repeat and verify, our observations in a careful and leisurely manner. The angle which, in the case before us, is in this predicament, is the altitude of the star, which attains its maximum or minimum on the meridian, and which is measured on the limb of the mural circle.

(146.) The purposes of astronomy, however, require that an observer should possess the means of observing any object not directly on the meridian, but at any point of its diurnal course, or wherever it may present itself in the heavens. Now, a point in the sphere is determined by reference to two great circles at right angles to each other; or of two circles one of which passes through the pole of the other. These, in the language of geometry, are *co-ordinates* by which its situation is ascertained: for instance,—on the earth, a place is known if we know its longitude and latitude; —in the starry heavens, if we know its right ascension and declination;—in the visible hemisphere, if we know its azimuth and altitude, &c.

(147.) To observe an object at any point of its diurnal course, we must possess the means of directing a telescope to it; which, therefore, must be capable of motion in two planes at right angles to each other; and the amount of its angular motion in each must be measured on two circles *co-ordinate* to each other, whose planes must be parallel to those in which the telescope moves. The practical accomplishment of this condition is effected by making the axis of one of the circles penetrate that of the other at right angles. The pierced axis turns on fixed supports, while the other has no connection with any external support, but is sustained entirely by that which it penetrates, which is strengthened and enlarged at the point of penetration to receive it. The annexed figure exhibits the simplest form of such a combination, though by no means the best in point of mechanism. The two circles are *read off* by verniers, or microscopes; the one attached to the fixed support which carries the principal axis, the other to an arm projecting from that axis. Both circles also are susceptible of being clamped, the clamps being attached to the same ultimate bearing with which the apparatus for reading off is connected.

(148.) It is manifest that such a combination, however its principal axis be pointed (provided that its direction be invariable), will enable us to ascertain the situation of any object with respect to the observer's station, by angles reckoned upon two great circles in the visible hemisphere, one of which has for its poles the prolongations of the principal axis or the vanishing points of a system of lines parallel to it, and the other passes always through these poles: for the former great circle is the vanishing line of all planes parallel to the circle A B, while the latter, in any position of the instrument, is the vanishing line of all the planes parallel to the circle G H; and these two planes being, by the construction of the instrument, at right angles, the great circles, which are their vanishing lines, must be so too. Now, if two great circles of a sphere be at right angles to each other, the one will always pass through the other's poles.

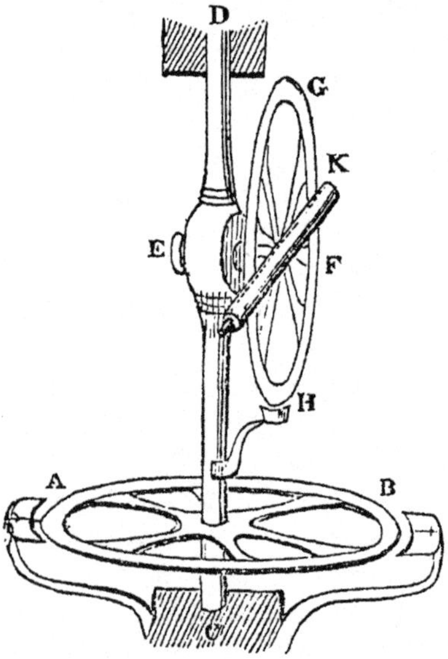

(149.) There are, however, but two positions in which such an apparatus can be mounted so as to be of any practical utility in astronomy. The first is, when the principal axis C D is parallel to the earth's axis, and therefore points to the poles of the heavens which are the vanishing points of all lines in this system of parallels; and when, of course, the plane of the circle A B is parallel to the earth's equator, and therefore has the equinoctial for its vanishing circle, and measures, by its arcs read off, hour angles, or differences of right ascension. In this case, the great circles in the heavens, corresponding to the various positions, which the circle G H can be made to assume, by the rotation of the instrument round its axis C D, are all hour-circles; and the arcs read off on this circle will be declinations, or polar distances, or their differences.

(150.) In this position the apparatus assumes the name of an *equatorial*, or, as it was formerly called, a *parallactic* instrument. It is one of the most convenient instruments for all such observations as require an object to be kept long in view, because, being once set upon the object, it can be followed as long as we please by a *single motion*, i.e. by merely turning the whole apparatus round on its polar axis. For since, when the telescope is set on a star, the angle between its direction and that of the polar axis is equal to the polar distance of the star, it follows, that when turned about its axis, without altering the position of the telescope on the circle GH, the point to which it is directed will always lie in the small circle of the heavens coincident with the star's diurnal path. In many observations this is an inestimable advantage, and one which belongs to no other instrument. The equatorial is also used for determining the place of an unknown by comparison with that of a known object, in a manner to be described in the fourth chapter. The adjustments of the equatorial are somewhat complicated and difficult. They are best performed by following the pole-star round the entire diurnal circle, and by observing, at proper intervals, other considerable stars whose places are well ascertained.*

(151.) The other position in which such a com-

* See Littrow on the Adjustment of the Equatorial. — *Mem. Astron. Soc.* vol. ii. p. 45.

pound apparatus as we have described in art. 147. may be advantageously mounted, is that in which the principal axis occupies a vertical position, and the one circle, A B, consequently corresponds to the celestial horizon, and the other, G H, to a vertical circle of the heavens. The angles measured on the former are therefore *azimuths*, or differences of azimuth, and those on the latter zenith distances, or altitudes, according as the graduation commences from the upper point of its limb, or from one 90° distant from it. It is therefore known by the name of an *azimuth and altitude instrument*. The vertical position of its principal axis is secured either by a plumb-line suspended from the upper end, which, however it be turned round, should continue always to intersect one and the same fiducial mark near its lower extremity, or by a level fixed directly across it, whose bubble ought not to shift its place, on moving the instrument in azimuth. The north or south point on the horizontal circle is ascertained by bringing the vertical circle to coincide with the plane of the meridian, by the same criterion by which the azimuthal adjustment of the transit is performed (art. 137.), and noting, in this position, the reading off of the lower circle, or by the following process.

(152.) Let a bright star be observed at a considerable distance to the *east* of the meridian, by bringing it on the cross wires of the telescope. In this position let the horizontal circle be read off, and the telescope securely clamped on the vertical one. When the star has passed the meridian, and is in the descending point of its daily course, let it be followed by moving the whole instrument round to the west, without, however, unclamping the telescope, until it comes into the field of view; and until, by continuing the horizontal motion, the star and the cross of the wires come once more to coincide. In this position it is evident the star must have the same precise altitude above the *western* horizon, that it had at the moment of the first observation above the *eastern*. At this point let the motion be arrested, and the horizontal circle be again read off. The difference of the readings will be the azimuthal arc described in the interval. Now, it is evident that when the altitudes of any star are equal on either side of the meridian, its *azimuths*, whether reckoned both from the north or both from the south point of the horizon, must also be equal,—consequently the north or south point of the horizon must bisect the azimuthal arc thus determined, and will therefore become known.

(153.) This method of determining the north and south points of a horizontal circle (by which, when known, we may draw a meridian line) is called the "method of equal altitudes," and is of great and constant use in practical astronomy. If we note, at the moments of the two observations, the time, by a clock or chronometer, the instant halfway between them will be the moment of the star's meridian passage, which may thus be determined without a transit; and, *vice versâ*, the error of a clock or chronometer may by this process be discovered. For this last purpose, it is not necessary that our instrument should be provided with a horizontal circle at all. Any means by which altitudes can be measured will enable us to determine the moments when the same star arrives at *equal* altitudes in the eastern and western halves of its diurnal course; and, these once known, the instant of meridian passage and the error of the clock become also known.

(154.) One of the chief purposes to which the altitude and azimuth circle is applicable is the investigation of the amount and laws of refraction. For, by following with it a circumpolar star which passes the zenith, and another which grazes the horizon, through their whole diurnal course, the exact *apparent* form of their diurnal orbits, or the ovals into which their circles are distorted by refraction, can be traced; and their deviation from circles, being at every moment given by the nature of the observation *in the direction in which the refraction itself takes place* (i. e. in altitude), is made a matter of direct observation.

(155.) The *zenith sector* and the *theodolite* are peculiar modifications of the altitude and azimuth instrument. The former is adapted for the very exact observation of stars in or near the zenith, by giving a great length to the vertical axis, and suppressing all the circumference of the vertical circle, except a few degrees of its lower part, by which a great length of radius, and a consequent proportional enlargement of the divisions of its arc, is obtained. The latter is especially devoted to the measure of horizontal angles between terrestrial objects, in which the telescope never requires to be elevated more than a few degrees, and in which, therefore, the vertical circle is either dispensed with, or executed on a smaller scale, and with less delicacy; while, on the other hand, great care is bestowed on securing the exact perpendicularity of the plane of the telescope's motion, by resting its horizontal axis on two supports like the piers of a transit-instrument, while themselves are firmly bedded on the spokes of the horizontal circle, and turn with it.

(156.) The last instrument we shall describe is one by whose aid the direct angular distance of any two objects may be measured, or the altitude of a single one determined, either by measuring its distance from the visible horizon (such as the sea-offing, allowing for its dip), or from its own reflexion on the surface of mercury. It is the sextant, or quadrant, commonly called *Hadley*'s, from its reputed inventor, though the priority of invention belongs undoubtedly to Newton, whose claims to the gratitude of the navigator are thus doubled, by his having furnished at once the only theory by which his vessel can be securely guided, and the only instrument which has ever been found to avail, in applying that theory to its nautical uses.*

* Newton communicated it to Dr. Halley, who suppressed it. The description of the instrument was found, after the death of Halley, among his papers, in Newton's own handwriting, by his executor, who communicated the papers to the Royal Society, twenty-five years after Newton's death, and eleven after the publication of Hadley's invention, which might be, and probably was, independent of any knowledge of Newton's, though Hutton insinuates the contrary.

(157.) The principle of this instrument is the optical property of reflected rays, thus announced:—"The

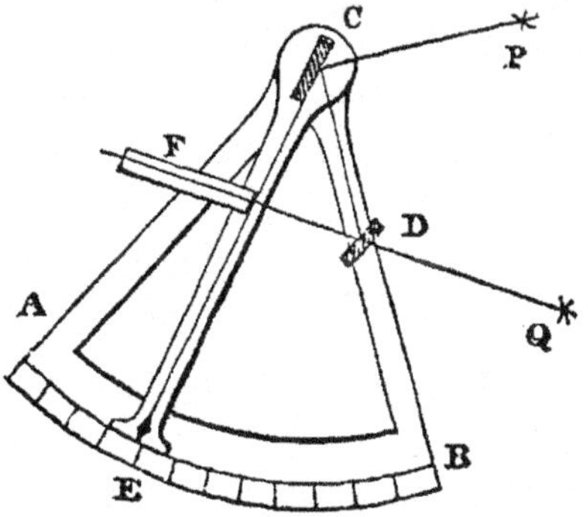

angle between the first and last directions of a ray which has suffered two reflexions in one plane is equal to twice the inclination of the reflecting surfaces to each other." Let A B be the limb, or graduated arc, of a portion of a circle 60° in extent, but divided into 120 equal parts. On the radius C B let a silvered plane glass D be fixed, at right angles to the plane of the circle, and on the moveable radius C E let another such silvered glass, C, be fixed. The glass D is permanently fixed parallel to A C, and only one half of it is silvered, the other half allowing objects to be seen through it. The glass C is wholly silvered, and its plane is parallel to the length of the moveable radius C E, at the extremity E, of which a vernier is placed to read off the divisions of the limb. On the radius A C is set a telescope F, through which any object, Q, may be seen by *direct* rays which pass through the unsilvered portion of the glass D, while another object, P, is seen through the same telescope by rays, which, after reflexion at C, have been thrown upon the silvered part of D, and are thence directed by a second reflexion into the telescope. The two images so formed will both be seen in the field of view at once, and by moving the radius C E will (if the reflectors be truly perpendicular to the plane of the circle) meet and pass over, without obliterating each other. The motion, however, is arrested when they meet, and at this point the angle included between the direction C P of one object, and F Q of the other, is twice the angle E C A included between the fixed and moveable radii C A, C E. Now, the graduations of the limb being purposely made only half as distant as would correspond to degrees, the arc A E, when read off, as if the graduations were *whole* degrees, will, in fact, read double its real amount, and therefore the numbers to read off will express not the angle E C A, but its double, the angle subtended by the objects.

(158.) To determine the exact distances between the stars by direct observation is comparatively of little service; but in nautical astronomy the measurement of their distances from the moon, and of their altitudes, is of essential importance; and as the sextant requires no fixed support, but can be held in the hand, and used on ship-board, the utility of the instrument becomes at once obvious. For altitudes at sea, as no level, plumb-line, or artificial horizon can be used, the sea-offing affords the only resource; and the image of the star observed, seen by reflexion, is brought to coincide with the boundary of the sea seen by direct rays. Thus the altitude above the sea-line is found; and this corrected for the *dip of the horizon* (art. 24.) gives the true altitude of the star. On land, an artificial horizon may be used (art. 139.), and the consideration of dip is rendered unnecessary.

(159.) The reflecting circle is an instrument destined for the same uses as the sextant, but more complete, the circle being entire, and the divisions carried all round. It is usually furnished with three verniers, so as to admit of three distinct readings off, by the average of which the error of graduation and of reading is reduced. This is altogether a very refined and elegant instrument.

(160.) We must not conclude this chapter without mention of the "principle of repetition;" an invention of Borda, by which the error of graduation may be diminished to any degree, and, practically speaking, annihilated. Let P Q be two objects which we may suppose fixed, for purposes of mere explanation, and let K L be a

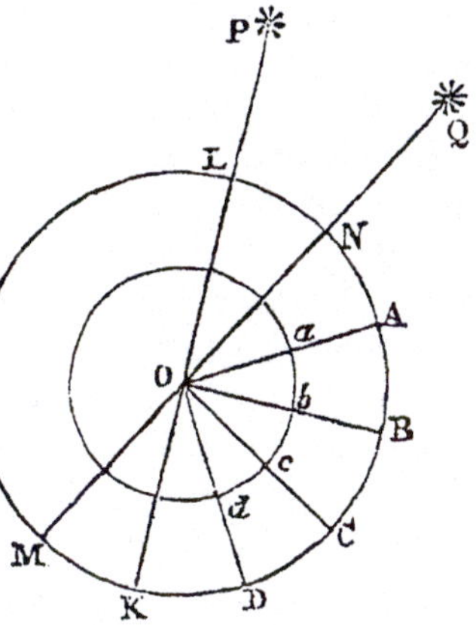

telescope moveable on O, the common axis of two circles, A M L and *a b c*, of which the former, A M L, is absolutely fixed in the plane of the objects, and carries the graduations, and the latter is freely moveable on the axis. The telescope is attached permanently to the latter circle, and moves with it. An arm O *a* A carries the index, or vernier, which reads off the graduated limb of the fixed circle. This arm is provided with two clamps, by which it can be temporarily connected with either circle, and detached at pleasure. Suppose, now, the telescope directed to P. Clamp the index arm O A to the *inner* circle, and unclamp it from the outer, and read off. Then carry the telescope round to the other object Q. In so doing, the inner circle, and the index-arm which is clamped to it, will also be carried round, over an arc A B, on the graduated limb of the outer, equal to the angle P O Q. Now clamp the index to the outer circle, and unclamp the inner, and read off: the difference of readings will of course measure the angle P O Q; but the result will be liable to two sources of error—that of *graduation* and that of observation, both which it is our object to get rid of. To this end transfer the telescope back to P, *without* unclamping the arm from the outer circle *then*, having made the bisection of P, clamp the arm to *b*, and unclamp it from B, and again transfer the telescope to Q, by which the arm will now be carried with it to C, over a second arc, B C, equal to the angle P O Q. Now again read off; then will the difference between this reading and the *original* one measure *twice* the angle P O Q, affected with *both* errors of observation, but only with *the same error of graduation as before.* Let this process be repeated as often as we please (suppose ten times); then will the final arc A B C D read off on the circle be ten times the required angle, affected by the joint errors of all the ten observations, but only by the same constant error of graduation, which depends on the initial and final readings off alone. Now the errors of observation, when numerous, tend to balance and destroy one another; so that, if sufficiently multiplied, their influence will disappear from the result. There remains, then, only the constant error of graduation, which comes to be divided in the final result by the number of observations, and is therefore diminished in its influence to one tenth of its possible amount, or to less if need be. The abstract beauty and advantage of this principle seem to be counterbalanced in practice by some unknown cause, which, probably, must be sought for in imperfect clamping.

CHAP. III.

OF GEOGRAPHY.

OF THE FIGURE OF THE EARTH. — ITS EXACT DIMENSIONS. — ITS FORM THAT OF EQUILIBRIUM MODIFIED BY CENTRIFUGAL FORCE. — VARIATION OF GRAVITY ON ITS SURFACE. — STATICAL AND DYNAMICAL MEASURES OF GRAVITY. — THE PENDULUM. — GRAVITY TO A SPHEROID. — OTHER EFFECTS OF EARTH'S ROTATION. — TRADE WINDS. — DETERMINATION OF GEOGRAPHICAL POSITIONS. — OF LATITUDES. — OF LONGITUDES. — CONDUCT OF A TRIGONOMETRICAL SURVEY. — OF MAPS. — PROJECTIONS OF THE SPHERE. — MEASUREMENT OF HEIGHTS BY THE BAROMETER.

(161.) GEOGRAPHY is not only the most important of the practical branches of knowledge to which astronomy is applied, but is also, theoretically speaking, an essential part of the latter science. The earth being the general station from which we view the heavens, a knowledge of the local situation of particular stations on its surface is of great consequence, when we come to enquire the distances of the nearer heavenly bodies from us, as concluded from observations of their parallax as well as on all other occasions, where a difference of locality can be supposed to influence astronomical results. We propose, therefore, in this chapter, to explain the principles, by which astronomical observation is applied to geographical determinations, and to give at the same time an outline of geography so far as it is to be considered a part of astronomy.

(162.) Geography, as the word imports, is a delineation or description of the earth. In its widest sense, this comprehends not only the delineation of the form of its continents and seas, its rivers and mountains, but their physical condition, climates, and products, and their appropriation by communities of men. With physical and political geography, however, we have no concern here. Astronomical geography has for its objects the exact knowledge of the form and dimensions of the earth, the parts of its surface occupied by sea and land, and the configuration of the surface of the latter, regarded as protuberant above the ocean, and broken into the various forms of mountain, table land, and valley; neither should the form of the bed of the ocean, regarded as a continuation of the surface of the land beneath the water, be left out of consideration; we know, it is true, very little of it; but this is an ignorance rather to be lamented, and, if possible, remedied, than acquiesced in, inasmuch as there are many very important branches of enquiry which would be greatly advanced by a better acquaintance with it.

(163.) With regard to the figure of the earth *as a whole*, we have already shown that, speaking loosely, it may be regarded as spherical; but the reader who has duly appreciated the remarks in art. 23. will not be at a loss to perceive that this result, concluded from observations not susceptible of much exactness, and embracing very small portions of the surface at once, can only be regarded as a first approximation, and may require to be materially modified by entering into minutiæ before neglected, or by increasing the delicacy of our observations, or by including in their extent larger areas of its surface. For instance, if it should turn out (as it will), on minuter enquiry, that the true figure is somewhat elliptical, or flattened, in the manner of an orange, having the diameter which coincides with the axis about $\frac{1}{300}$th part shorter than the diameter of its equatorial circle; — this is so trifling a deviation from the spherical form that, if a model of such proportions were turned in wood, and laid before us on a table, the nicest eye or hand would not detect the flattening, since the difference of diameters, in a globe of sixteen inches, would amount only to $\frac{1}{20}$th of an inch. In all common parlance, and for all ordinary purposes, then, it would still be called a globe; while, nevertheless, by careful measurement, the difference would not fail to be noticed, and, speaking strictly, it would be termed, not a globe, but an oblate ellipsoid, or spheroid, which is the name appropriated by geometers to the form above described.

(164.) The sections of such a figure by a plane are not circles, but ellipses; so that, on such a shaped earth, the horizon of a spectator would nowhere (except at the poles) be exactly circular, but somewhat elliptical. It is easy to demonstrate, however, that its deviation from the circular form, arising from so very slight an "*ellipticity*" as above supposed, would be quite imperceptible, not only to our eyesight but to the test of the dipsector; so that by that mode of observation we should never be led to notice so small a deviation from perfect sphericity. How we are led to this conclusion, as a practical result, will appear, when we have explained the means of determining with accuracy the dimensions of the whole, or any part of the earth.

(165.) As we cannot grasp the earth, nor recede from it far enough to view it at once as a whole, and compare it with a known standard of measure in any degree commensurate to its own size, but can only creep about upon it, and apply our diminutive measures to comparatively small parts of its vast surface in succession, it becomes necessary to supply, by geometrical reasoning, the defect of our physical powers, and from a delicate and careful measurement of such small parts to conclude the form and dimensions of the whole mass. This would present little difficulty, if we were sure the earth were strictly a sphere, for the proportion of the circumference of a circle to its diameter being known (viz. that of 3·1415926 to 1·0000000), we have only to ascertain the length of the entire circumference of any great circle, such as a meridian, in miles, feet or any other standard units, to know the diameter in units of the same kind. Now the circumference of the whole circle is known as soon as we know the exact length of any aliquot part of it, such as 1° or $\frac{1}{360}$th part; and this, being not more than about seventy miles in length, is not beyond the limits of very exact measurement, and could, in fact, be measured (if we knew its exact termination at each extremity) within a very few feet, or, indeed, inches, by methods presently to be particularized.

(166.) Supposing, then, we were to begin measuring with all due nicety from any station, in the exact direction of a meridian, and go measuring on, till by some indication we were informed that we had accomplished an exact *degree* from the point we set out from, our problem would then be at once resolved. It only remains, therefore, to enquire by what indications we can be sure, 1st, that we *have* advanced *an exact degree;* and, 2dly, that we have been measuring in the *exact direction of a great circle.*

(167.) Now, the earth has no landmarks on it to indicate degrees, nor traces inscribed on its surface to guide us in such a course. The compass, though it affords a tolerable guide to the mariner or the traveller, is far too uncertain in its indications, and too little known in its laws, to be of any use in *such* an operation. We must, therefore, look outwards, and refer our situation on the surface of our globe to natural marks, *external* to it, and which are of equal permanence and stability with the earth itself. Such marks are afforded by the stars. By observations of their meridian altitudes, performed at any station, and from their known polar distances, we conclude the height of the pole; and since the altitude of the pole is equal to the latitude of the place (art. 95.), the same observations give the latitudes of any stations where we may establish the requisite instruments. When our latitude, then, is found to have diminished a degree, we know that, *provided we have kept to the meridian,* we have described one three hundred and sixtieth part of the earth's circumference.

(168.) The direction of the meridian may be secured at every instant by the observations described in art. 137.; and although local difficulties may oblige us to deviate in our measurement from this exact direction, yet if we keep a strict account of the amount of this deviation, a very simple calculation will enable us to *reduce* our observed measure to its *meridional* value.

(169.) Such is the principle of that most important geographical operation, the measurement of an arc of the meridian. In its detail, however, a somewhat modified course must be followed. An observatory cannot be mounted and dismounted at every step; so that we cannot identify and measure an exact degree *neither more nor less*. But this is of no consequence, provided we know with equal precision *how much*, more or less, we have measured. In place, then, of measuring this precise aliquot part, we take the more convenient method of measuring from one good observing station to another, *about* a degree, or two or three degrees, as the case may be, apart, and determining by astronomical observation the precise difference of latitudes between the stations.

(170.) Again, it is of great consequence to avoid in this operation every source of uncertainty, because an error committed in the length of a single degree will be multiplied 360 times in the circumference, and nearly 115 times in the diameter of the earth concluded from it. Any error which may affect the astronomical determination of a star's altitude will be especially influential. Now there is still too much uncertainty and fluctuation in the amount of refraction at moderate altitudes, not to make it especially desirable to avoid this source of error. To effect this, we take care to select for observation, at the extreme stations, some star which passes through or near the zeniths of both. The amount of refraction, within a few degrees of the zenith, is very small, and its fluctuations and uncertainty, in point of quantity, so excessively minute as to be utterly inappreciable. Now, it is the same thing whether we observe the *pole* to be raised or depressed a degree, or the *zenith distance* of a star when on the meridian to have changed by the same quantity. If at one station we observe any star to pass through the zenith, and at the other to pass one degree south or north of the zenith, we are sure that the geographical latitudes, or the altitudes of the pole at the two stations, must differ by the same amount.

(171.) Granting that the terminal points of one degree can be ascertained, its *length* may be measured by the methods which will be presently described, as we have before remarked, to within a very few feet. Now, the error which may be committed in fixing each of these terminal points cannot exceed that which may be committed in the observation of the zenith distance of a star, properly situated for the purpose in question. This error, with proper care, can hardly exceed a single second. Supposing we grant the possibility of ten feet of error in the measured length of one degree, and of one second in each of the zenith distances of one star, observed at the northern and southern stations, and, lastly, suppose all these errors to conspire, so as to tend all of them to give a result greater or all less than the truth, it will appear, by a very easy proportion, that the whole amount of error which would be thus entailed on an estimate of the earth's diameter, as concluded from such a measure, would not exceed 544 yards, or about the third part of a mile, and this would be large allowance.

(172.) This, however, supposes that the form of the earth is that of a perfect sphere, and, in consequence, the lengths of its degrees in all parts precisely equal. But when we come to compare the measures of meridional arcs made in various parts of the globe, the results obtained, although they agree sufficiently to show that the supposition of a spherical figure is not *very* remote from the truth, yet exhibit discordances far greater than what we have shown to be attributable to error of observation, and which render it evident that the hypothesis, in strictness of its wording, is untenable. The following table exhibits the lengths of a degree of the meridian (astronomically determined as above described), expressed in British standard feet, as resulting from actual measurement made with all possible care and precision, by commissioners of various nations, men of the first eminence, supplied by their respective governments with the best instruments, and furnished with every facility which could tend to ensure a successful result of their important labours.*

Country.	Latitude of Middle of the Arc.	Arc measured.	Length of the Degree concluded.	Observers.
Sweden - -	66 20 10	1° 37′ 19″	365782	Svanberg.
Russia - -	58 17 37	3 35 5	365368	Struve.
England - -	52 35 45	3 57 13	364971	Roy, Kater.
France - -	46 52 2	8 20 0	364872	Lacaille, Cassini.
France - -	44 51 2	12 22 13	364535	Delambre, Mechain.
Rome - -	42 59 0	2 9 47	364262	Boscovich.
America, U. S. -	39 12 0	1 28 45	363786	Mason, Dixon.
Cape of Good Hope	33 18 30	1 13 17½	364713	Lacaille.
India - -	16 8 22	15 57 40	363044	Lambton, Everest.
India - -	12 32 21	1 34 56	363013	Lambton.
Peru - -	1 31 0	3 7 3	362808	Condamine, &c.

It is evident from a mere inspection of the second and fourth columns of this table that *the measured length of a degree increases with the latitude,* being greatest near the poles, and least near the equator. Let us now consider what interpretation is to be put upon this conclusion, as regards the form of the earth.

(173.) Suppose we held in our hands a model of the earth smoothly turned in wood, it would be, as already observed, so nearly spherical, that neither by the eye nor the touch, unassisted by instruments, could we detect any deviation from that form. Suppose, too, we were debarred from measuring directly across from surface to surface in different directions with any instrument, by which we might at once ascertain whether one diameter were longer than another; how, then, we may ask, are we to ascertain whether it is a true sphere or not? It is clear that we have no resource, but to endeavour to discover, by some nicer

* The first three columns of this table are extracted from among the data given in Professor's Airy's excellent paper "On the Figure of the Earth," in the Encyclopædia Metropolitana.

means than simple inspection or feeling, whether the convexity of its surface is the same in every part; and if not, where it is greatest, and where least. Suppose, then, a thin plate of metal to be cut into a concavity at its edge, so as exactly to fit the surface at A: let this

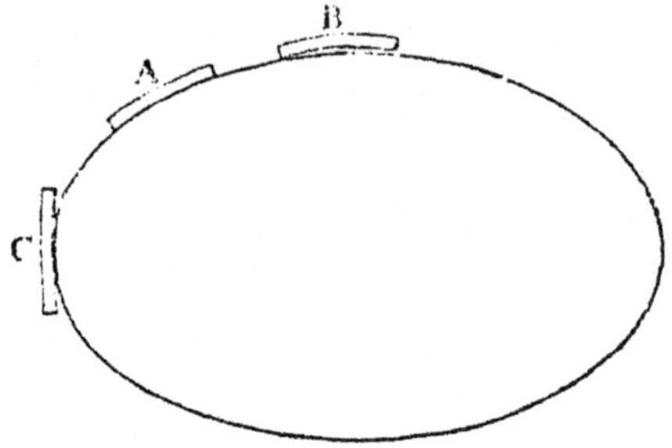

now be removed from A, and applied successively to several other parts of the surface, taking care to keep its plane always on a great circle of the globe, as here represented. If, then, we find any position, B, in which the light can enter in the middle between the globe and plate, or any other, C, where the latter tilts by pressure, or admits the light under its edges, we are sure that the *curvature* of the surface at B is less, and at C greater, than at A.

(174.) What we here do by the application of a metal plate of determinate length and curvature, we do on the earth by the measurement of a degree of variation in the altitude of the pole. Curvature of a surface is nothing but the continual deflection of its tangent from one fixed direction as we advance along it. When, in the *same measured distance of advance*, we find the tangent (which answers to our horizon) to have shifted its position with respect to a fixed direction in space, (such as the axis of the heavens, or the line joining the earth's centre and some given star,) *more* in one part of the earth's meridian than in another, we conclude, of necessity, that the curvature of the surface at the former spot is greater than at the latter; and, *vice versâ*, when, in order to produce the same change of horizon with respect to the pole (suppose 1°), we require to travel over a *longer* measured space at one point than at another, we assign to that point a less curvature. Hence we conclude that *the curvature of a meridional section of the earth is sensibly greater at the equator than towards the poles;* or, in other words, that the earth is not spherical, but *flattened* at the poles, or, which comes to the same, protuberant at the equator.

(175.) Let N A B D E F represent a meridional section of the earth, C its centre, and NA, BD, GE,

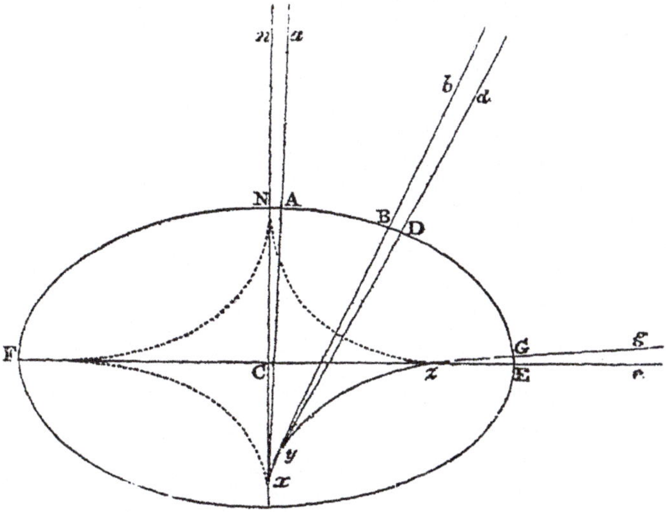

arcs of a meridian, each corresponding to one degree of difference of latitude, or to one degree of variation in the meridian altitude of a star, as referred to the horizon of a spectator travelling along the meridian. Let *n* N, *a* A, *b* B, *d* D, *g* G, *e* E, be the respective directions of the *plumb-line* at the stations N, A, B, D, G, E, of which we will suppose N to be at the pole and E at the equator; then will the tangents to the surface at these points respectively be perpendicular to these directions; and, consequently, if each pair, viz. *n* N and *a* A, *b* B and *d* D, *g* G and *e* E, be prolonged till they intersect each other (at the points *x*, *y*, *z*), the angles N *x* A, B *y* D, G *z* E, will each be one degree, and, therefore, all equal; so that the small curvilinear arcs N A, B D, G E, may be regarded as arcs of circles of one degree each, described about *x*, *y*, *z*, as centres. These are what in geometry are called *centres of curvature*, and the radii *x* N or *x* A, *y* B or *y* D, *z* G or *z* E, represent *radii of curvature*, by which the curvatures at those points are determined and measured. Now, as the arcs of different circles, which subtend equal angles at their respective centres, are in the direct proportion of their radii, and as the arc N A is greater than B D, and that again than G E, it follows that the radius N *x* must be greater than B *y*, and B *y* than E *z*. Thus it appears that the mutual intersections of the plumb-lines will not, as in the sphere, all coincide in one point C, the centre, but will be arranged along a certain curve, *xyz* (which will be rendered more evident by considering a number of intermediate stations). To this curve geometers have given the name of the *evolute* of the curve N A B D G E, from whose centres of curvature it is constructed.

(176.) In the flattening of a round figure at two opposite points, and its protuberance at points rectangularly situated to the former, we recognize the distinguishing feature of the elliptic form. Accordingly, the next and simplest supposition that we can make respecting the nature of the meridian, since it is proved not to be a circle, is, that it is an ellipse, or nearly so, having N S, the axis of the earth, for its shorter, and E F, the equatorial diameter, for its longer axis; and that the form of the earth's surface is that which would arise from making such a curve revolve about its shorter axis N S. This agrees well with the general course of the increase of the degree in going from the equator to the pole. In the ellipse, the radius of curvature at E, the extremity of the longer axis is the least, and at that of the shorter axis, the greatest it admits, and the form of its *evolute* agrees with that here represented.* Assuming, then, that it is an ellipse, the geometrical properties of that curve enable us to assign the proportion between the lengths of its axes which shall correspond to any proposed rate of variation in its curvature, as well as to fix upon their absolute lengths, corresponding to any assigned length of the degree in a given latitude. Without troubling the reader with the investigation, (which may be found in any work on the conic sections,) it will be sufficient to state that the lengths which agree on the whole best with the entire series of meridional arcs which have been satisfactorily measured, are as follow†:—

		Feet.		Miles.
Greater or equatorial diameter	=	41,847,426	=	7925·648
Lesser or polar diameter	=	41,707,620	=	7899·170
Difference of diameters, or polar compression	=	139,806	=	26·478

The proportion of the diameters is very nearly that of 298 : 299, and their difference $\frac{1}{299}$ of the greater, or a very little greater than $\frac{1}{300}$.

(177.) Thus we see that the rough diameter of 8000 miles we have hitherto used, is rather too great, the excess being about 100 miles, or $\frac{1}{80}$th part. We consider it extremely improbable that an error to the extent of five miles can subsist in the diameters, or an uncertainity to that of a tenth of its whole quantity in the compression just stated. As convenient numbers to remember, the reader may bear in mind, that in our latitude there are just as many thousands of feet in a degree of the meridian as there are days in the year (365): that, speaking loosely, a degree is about 70 British statute miles, and a second about 100 feet; and that the equatorial circumference of the earth is a little less than 25,000 miles (24,899).

(178.) The supposition of an elliptic form of the earth's section through the axis is recommended by its

* The dotted lines are the portions of the evolute belonging to the other quadrants.

† See Profess. Airy's Essay before cited.

simplicity, and confirmed by comparing the numerical results we have just set down with those of actual measurement. When this comparison is executed, discordances, it is true, are observed, which, although still too great to be referred to error of measurement, are yet so small, compared to the errors which would result from the spherical hypothesis, as completely to justify our regarding the earth as an ellipsoid, and referring the observed deviations to either local or, if general, to comparatively small causes.

(179.) Now, it is highly satisfactory to find that the general elliptical figure thus *practically* proved to exist, is precisely what *ought theoretically* to result from the rotation of the earth on its axis. For, let us suppose the earth a sphere, at rest, of uniform materials throughout, and externally covered with an ocean of equal depth in every part. Under such circumstances it would obviously be in a state of *equilibrium*; and the water on its surface would have no tendency to run one way or the other. Suppose, now, a quantity of its materials were taken from the polar regions, and piled up all around the equator, so as to produce that difference of the polar and equatorial diameters of 26 miles which we know to exist. It is not less evident that a mountain ridge or equatorial *continent, only,* would be thus formed, from which the water would run down to the excavated part at the poles. However solid matter might rest where it was placed, the liquid part, at least, would not remain there, any more than if it were thrown on the side of a hill. The consequence, therefore, would be the formation of two great polar seas, hemmed in all round by equatorial land. Now, this is by no means the case in nature. The ocean occupies, indifferently, all latitudes, with no more partiality to the polar than to the equatorial. Since, then, as we see, the water occupies an elevation above the centre no less than 13 miles greater at the equator than at the poles, and yet manifests no tendency to leave the former and run towards the latter, it is evident that it must be *retained* in that situation by some adequate *power*. No such power, however, would exist in the case we have supposed, which is therefore not conformable to nature. In other words, the spherical form is *not* the *figure of equilibrium;* and *therefore* the earth is either not at rest, or is so internally constituted as to *attract* the water to its equatorial regions, and retain it there. For the latter supposition there is no *primâ facie* probability, nor any analogy to lead us to such an idea. The former is in accordance with all the phenomena of the apparent diurnal motion of the heavens; and, therefore, if it will furnish us with the *power* in question, we can have no hesitation in adopting it as the true one.

(180.) Now, every body knows that when a weight is whirled round, it acquires thereby a tendency to recede from the centre of its motion; which is called the centrifugal force. A stone whirled round in a sling is a common illustration; but a better, for our present purpose, will be a pail of water, suspended by a cord, and made to *spin round*, while the cord hangs perpendicularly. The surface of the water, instead of remaining horizontal, will become concave, as in the figure. The centrifugal force generates a tendency in *all* the water to leave the axis, and press towards the circumference; it is, therefore, urged against the pail, and forced up its sides, till the excess of height, and consequent increase of pressure downwards, just counterbalances its centrifugal force, and a *state of equilibrium* is attained. The experiment is a very easy and instructive one, and is admirably calculated to show how the *form of equilibrium* accommodates itself to varying circumstances. If, for example, we

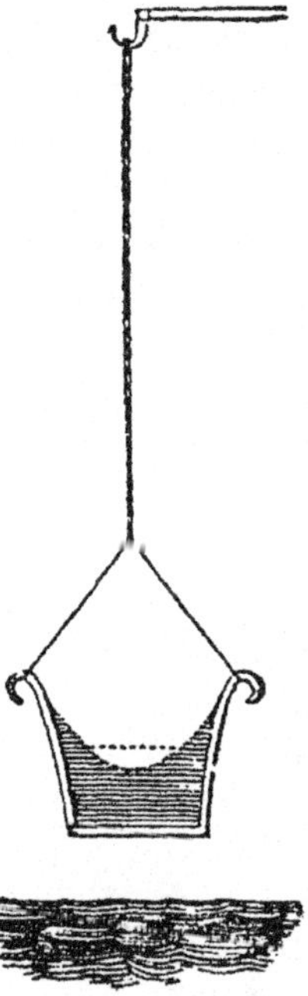

allow the rotation to cease by degrees, as it becomes slower we shall see the concavity of the water regularly diminish; the elevated outward portion will descend, and the depressed central rise, while all the time a perfectly *smooth* surface is maintained, till the rotation is exhausted, when the water resumes its horizontal state.

(181.) Suppose, then, a globe, of the size of the earth, at rest, and covered with a uniform ocean, were to be set in rotation about a certain axis, at first very slowly, but by degrees more rapidly, till it turned round once in twenty-four hours; a centrifugal force would be thus generated, whose general tendency would be to urge the water at every point of the surface to *recede* from the *axis*. A rotation might, indeed, be conceived so swift as to *flirt* the whole ocean from the surface, like water from a mop. But this would require a far greater velocity than what we now speak of. In the case supposed, the *weight* of the water would still keep it *on* the earth; and the tendency to recede from the axis *could* only be satisfied, therefore, by the water leaving the poles, and flowing towards the equator; there heaping itself up in a ridge, just as the water in our pail accumulates against the side; and being retained in opposition to its weight, or natural tendency towards the centre, by the pressure thus caused. This, however, could not take place without laying dry the polar portions of the land in the form of immensely protuberant continents; and the difference of our supposed cases, therefore, is this:— in the former, a great equatorial continent and polar seas would be formed; in the latter, protuberant land would appear at the poles, and a zone of ocean be disposed around the equator. This would be the first or immediate effect. Let us now see what would afterwards happen, in the two cases, if things were allowed to take their natural course.

(182.) The sea is constantly beating on the land, grinding it down, and scattering its worn off particles and fragments, in the state of mud and pebbles, over its bed. Geological facts afford abundant proof that the existing continents have all of them undergone this process, even more than once, and been entirely torn in fragments, or reduced to powder, and submerged and reconstructed. Land, in this view of the subject, loses its attribute of fixity. As a mass it might hold together in opposition to forces which the water freely obeys; but in its state of successive or simultaneous degradation, when disseminated through the water, in the state of sand or mud, it is subject to all the impulses of that fluid. In the lapse of time, then, the protuberant land in both cases would be destroyed, and spread over the bottom of the ocean, filling up the lower parts, and tending continually to remodel the surface of the solid nucleus, in correspondence with the *form of equilibrium* in both cases. Thus, after a sufficient lapse of time, in the case of an earth at rest, the equatorial continent, thus forcibly constructed, would again be levelled and transferred to the polar excavations, and the spherical figure be so at length restored. In that of an earth in rotation, the polar protuberances would gradually be cut down and disappear, being transferred to the equator (as being *then* the *deepest sea*), till the earth would assume by degrees the form we observe it to have — that of a flattened or *oblate* ellipsoid.

(183.) We are far from meaning here to trace the process *by which* the earth really assumed its actual form; all we intend is, to show that this is the form to which, under the condition of a rotation on its axis, it must *tend;* and which it would attain, even if originally and (so to speak) perversely constituted otherwise.

(184.) But, further, the dimensions of the earth and the time of its rotation being known, it is easy thence to calculate the exact amount of the centrifugal force *, which, at the equator, appears to be $\frac{1}{289}$th part of the force or weight by which all bodies, whether solid or liquid, tend to fall towards the earth. By this

* See Cab. Cyc., MECHANICS, ch. viii.

fraction of its weight, then, the sea at the equator is *lightened*, and thereby rendered susceptible of being supported at a higher level, or more remote from the centre than at the poles, where no such counteracting force exists; and where, in consequence, the water may be considered as *specifically heavier*. Taking this principle as a guide, and combining it with the laws of gravity (as developed by Newton, and as hereafter to be more fully explained), mathematicians have been enabled to investigate, *à priori*, what would be the figure of equilibrium of such a body, constituted internally as we have reason to believe the earth to be; covered wholly or partially with a fluid; and revolving uniformly in twenty-four hours; and the result of this enquiry is found to agree very satisfactorily with what experience shows to be the case. From their investigations it appears that the form of equilibrium is, in fact, no other than an oblate ellipsoid, of a degree of ellipticity very nearly identical with what is observed, and which would be no doubt accurately so, did we know the internal constitution and materials of the earth.

(185.) The confirmation thus incidentally furnished, of the hypothesis of the earth's rotation on its axis, cannot fail to strike the reader. A deviation of its figure from that of a sphere was not contemplated among the original reasons for adopting that hypothesis, which was assumed solely on account of the easy explanation it offers of the apparent diurnal motion of the heavens. Yet we see that, once admitted, it draws with it, as a necessary consequence, this other remarkable phenomenon, of which no other satisfactory account could be rendered. Indeed, so direct is their connection, that the ellipticity of the earth's figure was discovered and demonstrated by Newton to be a consequence of its rotation, and its amount actually calculated by him, long before any measurements had suggested such a conclusion. As we advance with our subject, we shall find the same simple principle branching out into a whole train of singular and important consequences, some obvious enough, others which at first seem entirely unconnected with it, and which, until traced by Newton up to this their origin, had ranked among the most inscrutable arcana of astronomy, as well as among its grandest phenomena.

(186.) Of its more obvious consequences, we may here mention one which falls in naturally with our present subject. If the earth really revolve on its axis, this rotation must generate a centrifugal force (see art. 184.), the effect of which must of course be to counteract a certain portion of the *weight* of every body situated at the equator, as compared with its weight at the poles, or in any intermediate latitudes. Now, this is fully confirmed by experience. There is actually observed to exist a difference in the *gravity*, or downward tendency, of one and the same body, when conveyed successively to stations in different latitudes. Experiments made with the greatest care, and in every accessible part of the globe, have fully demonstrated the fact of a regular and progressive increase in the weights of bodies corresponding to the increase of latitude, and fixed its amount and the law of its progression. From these it appears, that the extreme amount of this variation of gravity, or the difference between the equatorial and polar weights of one and the same mass of matter, is 1 part in 194 of its whole weight, the rate of increase in travelling from the equator to the pole being *as the square of the sine of the latitude.*

(187.) The reader will here naturally enquire, what is *meant* by speaking of the same body as having different weights at different stations; and, how such a fact, if true, can be ascertained. When we weigh a body by a balance or a steelyard we do but counteract its weight by the equal weight of another body under the very same circumstances; and if both the body weighed and its counterpoise be removed to another station, their gravity, if changed at all, will be changed equally, so that they will still continue to counterbalance each other. A difference in the intensity of gravity could, therefore, never be detected by these means; nor is it in *this* sense that we assert that a body weighing 194 pounds at the equator will weigh 195 at the pole. If counterbalanced in a scale or steelyard at the former station, an additional pound placed in one or other scale at the latter would inevitably sink the beam.

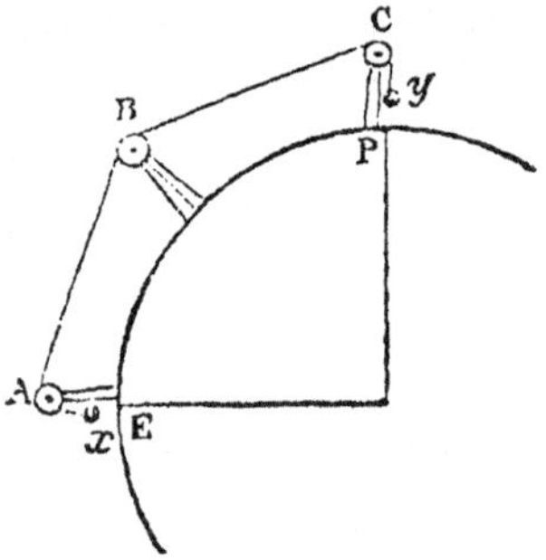

(188.) The meaning of the proposition may be thus explained:—Conceive a weight x suspended at the equator by a string without weight passing over a pulley, A, and conducted (supposing such a thing possible) over other pulleys, such as B, round the earth's convexity, till the other end hung down at the pole, and there sustained the weight y. If, then, the weights x and y were such as, at any one station, equatorial or polar, would exactly counterpoise each other on a balance, or when suspended side by side over a single pulley, they would not counterbalance each other in this supposed situation, but the polar weight y would preponderate; and to restore the equipoise the weight x must be increased by $\frac{1}{194}$th part of its quantity.

(189.) The means by which this variation of gravity may be shown to exist, and its amount measured, are twofold (like all estimations of mechanical power), statical and dynamical. The former consists in putting the gravity of a weight in equilibrium, not with that of another weight, but with a natural power of a different kind not liable to be affected by local situation. Such a power is the elastic force of a spring. Let ABC be a strong support of brass standing on the foot AED cast in one piece with it, into which is let a smooth plate of agate, D, which can be adjusted to perfect horizontality by a level. At C let a spiral spring G be attached, which carries at its lower end a weight F, polished and convex below. The length and strength of the spring must be so adjusted that the weight F shall be sustained by it just to swing clear of contact with the agate plate in the highest latitude at which it is intended to use the instrument. Then, if small weights be added cautiously, it may be made to descend till it *just grazes* the agate, a contact which can be made with the utmost imaginable delicacy. Let these weights be noted; the weight F detached; the spring G carefully lifted off its hook, and secured, for travelling, from rust, strain, or disturbance, and the whole apparatus conveyed to a station in a lower latitude. It will then

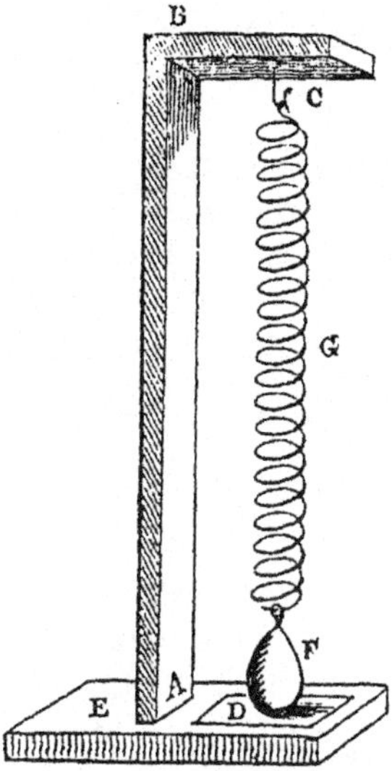

be found, on remounting it, that, although loaded with the same additional weights as before, the weight F will no longer have power enough to stretch the spring to the extent required for producing a similar contact. More weights will require to be added; and the additional quantity necessary will, it is evident, measure the difference of gravity between the two stations, as exerted on the whole quantity of pendent matter, *i.e.* the sum of the weight of F and *half* that of the spiral spring itself. Granting that a spiral spring can be constructed of such strength and dimensions that a weight of 10,000 grains, including its own, shall produce an elongation of 10 inches without permanently straining it *, one addi-

* Whether the process above described could ever be so far perfected and refined as to become a substitute for the use of the pendulum must depend on the degree of permanence and uniformity of action of springs, on the constancy or variability of the effect of temperature, on their elastic force, on the possibility of transporting them, absolutely unaltered, from place to place, &c. The great advantages, however, which such an apparatus and mode of observation would possess, in point of convenience, cheapness, portability, and expedition, over the present laborious, tedious, and expensive process, render the attempt well worth making.

tional grain will produce a further extension of $\frac{1}{1000}$th of an inch, a quantity which cannot possibly be mistaken in such a contact as that in question. Thus we should be provided with the means of measuring the power of gravity at any station to within $\frac{1}{10000}$th of its whole quantity.

(190.) The other, or dynamical process, by which the force urging any given weight to the earth may be determined, consists in ascertaining the velocity imparted by it to the weight when suffered to fall freely in a given time, as one second. This velocity cannot, indeed, be directly measured; but indirectly, the principles of mechanics furnish an easy and certain means of deducing it, and, consequently, the intensity of gravity, by observing the oscillations of a pendulum. It is proved in mechanics (see Cab. Cyc., MECHANICS, 216.), that, if one and the same pendulum be made to oscillate at different stations, or under the influence of different forces, and the numbers of oscillations made in the same time in each case be counted, the intensities of the forces will be to each other as the squares of the numbers of oscillations made, and thus their proportion becomes known. For instance, it is found that, under the equator, a pendulum of a certain form and length makes 86,400 vibrations in a mean solar day; and that, when transported to London, the same pendulum makes 86,535 vibrations in the same time. Hence we conclude, that the intensity of the force urging the pendulum downwards at the equator is to that at London as 86400 to 86535, or as 1 to 1 00315; or, in other words, that a mass of matter at the equator weighing 10,000 pounds exerts the same pressure on the ground, the same effort to crush a body placed below it, that 10,031½ of *the same pounds*, transported to London, would exert there.

(191.) Experiments of this kind have been made, as above stated, with the utmost care and minutest precaution to ensure exactness in all accessible latitudes; and their general and final result has been, to give $\frac{1}{194}$

for the fraction expressing the difference of gravity at the equator and poles. Now, it will not fail to be noticed by the reader, and will, probably, occur to him as an objection against the explanation here given of the fact by the earth's rotation, that this differs materially from the fraction $\frac{1}{289}$ expressing the centrifugal force at the equator. The difference by which the former fraction exceeds the latter is $\frac{1}{590}$, a small quantity in itself, but still far too large, compared with the others in question, not to be distinctly accounted for, and not to prove fatal to this explanation if it will not render a strict account of it.

(192.) The mode in which this difference arises affords a curious and instructive example of the indirect influence which mechanical causes often exercise, and of which astronomy furnishes innumerable instances. The rotation of the earth gives rise to the centrifugal force; the centrifugal force produces an ellipticity in the form of the earth itself; and this very ellipticity of form modifies its power of attraction on bodies placed at its surface, and thus gives rise to the difference in question. Here, then, we have the same cause exercising at once a direct and an indirect influence. The amount of the former is easily calculated, that of the latter with far more difficulty, by an intricate and profound application of geometry, whose steps we cannot pretend to trace in a work like the present, and can only state its nature and result.

(193.) The weight of a body (considered as undiminished by a centrifugal force) is the effect of the earth's attraction on it. This attraction, as Newton has demonstrated, consists, not in a tendency of all matter to any one particular centre, but in a disposition of every particle of matter in the universe to press towards, and if not opposed to approach to, every other. The attraction of the earth, then, on a body placed on its surface, is not a simple but a complex force, resulting from the separate attractions of all its parts. Now, it is evident, that if the earth were a perfect sphere, the attraction exerted by it on a body any where placed on its surface, whether at its equator or pole, must be exactly alike,— for the simple reason of the exact symmetry of the sphere in every direction. It is not less evident that, the earth being elliptical, and this symmetry or similitude of all its parts not existing, the same result cannot be expected. A body placed at the equator, and a similar one at the pole of a flattened ellipsoid, stand in a different geometrical relation to the mass as a whole. This difference, without entering further into particulars, may be expected to draw with it a difference in its forces of attraction on the two bodies. Calculation confirms this idea. It is a question of purely mathematical investigation, and has been treated with perfect clearness and precision by Newton, Maclaurin, Clairaut, and many other eminent geometers; and the result of their investigations is to show that, owing to the elliptic form of the earth alone, and independent of the centrifugal force, its attraction ought to increase the weight of a body in going from the equator to the pole by almost exactly $\frac{1}{590}$th part; which, together with $\frac{1}{289}$th due to the centrifugal force, make up the whole quantity, $\frac{1}{194}$th, observed.

(194.) Another great geographical phenomenon, which owes its existence to the earth's rotation, is that of the trade-winds. These mighty currents in our atmosphere, on which so important a part of navigation depends, arise from, 1st, the unequal exposure of the earth's surface to the sun's rays, by which it is unequally heated in different latitudes; and, 2dly, from that general law in the constitution of all fluids, in virtue of which they occupy a larger bulk, and become specifically lighter when hot than when cold. These causes, combined with the earth's rotation from west to east, afford an easy and satisfactory explanation of the magnificent phenomena in question.

(195.) It is a matter of observed fact, of which we shall give the explanation farther on, that the sun is constantly vertical over some one or other part of the earth between two parallels of latitude, called the tropics, respectively $23\frac{1}{2}°$ north, and as much south of the equator; and that the whole of that zone or belt of the earth's surface included between the tropics, and equally divided by the equator, is, in consequence of the great altitude attained by the sun in its diurnal course, maintained at a much higher temperature than those regions to the north and south which lie nearer the poles. Now, the heat thus acquired by the earth's surface is communicated to the incumbent air, which is thereby expanded, and rendered specifically lighter than the air incumbent on the rest of the globe. It is, therefore, in obedience to the general laws of hydrostatics, displaced and buoyed up from the surface, and its place occupied by colder, and therefore heavier air, which glides in, on both sides, along the surface, from the regions beyond the tropics; while the displaced air, thus raised above its due level, and unsustained by any lateral pressure, flows over, as it were, and forms an upper current in the contrary direction, or toward the poles; which, being cooled in its course, and also sucked down to supply the deficiency in the extra-tropical regions, keeps up thus a continual circulation.

(196.) Since the earth revolves about an axis passing through the poles, the equatorial portion of its surface has the greatest velocity of rotation, and all other parts less in the proportion of the radii of the circles of latitude to which they correspond. But as the air, when relatively and apparently at rest on any part of the earth's surface, is only so because in reality it participates in the motion of rotation proper to that part, it follows that when a mass of air near the poles is transferred to the region near the equator by any impulse urging it directly towards that circle, in every point of its progress towards its new situation it must be found deficient in rotatory velocity, and therefore unable to keep up with the speed of the new surface over which it is brought. Hence, the currents of air which set in towards the equator from the north and south must, as they glide along the surface, at the same time lag, or hang back, and *drag upon* it in the direction *opposite* to the earth's rotation, *i. e.* from east to west. Thus these currents, which but for the rotation would be simply northerly and southerly winds, acquire, from this cause, a *relative* direction towards the west, and assume the character of permanent north-easterly and south-easterly winds.

(197.) Were any considerable mass of air to be *suddenly* transferred from beyond the tropics to the equator, the difference of the rotatory velocities proper to the two situations would be so great as to produce not merely a wind, but a tempest of the most destructive violence. But this is not the case: the advance of the air from the north and south is gradual, and all the while the earth is continually acting on, and by the friction of its surface accelerating its rotatory velocity. Supposing its progress towards the equator to cease at any point, this cause would almost immediately communicate to it the deficient motion of rotation, after which it would revolve quietly with the earth, and be at relative rest. We have only to call to mind the comparative *thinness* of the coating which the atmosphere forms around the globe (art. 34.), and the immense *mass* of the latter, compared with the former (which it exceeds at least 100,000,000 times), to appreciate fully the absolute *command* of any extensive territory of the earth over the atmosphere immediately incumbent on it, in point of motion.

(198.) It follows from this, then, that as the winds on both sides approach the equator, their easterly tendency must diminish.* The lengths of the diurnal circles increase very slowly in the immediate vicinity of the equator, and for several degrees on either side of it hardly change at all. Thus the friction of the surface has more time to act in accelerating the velocity of the

* See Captain Hall's "Fragments of Voyages and Travels," 2d series, vol. i. p. 162., where this is very distinctly, and, so far as I am aware, for the first time, reasoned out. (*Author.*)

air, bringing it towards a state of *relative* rest, and diminishing thereby the relative set of the currents from east to west, which, on the other hand, is feebly, and, at length, not at all reinforced by the cause which originally produced it. Arrived, then, at the equator, the trades must be expected to lose their easterly character altogether. But not only this but the northern and southern currents here meeting and opposing, will mutually destroy each other, leaving only such preponderancy as may be due to a difference of local causes acting in the two hemispheres, — which in some regions around the equator may lie one way, in some another.

(199.) The result, then, must be the production of two great tropical belts, in the northern of which a constant north-easterly and in the southern a south-easterly, wind must prevail, while the winds in the equatorial belt, which separates the two former, should be comparatively calm and free from any steady prevalence of easterly character. All these consequences are agreeable to observed fact, and the system of aërial currents above described constitutes in reality what is understood by the regular *trade winds*.*

(200.) The constant friction thus produced between the earth and atmosphere in the regions near the equator must (it may be objected) by degrees reduce and at length destroy the rotation of the whole mass. The laws of dynamics, however, render such a consequence, generally, impossible; and it is easy to see, in the present case, where and how the compensation takes place. The heated equatorial air, while it rises and flows over towards the poles, carries with it the rotatory velocity due to its equatorial situation into a higher latitude, where the earth's surface has less motion. Hence, as it travels northward or southward, it will *gain* continually more and more on the surface of the earth in its diurnal motion, and assume constantly more and more a *westerly* relative direction; and when at length it returns to the surface, in its circulation, which it must

* See the work last cited.

do more or less in all the interval between the tropics and the poles, it will act on it by its friction as a powerful south-west wind in the northern hemisphere, and a north-west in the southern, and restore to it the impulse taken up from it at the equator. We have here the origin of the south-west and westerly gales so prevalent in our latitudes, and of the almost universal westerly winds in the North Atlantic, which are, in fact, nothing else than a part of the general system of the re-action of the trades, and of the process by which the equilibrium of the earth's motion is maintained under their action.*

(201.) In order to construct a map or model of the earth, and obtain a knowledge of the distribution of sea and land over its surface, the forms of the outlines of its continents and islands, the courses of its rivers and mountain chains, and the relative situations, with respect to each other, of those points which chiefly interest us, as centres of human habitation, or from other causes; it is necessary to possess the means of determining correctly the situation of any proposed station on its surface. For this two elements require to be known, the latitude and longitude, the former assigning its distance from the poles or the equator, the latter, the meridian on which that distance is to be reckoned. To these, in strictness, should be added, its height above the sea level; but the consideration of this had better be deferred, to avoid complicating the subject.

As it is our object merely to illustrate the mode in which the earth's rotation affects the atmosphere on the great scale, we omit all consideration of local periodical winds, such as monsoons, &c.

It seems worth enquiry, whether hurricanes in tropical climates may not arise from portions of the upper currents prematurely diverted downwards beforetheir relative velocity has been sufficiently reduced by friction on, and gradual mixing with, the lower strata; and so dashing upon the earth with that tremendous velocity which gives them their destructive character, and of which hardly any rational account has yet been given. Their course, generally speaking, is in opposition to the regular trade wind, as it ought to be, in conformity with this idea. (Young's Lectures, i. 704.) But it by no means follows that this must always be the case. In general, a rapid transfer, either way, in latitude, of any mass of air which local or temporary causes might carry *above the immediate reach of the friction of the earth's surface*, would give a fearful exaggeration to its velocity. Wherever such a mass should strike the earth, a hurricane might arise; and should two such masses encounter in mid-air, a tornado of any degree of intensity on record might easily result from their combination.—*Author.*

(202.) The latitude of a station on a sphere would be merely the length of an arc of the meridian, intercepted between the station and the nearest point of the equator, reduced into degrees. (See art. 86.) But as the earth is elliptic, this mode of conceiving latitudes becomes inapplicable, and we are compelled to resort for our definition of latitude to a generalization of that property, (art. 95.) which affords the readiest means of determining it by observation, and which has the advantage of being independent of the figure of the earth, which, after all, is not *exactly* an ellipsoid, or any known geometrical solid. The latitude of a station, then, is the altitude of the elevated pole, and is, therefore, astronomically determined by those methods already explained for ascertaining that important element. In consequence, it will be remembered that, to make a perfectly correct map of the whole, or any part of the earth's surface, equal differences of latitude are not represented by exactly equal intervals of surface.

(203.) To determine the latitude of a station, then, is easy. It is otherwise with its longitude, whose exact determination is a matter of more difficulty. The reason is this:—as there are no meridians marked upon the earth, any more than parallels of latitude, we are obliged in this case, as in the case of the latitude, to resort to marks external to the earth, *i. e.* to the heavenly bodies, for the objects of our measurement; but with this difference in the two cases—to observers situated at stations on the same *meridian* (*i. e.* differing in latitude) the heavens present different aspects at *all* moments. The portions of them which become visible in a complete diurnal rotation are not the same, and stars which are common to both describe circles differently inclined to their horizons, and differently divided by them, and attain different altitudes. On the other hand, to observers situated on the same *parallel* (*i. e.* differing only in longitude) the heavens present the same aspects. Their visible portions are the same; and the same stars describe circles equally inclined, and similarly divided by their horizons, and attain the same altitudes. In the former case there *is*, in the latter there is *not*, any thing in the appearance of the heavens, watched through a whole diurnal rotation, which indicates a difference of locality in the observer.

(204.) But no two observers, at different points of the earth's surface, can have at *the same instant* the same celestial hemisphere visible. Suppose, to fix our ideas, an observer stationed at a given point of the equator, and that at the moment when he noticed some bright star to be in his zenith, and therefore on his meridian, he should be suddenly transported, in an instant of time, round one quarter of the globe in a *westerly* direction, it is evident that he will no longer have the same star vertically above him: it will now appear to him to be just rising, and he will have to wait six hours before it again comes to his zenith, *i. e.* before the earth's rotation from west to east *carries him back again* to the line joining the star and the earth's centre from which he set out.

(205.) The difference of the cases, then, may be thus stated, so as to afford a key to the astronomical solution of the problem of the longitude. In the case of stations differing only in latitude, the same star comes to the meridian at the same *time*, but at different *altitudes*. In that of stations differing only in longitude, it comes to the meridian at the same *altitude* but at different *times*. Supposing, then, that an observer is in possession of any means by which he can certainly ascertain the *time* of a known star's transit across his meridian, he knows his longitude; or if he knows the difference between its times of transit across his meridian and across that of any other station, he knows their difference of longitudes. For instance, if the same star pass the meridian of a place A at a certain moment, and that of B exactly one hour of sidereal time, or one twenty-fourth part of the earth's diurnal period, later, then the difference of longitudes between A and B is one hour of time or 15°, and B is so much west of A.

(206.) In order to a perfectly clear understanding of the principle on which the problem of finding the longitude by astronomical observations is resolved, the reader must learn to distinguish between time, in the abstract, as common to the whole universe, and therefore reckoned from an epoch independent of local situation, and *local time*, which reckons, at each particular place, from an epoch, or initial instant, determined by local convenience. Of time reckoned in the former, or abstract manner, we have an example in what we have before defined as equinoctial time, which dates from an epoch determined by the sun's motion among the stars. Of the latter, or *local* reckoning, we have instances in every sidereal clock in an observatory, and in every town clock for common use. Every astronomer regulates, or aims at regulating, his sidereal clock, so that it shall indicate $0^h\ 0^m\ 0^s$, when a certain point in the heavens, called the equinox, is on the meridian of his station. This is the *epoch* of his sidereal time; which is, therefore, entirely a *local* reckoning. It gives no information to say that an event happened at such and such an hour of sidereal time, unless we particularize the station to which the sidereal time meant appertains. Just so it is with mean or common time. This is also a local reckoning, having for its epoch *mean noon*, or the average of all the times throughout the year, when the sun is on the meridian *of that particular place to which it belongs;* and, therefore, in like manner, when we date any event by mean time, it is necessary to name the place, or particularize *what* mean time we intend. On the other hand, a date by equinoctial time is absolute, and requires no such explanatory addition.

(207.) The astronomer sets and regulates his sidereal clock by observing the meridian passages of the more conspicuous and well known stars. Each of these holds in the heavens a certain determinate and known place with respect to that imaginary point called the equinox, and by noting the times of their passage in succession by his clock he knows when the equinox passed. At that moment his clock ought to have marked $0^h\ 0^m\ 0^s$; and if it did not, he knows and can correct its error, and by the agreement or disagreement of the errors assigned by each star he can ascertain whether his clock is correctly regulated to go twenty-four hours in one diurnal period, and if not, can ascertain and allow for its rate. Thus, although his clock may not, and indeed cannot, either be set correctly, or go truly, yet by applying its *error* and *rate* (as they are technically termed), he can correct its indications, and ascertain the exact sidereal times corresponding to them, and proper to his locality. This indispensable operation is called getting his *local time*. For simplicity of explanation, however, we shall suppose the clock a perfect instrument; or, which comes to the same thing, its error and rate applied at every moment it is consulted, and included in its indications.

(208.) Suppose, now, two observers, at distant stations, A and B, each independently of the other, to set and regulate his clock to the true sidereal time of his station. It is evident that if one of these clocks could be taken up without deranging its going, and set down by the side of the other, they would be found, on comparison, to differ by the exact difference of their local epochs; that is, by the time occupied by the equinox, or by any star, in passing from the meridian of A to that of B: in other words, by their difference of longitude, expressed in sidereal hours, minutes, and seconds.

(209.) A pendulum clock cannot be thus taken up and transported from place to place without derangement, but a chronometer may. Suppose, then, the observer at B to use a chronometer instead of a clock, he may, by bodily transfer of the instrument to the other station, procure a direct comparison of sidereal times, and thus obtain his longitude from A. And even if he employ a clock, yet by comparing it first with a good chronometer, and then transferring the latter instrument for comparison with the other clock, the same end will be accomplished, provided the going of the chronometer can be depended on.

(210.) Were chronometers perfect, nothing more complete and convenient than this mode of ascertaining differences of longitude could be desired. An observer, provided with such an instrument, and with a portable transit, or some equivalent method of determining the local time at any given station, might, by journeying from place to place, and observing the meridian passages of stars at each, (taking care not to alter his chronometer, or let it run down,) ascertain their differences of longitude with any required precision. In this case, the same time-keeper being used at every station, if, at one of them, A, it mark true sidereal time, at any other, B, it will be just so much sidereal time in error as the difference of longitudes of A and B is equivalent to: in other words, the longitude of B from A will appear as the error of the time-keeper on the local time of B. If he travel westward, then his chronometer will appear continually to gain, although it really goes correctly. Suppose, for instance, he set out from A, when the equinox was on the meridian, or his chronometer at 0^h, and in twenty-four hours (sid. time) had travelled 15° westward to B. At the moment of arrival there, his chronometer will again point to 0^h; but the equinox will be, not on his new meridian, but on that of A, and he must wait one hour more for its arrival at that of B. When it does arrive there, then his watch will point not to 0^h but to 1^h, and will therefore be 1^h *fast* on the local time of B If he travel eastward, the reverse will happen

(211.) Suppose an observer now to set out from any station as above described, and constantly travelling westward to make the tour of the globe, and return to the point he set out from. A singular consequence will happen: he will have lost a day in his reckoning of time. He will enter the day of his arrival in his diary, as Monday, for instance, when, in fact, it is Tuesday.

The reason is obvious. Days and nights are caused by the alternate appearance of the sun and stars, as the rotation of the earth carries the spectator round to view them in succession. So many turns as he makes round the centre, so many days and nights will he experience. But if he travel once round the globe in the direction of its motion, he will, on his arrival, have really made one turn *more* round its centre; and if in the opposite direction, one turn *less* than if he had remained stationary at one point of its surface: in the former case, then, he will have witnessed one alternation of day and night more, in the latter one less, than if he had trusted to the rotation of the earth alone to carry him round. As the earth revolves from west to east, it follows that a westward direction of his journey, by which he counteracts its rotation, will cause him to lose a day, and an eastward direction, by which he conspires with it, to gain one. In the former case, all his days will be longer; in the latter, shorter than those of a stationary observer. This contingency has actually happened to circumnavigators. Hence, also, it must necessarily happen that distant settlements, *on the same meridian*, will differ a day in their usual reckoning of time, according as they have been colonized by settlers arriving in an eastward or in a westward direction,—a circumstance which may produce strange confusion when they come to communicate with each other. The only mode of correcting the ambiguity, and settling the disputes which such a difference may give rise to, consists in having recourse to the equinoctial date, which can never be ambiguous.

(212.) Unfortunately for geography and navigation, the chronometer, though greatly and indeed wonderfully improved by the skill of modern artists, is yet far too imperfect an instrument to be relied on implicitly. However such an instrument may preserve its uniformity of rate for a few hours, or even days, yet in long absences from home the chances of error and accident become so multiplied, as to destroy all security of reliance on even the best. To a certain extent this may, indeed, be remedied by carrying out several, and using them as checks on each other; but, besides the expense and trouble, this is only a palliation of the evil — the great and fundamental, — as it is the only one to which *the determination of longitudes by time-keepers* is liable. — It becomes necessary, therefore, to resort to other means of communicating from one station to another a knowledge of its local time, or of propagating from some principal station, as a centre, its local time as a universal standard with which the local time at any other, however situated, may be at once compared, and thus the longitudes of all places be referred to the meridian of such central point.

(213.) The simplest and most accurate method by which this object can be accomplished, when circumstances admit of its adoption, is that by telegraphic signal. Let A and B be two observatories, or other stations, provided with accurate means of determining *their respective local times*, and let us first suppose them visible from each other. Their clocks being regulated, and their errors and rates ascertained and applied, let a signal be made at A, of some sudden and definite kind. such as the flash of gunpowder, the explosion of a rocket, the sudden extinction of a bright light, or any other which admits of no mistake, and can be seen at great distances. The moment of the signal being made must be noted by *each* observer at his respective clock or watch, as if it were the transit of a star, or any astronomical phenomenon, and the error and rate of the clock at each station being applied, the local time of the signal at each is determined. Consequently, when the observers communicate their observations of the signal to each other, since (owing to the almost instantaneous transmission of light) it must have been seen at the same *absolute* instant by both, the difference of their local times, and therefore of their longitudes, becomes known, For example; at A the signal is observed to happen at $5^h\ 0^m\ 0^s$ sid. time at A, as obtained by applying the error and rate to the time shown by the clock at A.

when the signal was seen there. At B the same signal was seen at $5^h\ 4^m\ 0^s$, sid. time at B, similarly deduced from the time noted by the clock at B, by applying *its* error and rate. Consequently, the difference of their local epochs is $4^m\ 0^s$, which is also their difference of longitudes in time, or $1°\ 0'\ 0''$ in hour angle.

(214.) The accuracy of the final determination may be increased by making and observing several signals at stated intervals, each of which affords a comparison of times, and the mean of all which is, of course, more to be depended on than the result of any single comparison. By this means, the error introduced by the comparison of clocks may be regarded as altogether destroyed.

(215.) The distances at which signals can be rendered visible must of course depend on the nature of the interposed country. Over sea the explosion of rockets may easily be seen at fifty or sixty miles; and in mountainous countries the flash of gunpowder in an open spoon may be seen, if a proper station be chosen for its exhibition, at much greater distances. The interval between the stations of observation may also be increased by causing the signals to be made not at one of them, but at an intermediate point; for, provided they are seen by both parties, it is a matter of indifference where they are exhibited. Still the interval which could be thus embraced would be very limited, and the method in consequence of little use, but for the following ingenious contrivance, by which it can be extended to any distance, and carried over any tract of country, however difficult.

(216.) This contrivance consists in establishing, between the extreme stations, whose difference of longitude is to be ascertained, and at which the local times are observed, a chain of intermediate stations, alternately destined for signals and for observers. Thus, let A and Z be the extreme stations. At B let a signal station be established, at which rockets, &c. are fired at stated intervals. At C let an observer be placed, provided with a chronometer; at D, another signal station; at E, another observer and chronometer; and so on till

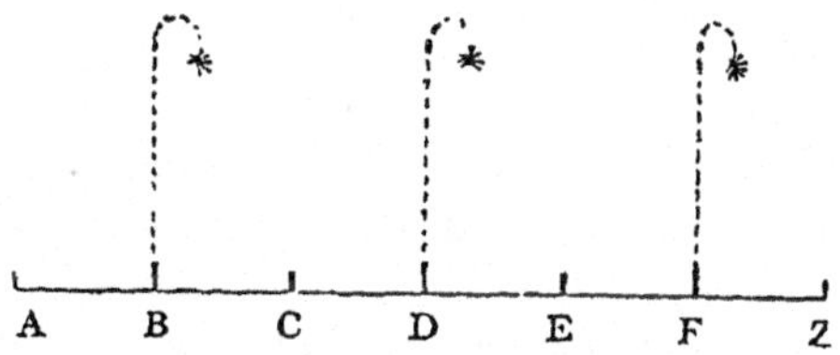

the whole line is occupied by stations so arranged, that the signals at B can be seen from A and C; those at D, from C and E; and so on. Matters being thus arranged, and the errors and rates of the clocks at A and Z ascertained by astronomical observation, let a signal be made at B, and observed at A and C, and the times noted. Thus the difference between A's clock and C's chronometer becomes known. After a short interval (five minutes for instance) let a signal be made at D, and observed by C and E. Then will the difference between their respective chronometers be determined; and the difference between the former and the clock at A being already ascertained, the difference between the clock A and chronometer E is therefore known. This, however, supposes that the intermediate chronometer C has kept true sidereal time, or at least a known rate, in the interval between the signals. Now this interval is purposely made so very short, that no instrument of any pretension to character can possibly produce an appreciable amount of error in its lapse. Thus the time propagated from A to C may be considered as handed over, without gain or loss (save from error of observation), to E. Similarly, by the signal made at F, and observed at E and Z, the time so transmitted to E is forwarded on to Z; and thus at length the clocks at A and Z are compared. The process may be repeated as often as is necessary to destroy error by a mean of results; and when the line of stations is numerous, by keeping up a succession of signals, so as to allow each observer to note alternately those on either side, which is easily pre-arranged, many comparisons may be kept running along the line at once, by which time is saved, and other advantages obtained.* In important cases the process is usually repeated on several nights in succession.

(217.) In place of artificial signals, natural ones, when they occur sufficiently definite for observation, may be equally employed. In a clear night the number of those singular meteors, called shooting stars, which may be observed, is usually very great; and as they are sudden in their appearance and disappearance, and from the great height at which they have been ascertained to take place are visible over extensive regions of the earth's surface, there is no doubt that they may be resorted to with advantage, by previous concert and agreement between distant observers to watch and note them.†

(218.) Another species of natural signal, of still greater extent and universality (being visible at once over a whole terrestrial hemisphere), is afforded by the eclipses of Jupiter's satellites, of which we shall speak more at large when we come to treat of those bodies. Every such eclipse is an event which possesses one great advantage in its applicability to the purpose in question, viz. that the time of its happening, at any fixed station, such as Greenwich, can be *predicted* from a long course of previous recorded observation and calculation thereon founded, and that this prediction is sufficiently precise and certain, to stand in the place of a corresponding observation. So that an observer at any other station wherever, who shall have observed one or more of these eclipses, and ascertained his local time, instead of waiting for a communication with Greenwich, to inform him at what moment the eclipse took place there, may use the *predicted Greenwich time* instead, and thence, at once, and on the spot, determine his lon-

* For a complete account of this method, and the mode of deducing the most advantageous result from a combination of all the observations, see a paper on the difference of longitudes of Greenwich and Paris, Phil. Trans. 1826; by the author of this volume.

† This idea was first suggested by the late Dr. Maskelyne.

gitude. This mode of ascertaining longitudes is, however, as will hereafter appear, not susceptible of great exactness, and should only be resorted to when others cannot be had. The nature of the observation also is such that it cannot be made at sea; so that, however useful to the geographer, it is of no advantage to navigation.

(219.) But such phenomena as these are of only occasional occurrence; and in their intervals, and when cut off from all communication with any fixed station, it is indispensable to possess some means of determining longitudes, on which not only the geographer may rely for a knowledge of the exact position of important stations on land in remote regions, but on which the navigator can securely stake, at every instant of his adventurous course, the lives of himself and comrades, the interests of his country, and the fortunes of his employers. Such a method is afforded by LUNAR OBSERVATIONS. Though we have not yet introduced the reader to the phenomena of the moon's motion, this will not prevent us from giving here the exposition of the principle of the lunar method; on the contrary, it will be highly advantageous to do so, since by this course we shall have to deal with the naked principle, apart from all the peculiar sources of difficulty with which the lunar theory is encumbered, but which are, in fact, completely extraneous to the *principle* of its application to the problem of the longitudes, which is quite elementary.

(220.) If there were in the heavens a clock furnished with a dial-plate and hands, which always marked Greenwich time, the longitude of any station would be at once determined, so soon as the *local time* was known, by comparing it with this clock. Now, the offices of the dial-plate and hands of a clock are these:—the former carries a set of marks upon it, whose position is known; the latter, by passing over and among these marks, informs us, by the place it holds with respect to them, what it is o'clock, or what time has elapsed since a certain moment when it stood at one particular spot.

(221.) In a clock the marks on the dial-plate are uniformly distributed all around the circumference of a circle, whose centre is that on which the hands revolve with a uniform motion. But it is clear that we should, with equal certainty, though with much more trouble, tell what o'clock it were, if the marks on the dial-plate were *un*equally distributed,—if the hands were *ex*-centric, and their motion *not* uniform,—provided we knew, 1st, the exact intervals round the circle at which the hour and minute marks were placed; which would be the case if we had them all registered in a table, from the results of previous careful measurement:—2dly, if we knew the exact amount and direction of excentricity of the centre of motion of the hands;—and, 3dly, if we were fully acquainted with all the mechanism which put the hands in motion, so as to be able to say at every instant what were their velocity of movement, and so as to be able to calculate, without fear of error, HOW MUCH *time* should correspond to so MUCH *angular movement.*

(222.) The visible surface of the starry heavens is the dial-plate of our clock, the stars are the fixed marks distributed around its circuit, the moon is the moveable hand, which, with a motion that, superficially considered, seems uniform, but which, when carefully examined, is found to be far otherwise, and regulated by mechanical laws of astonishing complexity and intricacy in result, though beautifully simple in principle and design, performs a monthly circuit among them, passing visibly over and hiding, or, as it is called, occulting, some, and gliding beside and between others; and whose position among them can, at any moment when it is visible, be exactly measured by the help of a sextant, just as we might measure the place of our clock-hand among the marks on its dial-plate with a pair of compasses, and thence, from the known and calculated laws of its motion, deduce the time. That the moon *does* so move *among the stars*, while the latter hold constantly, with respect to each other, the same relative position, the notice of a few nights, or even hours, will satisfy the commencing student, and this is all that at present we require.

(223.) There is only one circumstance wanting to make our analogy complete. Suppose the hands of our clock, instead of moving *quite close* to the dial-plate, were considerably elevated above, or distant in front of it. Unless, then, in viewing it, we kept our eye just in the line of their center, we should not see them exactly thrown or *projected* upon their proper places on the dial. And if we were either unaware of this cause of optical change of place, this *parallax* — or negligent in not taking it into account — we might make great mistakes in reading the time, by referring the hand to the wrong mark, or incorrectly appreciating its distance from the right. On the other hand, if we took care to note, in every case, when we had occasion to observe the time, the exact position of the eye, there would be no difficulty in ascertaining and allowing for the precise influence of this cause of apparent displacement. Now, this is just what obtains with the apparent motion of the moon among the stars. The former (as will appear) is comparatively near to the earth — the latter immensely distant; and in consequence of our not occupying the center of the earth, but being carried about on its surface, and constantly changing place, there arises a *parallax*, which displaces the moon apparently among the stars, and must be allowed for before we can tell the true place she would occupy if seen from the center.

(224.) Such a clock as we have described might, no doubt, be considered a very bad one; but if it were our *only* one, and if incalculable interests were at stake on a perfect knowledge of time, we should justly regard it as most precious, and think no pains ill bestowed in studying the laws of its movements, or in facilitating the means of *reading* it correctly. Such, in the parallel we are drawing, is the lunar theory, whose object is to reduce to regularity, the indications of this strangely irregular-going clock, to enable us to predict, long beforehand, and with absolute certainty, *whereabouts* among the stars, at every hour, minute, and second, in every day of every year, in Greenwich local time, the moon *would* be seen from the earth's center, and *will* be seen from every accessible point of its surface; and such is the *lunar method* of longitudes. The moon's apparent angular distances from all those principal and conspicuous stars which lie in its course, as seen from the earth's center, are computed and tabulated with the utmost care and precision in almanacks published under national control. No sooner does an observer, in any part of the globe, at sea or on land, measure its actual distance from any one of those *standard stars* (whose places in the heavens have been ascertained for the purpose with the most anxious sollicitude), than he has, in fact, performed that comparison of his local time with the local times of every observatory in the world, which enables him to ascertain his difference of longitude from one or all of them.

(225.) The latitudes and longitudes of any number of points on the earth's surface may be ascertained by the methods above described; and by thus laying down a sufficient number of principal points, and filling in the intermediate spaces by local surveys, might maps of counties be constructed, the outlines of continents and islands ascertained, the courses of rivers and mountain chains traced, and cities and towns referred to their proper localities. In practice, however, it is found simpler and easier to divide each particular nation into a series of great triangles, the angles of which are stations conspicuously visible from each other. Of these triangles, the *angles* only are measured by means of the *theodolite*, with the exception of *one side* only of *one triangle*, which is called *a base*, and which is measured with every refinement which ingenuity can devise or expense command. This *base* is of moderate extent, rarely surpassing six or seven miles, and purposely selected in a perfectly horizontal plane, otherwise conveniently adapted for purposes of measurement. Its length between its two extreme points (which are dots on plates of gold or platina let into massive blocks of stone, and which are, or at least *ought to be*, in all cases preserved with almost religious care, as monumental records of the highest importance), is then measured, with every precaution to ensure precision*, and its position with respect to the meridian, as well as the geographical positions of its extremities, carefully ascertained.

(226.) The annexed figure represents such a chain of triangles. A B is the base, O, C, stations visible from

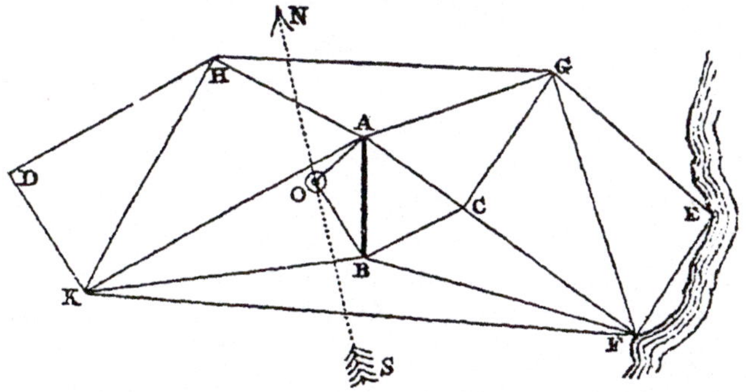

both its extremities (one of which, O, we will suppose to be a national observatory, with which it is a principal object that the base should be as closely and immediately connected as possible); and D, E, F, G, H, K, other stations, remarkable points in the county, by whose connection its whole surface may be covered, as it were, with a network of triangles. Now, it is evident that the angles of the triangle A, B, C being observed, and one of its sides, A B, measured, the other two sides, A C, B C, may be calculated by the rules of trigonometry; and thus each of the sides A C and B C becomes in its turn a *base* capable of being employed as known sides of other triangles. For instance, the angles of the triangles A C G and B C F being known by ob-

* The greatest *possible* error in the Irish base of between seven and eight miles, near Londonderry, is supposed not to exceed two inches.

servation, and their sides A C and B C, we can thence calculate the lengths A G, C G, and B F, C F. Again, C G and C F being known, and the included angle G C F, G F may be calculated, and so on. Thus may all the stations be accurately determined and laid down, and as this process may be carried on to any extent, a map of the whole county may be thus constructed, and filled in to any degree of detail we please.

(227.) Now, on this process there are two important remarks to be made. The first is, that it is necessary to be careful in the selection of stations, so as to form triangles free from any *very* great inequality in their angles. For instance, the triangle K B F would be a very improper one to determine the situation of F from observations at B and K, because the angle F being very acute, a small error in the angle K would produce a great one in the place of F *upon the line B F.* Such *ill-conditioned* triangles, therefore, must be avoided. But if this be attended to, the accuracy of the determination of the calculated sides will not be much short of that which would be obtained by actual measurement (were it practicable); and, therefore, as we recede from the base on all sides as a center, it will speedily become practicable to use *as bases*, the sides of *much larger* triangles, such as G F, G H, H K, &c.; by which means the next step of the operation will come to be carried on on a much larger scale, and embrace far greater intervals, than it would have been safe to do (for the above reason) in the immediate neighbourhood of the base. Thus it becomes easy to divide the whole face of a country into *great triangles* of from 30 to 100 miles in their sides (according to the nature of the ground), which, being once well determined, may be afterwards, by a second series of subordinate operations, broken up into smaller ones, and these again into others of a still minuter order, till the final filling in is brought within the limits of personal survey and draftsmanship, and till a map is constructed, with any required degree of detail.

(228.) The next remark we have to make is, that all the triangles in question are not, rigorously speaking, *plane*, but *spherical* — existing on the surface of a sphere, or rather, to speak correctly, of an ellipsoid. In very small triangles, of six or seven miles in the side, this may be neglected, as the difference is imperceptible; but in the larger ones it must be taken into consideration.

It is evident that, as every object used for pointing the telescope of a theodolite has some certain *elevation*, not only above the *soil*, but above the level of the *sea*, and as, moreover, these elevations differ in every instance, a *reduction to the horizon* of all the measured angles would appear to be required. But, in fact, by the construction of the theodolite (art. 155.) which is nothing more than an altitude and azimuth instrument, this reduction *is made* in the very act of reading off the horizontal angles. Let E be the center of the earth; A, B, C, the places on its *spherical surface*, to which three stations, A, P, Q, in a country are referred by radii E A, E B P, E C Q. If a theodolite be stationed at A, the axis of its horizontal circle will point to E when truly adjusted, and its plane will be a tangent to the sphere at A, intersecting the radii E B P, E C Q, at M and N, *above* the spherical surface. The telescope of the theodolite, it is true, is pointed in succession to P, and Q; but the readings off of its azimuth circle give — *not* the angle P A Q between the directions of the telescope, or between the objects P, Q, as seen from A; *but the azimuthal angle* M A N, which is the measure of the angle A of the spherical triangle B A C. Hence arises this remarkable circumstance, — that the

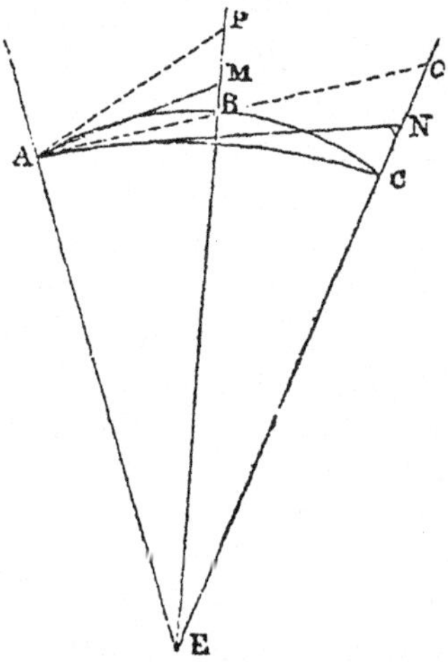

sum of the three observed angles of any of the great triangles in geodesical operations is always found to be rather *more* than 180°: were the earth's surface a *plane*, it ought to be exactly 180°; and this *excess*, which is called the *spherical excess*, is so far from being a proof of incorrectness in the work, that it is essential to its accuracy, and offers at the same time another palpable proof of the earth's sphericity.

(229.) The true way, then, of conceiving the subject of a trigonometrical survey, when the spherical form of the earth is taken into consideration, is to regard the network of triangles with which the country is covered, as the bases of an assemblage of pyramids converging to the center of the earth. The theodolite gives *us the true measures of the angles included by the planes of these pyramids;* and the surface of an imaginary sphere on the level of the sea intersects them in an assemblage of spherical triangles, above whose angles, in the radii prolonged, the real stations of observation are raised, by the superficial inequalities of mountain and valley. The operose calculations of spherical trigonometry which this consideration would seem to render necessary for the reductions of a survey, are dispensed with in practice by a very simple and easy rule, called the *rule for the spherical excess,* which is to be found in most works on trigonometry.* If we would take into account the ellipticity of the earth, it may also be done by appropriate processes of calculation, which, however, are too abstruse to dwell upon in a work like the present.

(230.) Whatever process of calculation we adopt, the result will be a reduction to the level of the sea, of all the triangles, and the consequent determination of the geographical latitude and longitude of every station observed. Thus we are at length enabled to construct maps of countries; to lay down the outlines of continents and islands; the courses of rivers; the direction of mountain ridges, and the places of their principal

* Lardner's Trigonometry, prop. 94. Woodhouse's ditto, p. 148. 1st edition.

summits; and all those details which, as they belong to physical and statistical, rather than to astronomical geography, we need not here dilate on. A few words, however, will be necessary respecting maps, which are used as well in astronomy as in geography.

(231.) A map is nothing more than a representation, upon a plane, of some portion of the surface of a sphere, on which are traced the particulars intended to be expressed, whether they be continuous outlines or points. Now, as a spherical surface * can by no contrivance be extended or projected into a plane, without undue enlargement or contraction of some parts in proportion to others; and as the system adopted in so extending or projecting it will decide *what* parts shall be enlarged or relatively contracted, and in what proportions; it follows, that when large portions of the sphere are to be mapped down, a great difference in their representations may subsist, according to the system of projection adopted.

(232.) The projections chiefly used in maps, are the *orthographic*, *stereographic*, and *Mercator's*. In the *orthographic* projection, every point of the hemisphere is referred to its diametral plane or base, by a perpendicular let fall on it, so that the representation of the hemisphere thus mapped on its base, is such as it would actually appear to an eye placed at an infinite distance from it. It is obvious, from the annexed figure, that in this projection only the central portions are represented of their true forms, while all the exterior is more and more distorted and crowded together as we approach the edges of the map. Owing to this cause, the orthographic projection, though very good for small portions of the globe, is of little service for large ones.

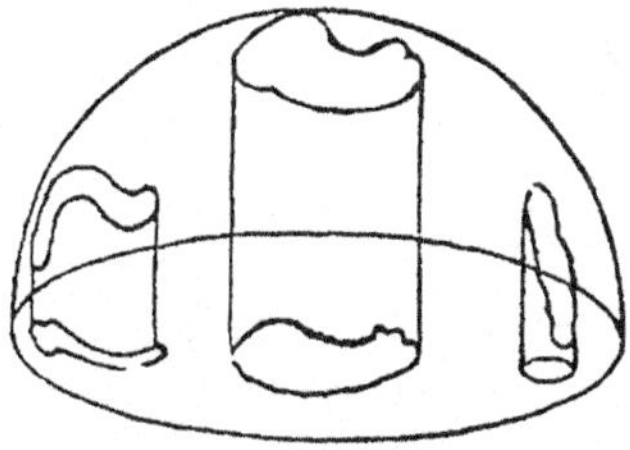

(233.) The *stereographic* projection is in great mea-

* We here neglect the ellipticity of the earth, which, for such a purpose as map-making, is too trifling to have any material influence.

sure free from this defect. To understand this projection, we must conceive an eye to be placed at E, one extremity of a diameter, E C B, of the sphere, and to view the concave surface of the sphere, every point of which, as P, is referred to the diametral plane A D F,

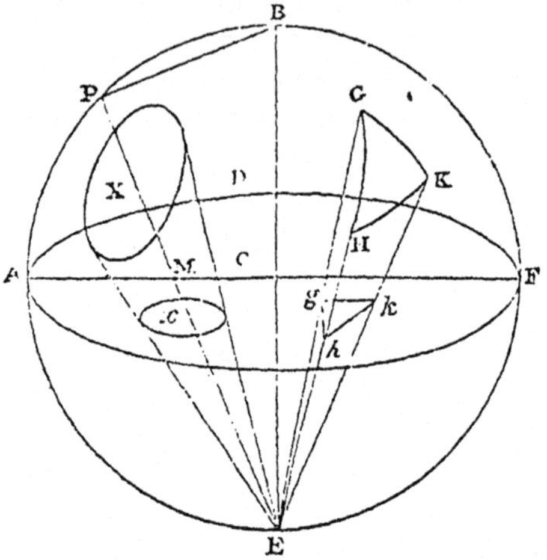

perpendicular to E B by the visual line P M E. The stereographic projection of a sphere, then, is a true perspective representation of its concavity on a diametral plane; and, as such, it possesses some singularly elegant geometrical properties, of which we shall state one or two of the principal.

(234.) And first, then, all circles on the sphere are represented by circles in the projection. Thus the circle X is projected into *x*. Only great circles passing through the vertex B are projected into straight lines traversing the center C: thus, B P A is projected into C A.

2dly. Every very small triangle, G H K, on the sphere, is represented by a *similar* triangle, *g h k*, in the projection. This is a very valuable property, as it insures a general similarity of appearance in the map to the reality in all its parts, and enables us to project at least a hemisphere in a single map, without any violent distortion of the configurations on the surface from their real forms. As in the orthographic projection, the borders of the hemisphere are unduly crowded together; in the stereographic, their projected dimensions are, on the contrary, somewhat enlarged in receding from the center.

(235.) Both these projections may be considered *natural* ones, inasmuch as they are really perspective representations of the surface on a plane. Mercator's is entirely an artificial one, representing the sphere as it cannot be seen from any one point, but as it might be seen by an eye carried successively over every part

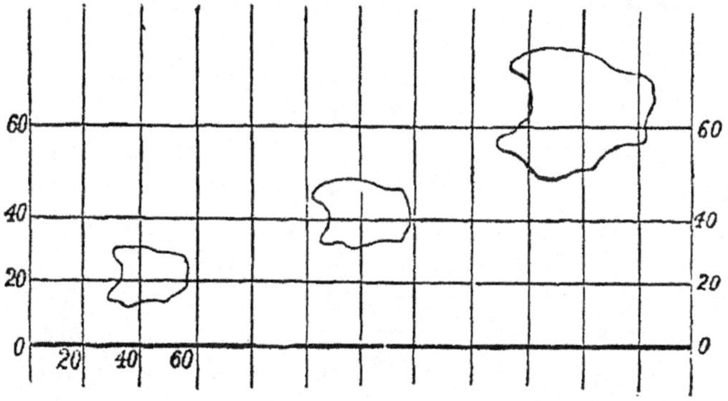

of it. In it, the degrees of *longitude*, and those of *latitude*, bear always to each other their due proportion: the equator is conceived to be extended out into a straight line, and the meridians are straight lines at right angles to it, as in the figure. Altogether, the general character of maps on this projection is not very dissimilar to what would be produced by referring every point in the globe to a circumscribing cylinder, by lines drawn from the center, and then unrolling the cylinder into a plane. Like the stereographic projection, it gives a true representation, as to *form*, of every particular small part, but varies greatly in point of *scale* in its different regions; the polar portions in particular being extravagantly enlarged; and the whole map, even of a single hemisphere, not being comprizable within any finite limits.

(236.) We shall not, of course, enter here into any geographical details; but one result of maritime discovery on the great scale is, so to speak, *massive* enough to call for mention as an astronomical feature. When the continents and seas are laid down on a globe (and since the discovery of Australia we are sure that no very extensive tracts of land remain unknown, except perhaps at the south pole), we find that it is possible so to divide the globe into two hemispheres, that one shall contain *nearly all the land;* the other being almost entirely sea. It is a fact, not a little interesting to Englishmen, and, combined with our insular station in that great highway of nations, the Atlantic, not a little explanatory of our commercial eminence, that London occupies nearly the center of the terrestrial hemisphere. Astronomically speaking, the fact of this divisibility of the globe into an oceanic and a terrestrial hemisphere is important, as demonstrative of a want of absolute equality in the density of the solid material of the two hemispheres. Considering the whole mass of land and water as in a state of *equilibrium,* it is evident that the half which protrudes must of necessity be *buoyant;* not, of course, that we mean to assert it to be lighter than *water,* but, as compared with the whole globe, *in a less degree heavier* than that fluid. We leave to geologists to draw from these premises their own conclusions (and we think them obvious enough) as to the internal constitution of the globe, and the immediate nature of the forces which sustain its continents at their actual elevation; but in any future investigations which may have for their object to explain the local deviations of the intensity of gravity, from what the hypothesis of an exact elliptic figure would require, this, as a general fact, ought not to be lost sight of.

(237.) Our knowledge of the surface of our globe is incomplete, unless it include the heights above the sea level of every part of the land, and the depression of the bed of the ocean below the surface over all its extent. The latter object is attainable (with whatever difficulty and however slowly) by direct sounding; the former by two distinct methods: the one consisting in trigonometrical measurement of the differences of level of all the stations of a survey; the other, by the use of the barometer, the principle of which is, in fact, identical with that of the sounding line. In both cases we measure the distance of the point whose level we would know from the surface of an equilibrated ocean: only in the one case it is an ocean of water; in the other, of air. In the one case our sounding line is real and tangible; in the other, an imaginary one, measured by the length of the column of quicksilver the superincumbent air is capable of counterbalancing.

(238.) Suppose that instead of air, the earth and ocean were covered with oil, and that human life could subsist under such circumstances. Let A B C D E be a

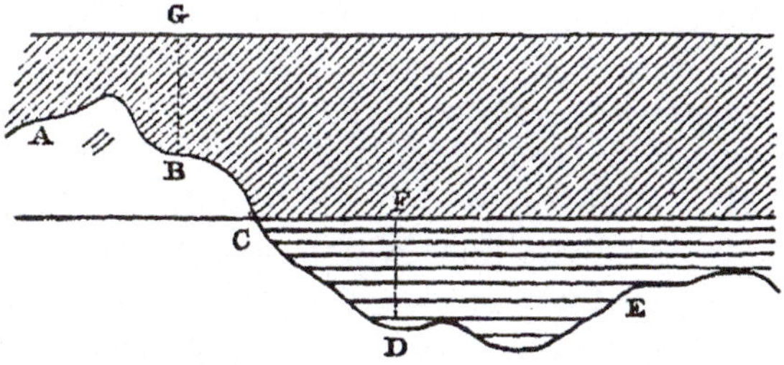

continent, of which the portion A B C projects above the water, but is covered by the oil, which also floats at an uniform depth on the whole ocean. Then if we would know the depth of any point D below the sea level, we let down a plummet from F. But if we would know the height of B above the same level, we have only to send up a float from B to the surface of the oil; and *having done the same at C, a point at the sea level, the difference of the two float lines gives the height in question.*

(239.) Now, though the atmosphere differs from oil in not having a positive *surface* equally definite, and in not being capable of carrying up any float adequate to such an use, yet it possesses all the properties of a fluid really essential to the purpose in view, and this in particular,—that, over the whole surface of the globe, its *strata of equal density* are parallel to the surface of equilibrium, or to what *would be* the surface of the sea, if *prolonged under the continents*, and therefore each or any of them has all the characters of a definite surface to measure from, provided it can be ascertained and identified. Now the height at which, at any station B, the mercury in a barometer is supported, informs us *at once* how much of the atmosphere is incumbent on B, or, in other words, *in what stratum* of the general atmosphere (indicated by its density) B is situated: whence we are enabled finally to conclude, by mechanical reasoning*, at what height above the sea-level *that degree of density* is to be found over the whole surface of the globe. Such is the principle of the application of the barometer to the measurement of heights. For details, the reader is referred to other works.†

(240.) Possessed of a knowledge of the heights of stations above the sea, we may connect all stations at the same altitude by level lines, the lowest of which will be the outline of the sea-coast; and the rest will mark out the successive coast-lines which would take place were the sea to rise by regular and equal accessions of level over the whole world, till the highest mountains were submerged. The bottoms of valleys and the ridge-lines of hills are determined by their property of intersecting all these level lines at right angles, and being, subject to that condition, the shortest and longest courses respectively which can be pursued from the summit to the sea. The former constitute the water-courses of a country; the latter divide it into drainage basins: and thus originate natural districts of the most ineffaceable character, on which the distribution, limits, and peculiarities of human communities are in great measure dependent.

* See Cab. Cycl. PNEUMATICS, art. 143.

† Biot, Astronomie Physique, vol. 3. For tables, see the work of Biot cited. Also those of Oltmann, annually published by the French board of longitudes in their Annuaire; and Mr. Baily's Collection of Astronomical Tables and Formulæ.

CHAP. IV.

OF URANOGRAPHY.

CONSTRUCTION OF CELESTIAL MAPS AND GLOBES BY OBSERVATIONS OF RIGHT ASCENSION AND DECLINATION. — CELESTIAL OBJECTS DISTINGUISHED INTO FIXED AND ERRATIC. — OF THE CONSTELLATIONS. — NATURAL REGIONS IN THE HEAVENS. — THE MILKY WAY. — THE ZODIAC. — OF THE ECLIPTIC. — CELESTIAL LATITUDES AND LONGITUDES. — PRECESSION OF THE EQUINOXES. — NUTATION. — ABERRATION. — URANOGRAPHICAL PROBLEMS.

(241.) THE determination of the relative situations of objects in the heavens, and the construction of maps and globes which shall truly represent their mutual configurations, as well as of catalogues which shall preserve a more precise numerical record of the position of each, is a task at once simpler and less laborious than that by which the surface of the earth is mapped and measured. Every star in the great constellation which appears to revolve above us, constitutes, so to speak, a celestial station; and among these stations we may, as upon the earth, triangulate, by measuring with proper instruments their angular distances from each other, which, cleared of the effect of refraction, are then in a state for laying down on charts, as we would the towns and villages of a country: and this without moving from our place, at least for all the stars which rise above our horizon.

(242.) Great exactness might, no doubt, be attained by this means, and excellent celestial charts constructed; but there is a far simpler and easier, and, at the same time, infinitely *more* accurate course laid open to us, if we take advantage of the earth's rotation on its axis, and by observing each celestial object as it passes our meridian, refer it separately and independently to the celestial equator, and thus ascertain its place on the surface of an imaginary sphere, which may be conceived to revolve with it, and on which it may be considered as projected.

(243.) The right ascension and declination of a point in the heavens correspond to the longitude and latitude of a station on the earth; and the place of a star on a celestial sphere is determined, when the former elements are known, just as that of a town on a map, by knowing the latter. The great advantages which the method of meridian observation possesses over that of triangulation from star to star, are, then, 1st, That in it every star is observed in that point of its diurnal course, when it is best seen and least displaced by refraction. 2dly, That the instruments required (the transit and mural circle) are the simplest and least liable to error or derangement of any used by astronomers. 3dly, That all the observations can be made systematically, in regular succession, and with equal advantages; there being here no question about advantageous or disadvantageous triangles, &c. And, lastly, That, by adopting this course, the very quantities which we should otherwise have to calculate by long and tedious operations of spherical trigonometry, and which are essential to the formation of a catalogue, are made the objects of immediate measurement. It is almost needless to state, then, that this is the course adopted by astronomers.

(244.) To determine the right ascension of a celestial object, all that is necessary is to observe the moment of its meridian passage with a transit instrument, by a clock regulated to exact sidereal time, or reduced to such by applying its known error and rate. The *rate* may be obtained by repeated observations of the same star at its successive meridian passages. The *error*, however, requires a knowledge of the *equinox*, or initial point from which all right ascensions in the heavens reckon, as longitudes do on the earth from a first meridian.

(245.) The nature of this point will be explained presently; but for the purposes of uranography, in so far as they concern only the actual configurations of the stars *inter se*, a knowledge of the equinox is not necessary. The choice of the equinox, as a zero point of right ascensions, is purely artificial, and a matter of convenience: but as on the earth, any station (as a national observatory) may be chosen for an origin of longitudes; so in uranography, any conspicuous star may be selected as an initial point from which hour angles may be reckoned, and from which, by merely observing *differences* or *intervals* of time, the situation of all others may be deduced. In practice, these intervals are affected by certain minute causes of inequality, which must be allowed for, and which will be explained in their proper places.

(246.) The declinations of celestial objects are obtained, 1st, By observation of their *meridian altitudes*, with the mural circle, or other proper instruments. This requires a knowledge of the geographical latitude of the station of observation, which itself is only to be obtained by celestial observation. 2dly, And more directly by observation of their *polar distances* on the mural circle, as explained in art. 136., which is independent of any previous determination of the latitude of the station; neither, however, in this case, does observation give directly and immediately the *exact* declinations. The observations require to be corrected, first for refraction, and moreover for those minute causes of inequality which have been just alluded to in the case of right ascensions.

(247.) In this manner, then, may the places, one among the other, of all celestial objects be ascertained, and maps and globes constructed. Now here arises a very important question. How far are these places permanent? Do these stars and the greater luminaries of heaven preserve for ever one invariable connection and relation of place *inter se*, as if they formed part of a solid though invisible firmament; and, like the great natural land-marks on the earth, preserve immutably the same distances and bearings each from the other? If so, the most rational idea we could form of the universe would be that of an earth at absolute rest in the centre, and a hollow crystalline sphere circulating round it, and carrying sun, moon, and stars along in its diurnal motion. If not, we must dismiss all such notions, and enquire individually into the distinct history of each object, with a view to discovering the laws of its peculiar motions, and whether any and what other connection subsists between them.

(248.) So far is this, however, from being the case, that observations, even of the most cursory nature, are sufficient to show that some, at least, of the celestial bodies, and those the most conspicuous, are in a state of continual change of place among the rest. In the case of the moon, indeed, the change is so rapid and remarkable, that its alteration of situation with respect to such bright stars as may happen to be near it may be noticed any fine night in a few hours; and if noticed on two successive nights, cannot fail to strike the most careless observer. With the sun, too, the change of place among the stars is constant and rapid; though, from the invisibility of stars to the naked eye in the day-time, it is not so readily recognized, and requires either the use of telescopes and angular instruments to measure it, or a longer continuance of observation to be struck with it. Nevertheless, it is only necessary to call to mind its greater meridian altitude in summer than in winter, and the fact that the stars which come into view at night vary with the season of the year, to perceive that a great change must have taken place in that interval in its relative situation with respect to all the stars. Besides the sun and moon, too, there are several other bodies, called planets, which, for the most part, appear to the naked eye only as the largest and most brilliant stars, and which offer the same phenomenon of a constant change of place among the stars; now approaching, and now receding from, such of them as we may refer them to as marks; and, some in longer, some in shorter periods, making, like the sun and moon, the complete tour of the heavens.

(249.) These, however, are exceptions to the general rule. The innumerable multitude of the stars which are distributed over the vault of the heavens form a constellation, which preserves, not only to the eye of the casual observer, but to the nice examination of the astronomer, a uniformity of aspect which, when contrasted with the perpetual change in the configurations of the sun, moon, and planets, may well be termed invariable. It is not, indeed, that, by the refinement of exact measurements prosecuted from age to age, some small changes of apparent place, attributable to no illusion and to no *terrestrial* cause, cannot be detected in some of them;—such are called, in astronomy, the *proper motions* of the stars;—but these are so excessively slow, that their accumulated amount (even in those stars for which they are greatest) has been insufficient, in the whole duration of astronomical history, to produce any obvious or material alteration in the appearance of the starry heavens.

(250.) This circumstance, then, establishes a broad distinction of the heavenly bodies into two great classes; — the fixed, among which (unless in a course of observations continued for many years) no change of mutual situation can be detected; and the erratic, or wandering—(which is implied in the word planet*)— including the sun, moon, and planets, as well as the singular class of bodies termed comets, in whose apparent places among the stars, and among each other, the observation of a few days, or even hours, is sufficient to exhibit an indisputable alteration.

(251.) Uranography, then, as it concerns the fixed celestial bodies (or, as they are usually called, the *fixed stars*), is reduced to a simple marking down of their relative places on a globe or on maps; to the insertion on that globe, in its due place in the great constellation

* Πλανητης, a wanderer.

of the stars, of the pole of the heavens, or the vanishing point of parallels to the earth's axis; and of the equator and place of the equinox: points and circles these, which, though artificial, and having reference entirely to our earth, and therefore subject to all changes (if any) to which the earth's axis may be liable, are yet so convenient in practice, that they have obtained an admission (with some other circles and lines), sanctioned by usage, in all globes and planispheres. The reader, however, will take care to keep them separate in his mind, and to familiarize himself with the idea rather of *two* or more celestial globes, superposed and fitting on each other, on one of which—a real one—are inscribed the stars; on the others those imaginary points, lines, and circles which astronomers have devised for their own uses, and to aid their calculations; and to accustom himself to conceive in the latter, or artificial, spheres a capability of being shifted in any manner upon the surface of the other; so that, should experience demonstrate (as it does) that these artificial points and lines are brought, by a slow motion of the earth's axis, or by other *secular variations* (as they are called), to coincide, at very distant intervals of time, with different stars, he may not be unprepared for the change, and have no confusion to correct in his notions.

(252.) Of course we do not here speak of those uncouth figures and outlines of men and monsters, which are usually scribbled over celestial globes and maps, and serve, in a rude and barbarous way, to enable us to talk of groups of stars, or districts in the heavens, by names which, though absurd or puerile in their origin, have obtained a currency from which it would be difficult, and perhaps wrong, to dislodge them. In so far as they have really (as some have) any slight resemblance to the figures called up in imagination by a view of the more splendid "constellations," they have a certain convenience; but as they are otherwise entirely arbitrary, and correspond to no *natural* subdivisions or groupings of the stars, astronomers treat them lightly, or altogether disregard them *, except for briefly *naming* remarkable stars, as α Leonis, β Scorpii, &c. &c., by letters of the Greek alphabet attached to them. The reader will find them on any celestial charts or globes, and may compare them with the heavens, and there learn for himself their position.

(253.) There are not wanting, however, *natural* districts in the heavens, which offer great peculiarities of character, and strike every observer: such is the *milky way*, that great luminous band, which stretches, every evening, all across the sky, from horizon to horizon, and which, when traced with diligence, and mapped down, is found to form a zone completely encircling *the whole sphere*, almost in a great circle, which is neither an *hour* circle, nor coincident with any other of our astronomical *grammata*. It is divided in one part of its course, sending off a kind of branch, which unites again with the main body, after remaining distinct for about 150 degrees. This remarkable belt has maintained, from the earliest ages, the same relative situation among the stars; and, when examined through powerful telescopes, is found (wonderful to relate!) *to consist entirely of stars scattered by millions*, like glittering dust, on the black ground of the general heavens.

(254.) Another remarkable region in the heavens is the *zodiac*, not from any thing peculiar in its own constitution, but from its being the area within which the apparent motions of the sun, moon, and all the greater planets are confined. To trace the path of any one of these, it is only necessary to ascertain, by continued observation, its places at successive epochs, and entering these upon our map or sphere in sufficient number to form a series, not too far disjoined, to connect them by lines from point to point, as we mark out the course of

* This disregard is neither supercilious nor causeless. The constellations seem to have been almost purposely named and delineated to cause as much confusion and inconvenience as possible. Innumerable snakes twine through long and contorted areas of the heavens, where no memory can follow them; bears, lions and fishes, large and small, northern and southern, confuse all nomenclature, &c. A better system of constellations might have been a material help as an artificial memory.

a vessel at sea by mapping down its place from day to day. Now when this is done, it is found, first, that the apparent path, or track, of the sun on the surface of the heavens, is no other than an exact great circle of the sphere which is called the *ecliptic*, and which is inclined to the equinoctial at an angle of about 23° 28′, intersecting it at two opposite points, called the equinoctial points, or equinoxes, and which are distinguished from each other by the epithets vernal and autumnal; the vernal being that at which the sun crosses the equinoctial from south to north; the autumnal, when it quits the northern and enters the southern hemisphere. Secondly, that the moon and all the planets pursue paths which, in like manner, encircle the whole heavens, but are not, like that of the sun, great circles exactly returning into themselves and bisecting the sphere, but rather spiral curves of much complexity, and described with very unequal velocities in their different parts. They have all, however, this in common, that the *general direction* of their motions is the same with that of the sun, viz. from *west to east*, that is to say, the contrary to that in which both they and the stars appear to be carried by the diurnal motion of the heavens; and, moreover, that they never deviate far from the ecliptic on either side, crossing and recrossing it at regular and equal intervals of time, and confining themselves within a zone, or belt (the *zodiac* already spoken of), extending 9° on either side of the ecliptic.

(255.) It would manifestly be useless to map down on globes or charts the apparent paths of any of those bodies which never retrace the same course, and which, therefore, demonstrably, must occupy at some one moment or other of their history, every point in the area of that zone of the heavens within which they are circumscribed. The apparent complication of their movements arises (that of the moon excepted) from our viewing them from a station which is itself in motion, and would disappear, could we shift our point of view and observe them from the sun. On the other hand the apparent motion of the sun is presented to us under its least involved form, and is studied, from the station we occupy, to the greatest advantage. So that, independent of the importance of that luminary to us in other respects, it is by the investigation of the laws of its motions in the first instance that we must rise to a knowledge of those of all the other bodies of our system.

(256.) The ecliptic, which is its apparent path among the stars, is traversed by it in the period called the *sidereal year*, which consists of $365^d\ 6^h\ 9^m\ 9^s{\cdot}6$, reckoned in mean solar time, or $366^d\ 6^h\ 9^m\ 9^s{\cdot}6$ reckoned in sidereal time. The reason of this difference (and it is this which constitutes the origin of the difference between solar and sidereal time) is, that as the sun's apparent annual motion *among* the stars is performed in a contrary direction to the apparent *diurnal* motion of both sun and stars, it comes to the same thing as if the diurnal motion of the sun were so much *slower* than that of the stars, or as if the sun lagged behind them in its daily course. Where this has gone on for a whole year, the sun will have fallen behind the stars by a whole circumference of the heavens — or, in other words — in a year, the sun will have made fewer diurnal revolutions, by one, than the stars. So that the same interval of time which is measured by $366^d\ 6^h$, &c. of sidereal time, if reckoned in mean solar days, hours, &c. will be called $365^d\ 6^h$, &c. Thus, then, is the proportion between the mean solar and sidereal day established, which, reduced into a decimal fraction, is that of 1·00273791 to 1. The measurement of time by these different standards may be compared to that of space by the standard feet, or ells of two different nations; the proportion of which, once settled, can never become a source of error.

(257.) The position of the ecliptic among the stars may, for our present purpose, be regarded as invariable. It is true that this is not strictly the case; and on comparing together its position at present with that which it held at the most distant epoch at which we possess observations, we find evidences of a small change, which theory accounts for, and whose nature will be hereafter explained; but that change is so excessively slow, that for a great many successive years, or even for whole centuries, this circle may be regarded as holding the same position in the sidereal heavens.

(258.) The *poles of the ecliptic*, like those of any other great circle of the sphere, are opposite points on its surface, equidistant from the ecliptic in every direction. They are of course not coincident with those of the equinoctial, but removed from it by an angular interval equal to the inclination of the ecliptic to the equinoctial (23° 28′), which is called the *obliquity of the ecliptic*. In the annexed figure, if P *p* represent the north and south poles (by which, when used without qualification we always mean the poles of *the equinoctial*), and E Q A V the equinoctial, V S A W the ecliptic, and K *k*, its poles — the spherical angle Q V S is the obliquity of the ecliptic, and is equal in angular measure to P K or S Q. If we suppose the sun's apparent motion to be in the direction V S A W, V will be the *vernal* and A the *autumnal equinox*. S and W, the two points at which the ecliptic is most distant from the equinoctial, are termed *solstices*, because, when arrived there, the sun ceases to recede from the equator, and (in that sense, so far as its motion in declination is concerned) to stand still in the heavens. S, the point where the sun has the greatest *northern* declination, is called the summer solstice, and W, that where it is farthest south, the *winter*. These epithets obviously have their origin in the dependence of the seasons on the sun's declination, which will be explained in the next chapter. The circle E K P Q *k p*, which passes through the poles of the ecliptic and equinoctial, is called the solstitial colure; and a meridian drawn through the equinoxes, P V *p* A, the equinoctial colure.

(259.) Since the ecliptic holds a determinate situation in the starry heavens, it may be employed, like the equinoctial, to refer the positions of the stars to, by circles drawn through them from *its* poles, and therefore perpendicular to it. Such circles are termed, in astronomy, circles of latitude — the distance of a star from the ecliptic, reckoned on the circle of latitude passing through it, is called the *latitude* of the stars — and the arc of the ecliptic intercepted between the vernal equinox and this circle, its *longitude*. In the figure X is a

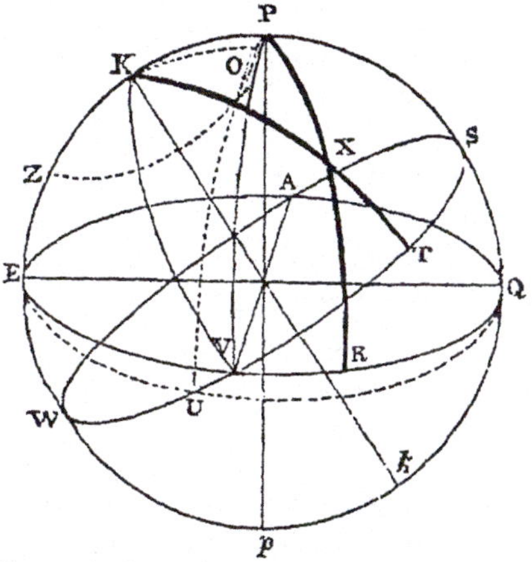

star, P X R a circle of declination drawn through it, by which it is referred to the equinoctial, and K X T a circle of latitude referring it to the ecliptic — then, as V R is the right ascension, and R X the declination, of X, so also is V T its longitude, and T X its latitude. The use of the terms longitude and latitude, in this sense, seems to have originated in considering the ecliptic as forming a kind of natural equator to the heavens, as the terrestrial equator does to the earth — the former holding an invariable position with respect to the stars, as the latter does with respect to stations on the earth's surface. The force of this observation will presently become apparent.

(260.) Knowing the right ascension and declination of an object, we may find its longitude and latitude, and *vice versâ*. This is a problem of great use in physical astronomy — the following is its solution: — In our last figure, E K P Q, the solstitial colure is of course 90° distant from V, the vernal equinox, which is one of its poles — so that V R (the right ascension) being given, and also V E, the arc E R, and its measure, the spherical angle E P R, or K P X, is known. In the spherical triangle K P X, then, we have given, 1st, The side P K, which, being the distance of the poles of the ecliptic and equinoctial, is equal to the obliquity of the ecliptic; 2d, The side P X, the *polar distance*, or the complement of the declination R X; and, 3d, the included angle K P X; and therefore, by spherical trigonometry, it is easy to find the other side K X, and the remaining angles. Now K X is the complement of the required latitude X T, and the angle P K X being known, and P K V being a right angle (because S V is 90°), the angle X K V becomes known. Now this is no other than the measure of the longitude V T of the object. The inverse problem is resolved by the same triangle, and by a process exactly similar.

(261.) The same course of observations by which the path of the sun among the fixed stars is traced, and the ecliptic marked out among them, determines, of course, the place of the equinox V upon the starry sphere, at that time — a point of great importance in practical astronomy, as it is the origin or zero point of right ascension. Now, when this process is repeated at considerably distant intervals of time, a very remarkable phenomenon is observed; viz. that the equinox does *not* preserve a constant place among the stars, but shifts its position, travelling continually and regularly, although with extreme lowness, *backwards*, along the ecliptic, in the direction V W from east to west, or the *contrary* to that in which the sun appears to move in that circle. As the ecliptic and equinoctial are not *very* much inclined, this motion of the equinox from east to west along the former, conspires (speaking generally) with the diurnal motion, and carries it, with reference to that motion, continually in advance upon the stars: hence it has acquired the name of the *precession of the equinoxes*, because the place of the equinox among the stars, at every subsequent moment, *precedes* (with reference to the diurnal motion) that which is held the moment before. The amount of this motion by which the equinox travels backward, or retrogrades (as it is called), on the ecliptic, is 0° 0′ 50″·10 *per annum*, an extremely minute quantity, but which, by its continual accumulation from year to year, at last makes itself very palpable, and that in a way highly inconvenient to practical astronomers, by destroying, in the lapse of a moderate number of years, the arrangement of their catalogues of stars, and making it necessary to reconstruct them. Since the formation of the earliest catalogue on record, the place of the equinox has retrograded already about 30°. The period in which it performs a complete tour of the ecliptic, is 25,868 years.

(262.) The immediate uranographical effect of the precession of the equinoxes is to produce a uniform *increase of longitude* in all the heavenly bodies, whether fixed or erratic. For the vernal equinox being the initial point of longitudes, as well as of right ascension, a retreat of this point on the ecliptic *tells* upon the longitudes of all alike, whether at rest or in motion, and produces, so far as its amount extends, the *appearance* of a motion in longitude common to all, *as if* the whole heavens had a slow rotation round the poles of the ecliptic in the long period above mentioned, similar to what they have in twenty-four hours round those of the equinoctial.

(263.) To form a just idea of this curious astronomical phenomenon, however, we must abandon, for a time, the consideration of the ecliptic, as tending to produce confusion in our ideas; for this reason, that the stability of the ecliptic itself among the stars is (as already hinted, art. 257.) only approximate, and that in consequence its intersection with the equinoctial is liable to a certain amount of change, arising from *its* fluctuation, which mixes itself with what is due to the principal uranographical cause of the phenomenon. This cause will become at once apparent, if, instead of regarding the equinox, we fix our attention on the pole of the equinoctial, or the vanishing point of the earth's axis.

(264.) The place of this point among the stars is easily determined at any epoch, by the most direct of all astronomical observations, — those with the mural circle. By this instrument we are enabled to ascertain at every moment the exact distance of the polar point from any three or more stars, and therefore to lay it down, by triangulating from these stars, with unerring precision, on a chart or globe, without the least reference to the position of the ecliptic, or to any other circle not naturally connected with it. Now, when this is done with proper diligence and exactness, it results that, although for short intervals of time, such as a few days, the place of the pole may be regarded as not sensibly variable, yet in reality it is in a state of constant, although extremely slow motion; and, what is still more remarkable, this motion is not uniform, but compounded of one principal, uniform, or nearly uniform, part, and other smaller and subordinate periodical fluctuations: the former giving rise to the phenomena of *precession;* the latter to another distinct phenomenon called *nutation.* These two phenomena, it is true, belong, theoretically speaking, to one and the same general head, and are intimately connected together, forming part of a great and complicated chain of consequences flowing from the earth's rotation on its axis: but it will be of advantage to present clearness to consider them separately.

(265.) It is found, then, that in virtue of the uniform part of the motion of the pole, it describes a circle in the heavens around the pole of the ecliptic as a centre, keeping constantly at the same distance of 23° 28′ from it, in a direction from east to west, and with such a velocity, that the annual angle described by it, in this its imaginary orbit, is $50''\cdot10$; so that the whole circle would be described by it in the above-mentioned period of 25,868 years. It is easy to perceive how such a motion of the pole will give rise to the retrograde motion of the equinoxes; for in the figure, art. 259., suppose the pole P in the progress of its motion in the small circle P O Z round K to come to O, then, as the situation of the equinoctial E V Q is determined by that of the pole, this, it is evident, must cause a displacement of the equinoctial, which will take a new situation, E U Q, 90° distant in every part from the new position O of the pole. The point U, therefore, in which the displaced equinoctial will intersect the ecliptic, *i. e.* the displaced equinox, will lie on that side of V, its original position, towards which the motion of the pole is directed, or to the westward.

(266.) The precession of the equinoxes thus conceived, consists, then, in a real but very slow motion of the pole of the heavens among the stars, in a small circle round the pole of the ecliptic. Now this cannot happen without producing corresponding changes in the apparent diurnal motion of the sphere, and the aspect which the heavens must present at very remote periods of history. The pole is nothing more than the vanishing point of the earth's axis. As this point, then, has such a motion as described, it necessarily follows that the earth's *axis* must have a conical motion, in virtue of which it points successively to every part of the small circle in question. We may form the best idea of such a motion by noticing a child's peg-top, when it spins not upright, or that amusing toy the te-to-tum, which, when delicately executed, and nicely balanced, becomes an elegant philosophical instrument, and exhibits, in the most beautiful manner, the whole phenomenon, in a way calculated to give at once a clear conception of it as a fact, and a considerable insight into its physical cause as a dynamical effect. The reader will take care not to confound the variation of the *position of the earth's* axis *in space* with a mere shifting of the imaginary line about which it revolves, in its interior. The whole earth participates in the motion, and goes along with the axis as if it were really a bar of iron driven through it. That such is the case is proved by the two great facts: 1st, that the latitudes of places on the earth, or their geographical situation with respect to the poles, have undergone no perceptible change from the earliest ages. 2dly, that the sea maintains its level, which could not be the case if the motion of the axis were not accompanied with a motion of the whole mass of the earth.

(267.) The visible effect of precession on the aspect of the heavens consists in the *apparent* approach of some stars and constellations to the pole and recess of others. The bright star of the Lesser Bear, which we call the pole star, has not always been, nor will always continue to be, our cynosure: at the time of the construction of the earliest catalogues it was 12° from the pole—it is now only 1° 24′, and will approach yet nearer, to within half a degree, after which it will again recede, and slowly give place to others, which will succeed it in its companionship to the pole. After a lapse of about 12,000 years, the star α Lyræ, the brightest in the northern hemisphere, will occupy the remarkable situation of a pole star, approaching within about 5° of the pole.

(268.) The *nutation* of the earth's axis is a small and slow subordinate gyratory movement, by which, if subsisting alone, the pole would describe among the stars, in a period of about nineteen years, a minute ellipsis, having its longer axis equal to 18″·5, and its shorter to 13″·74; the longer being directed towards the pole of the ecliptic, and the shorter, of course, at right angles to it. The consequence of this real motion of the pole is an *apparent* approach and recess of all the stars in the heavens to the pole in the same period. Since, also, the place of the equinox on the ecliptic is determined by the place of the pole in the heavens, the same cause will give rise to a small alternate advance and recess of the equinoctial points, by which, in the same period, both the longitudes and the right ascensions of the stars will be also alternately increased and diminished.

(269.) Both these motions, however, although here considered separately, subsist jointly; and since, while in virtue of the nutation, the pole is describing its little ellipse of $18''\cdot5$ in diameter, it is carried by the greater and regularly progressive motion of precession over so much of its circle round the pole of the ecliptic as corresponds to nineteen years, — that is to say, over an angle of nineteen times $50''\cdot1$ round the centre (which, in a small circle of $23° 28'$ in diameter, corresponds to $6' 20''$, as seen from the centre of the sphere): the path which it will pursue in virtue of the two motions, subsisting jointly, will be neither an ellipse nor an exact circle, but a gently undulated ring like that in the figure (where, however, the undulations are much exaggerated). (See *fig.* to art. 272.)

(270.) These movements of precession and nutation are common to all the celestial bodies both fixed and erratic; and this circumstance makes it impossible to attribute them to any other cause than a real motion of the earth's axis, such as we have described. Did they only affect the stars, they might, with equal plausibility, be urged to arise from a *real* rotation of the starry heavens, as a solid shell round an axis passing through the poles of the ecliptic in 25,868 years, and a real elliptic gyration of *that* axis in nineteen years: but since they also affect the sun, moon, and planets, which, having motions independent of the general body of the stars, cannot without extravagance be supposed *attached to* the celestial concave*, this idea falls to the ground; and there only remains, then, a real motion in the earth by which they *can* be accounted for. It will be shown in a subsequent chapter that they are necessary consequences of the rotation of the earth, combined with its elliptical figure, and the unequal attraction of the sun and moon on its polar and equatorial regions.

(271.) Uranographically considered, as affecting the apparent places of the stars, they are of the utmost

* This argument, cogent as it is, acquires additional and decisive force from the *law* of nutation, which is dependent on the position, for the time, of the *lunar* bit. If we attribute it to a real motion of the celestial sphere, we must ther maintain that sphere to be kept in a constant state of *tremor* by the motion of the moon!

importance in practical astronomy. When we speak of the right ascension and declination of a celestial object, it becomes necessary to state what *epoch* we intend, and whether we mean the *mean* right ascension—cleared, that is, of the periodical fluctuation in its amount, which arises from nutation, or the *apparent* right ascension, which, being reckoned from the actual place of the vernal equinox, is affected by the periodical advance and recess of the equinoctial point thence produced—and so of the other elements. It is the practice of astronomers to *reduce*, as it is termed, all their observations, both of right ascension and declination, to some common and convenient epoch—such as the beginning of the year for temporary purposes, or of the decade, or the century for more permanent uses, by subtracting from them the whole effect of precession in the interval; and, moreover, to divest them of the influence of nutation by investigating and subducting the amount of change, both in right ascension and declination, due to the displacement of the pole from the centre to the circumference of the little ellipse above mentioned. This last process is technically termed correcting or *equating* the observation for nutation; by which latter word is always understood, in astronomy, the getting rid of a periodical cause of fluctuation, and presenting a result, not as it *was* observed, but as it would have been observed, had that cause of fluctuation had no existence.

(272.) For these purposes, in the present case, very convenient formulæ have been derived, and tables constructed. They are, however, of too technical a character for this work; we shall, however, point out the manner in which the investigation is conducted. It has been shown in art. 260. by what means the right ascension and declination of an object are derived from its longitude and latitude. Referring to the figure of that article, and supposing the triangle K P X orthographically projected on the plane of the ecliptic as in the annexed figure: in the triangle K P X,

K P is the obliquity of the ecliptic, K X the *co-latitude* (or complement of latitude), and the angle P K X the *co-longitude* of the object X. These are the *data* of our question, of which the first is constant, and the two latter are varied by the effect of precession and nutation; and their variations (considering the minuteness of the latter effect generally, and the small number of years in comparison of the whole period of 25,868, for which we ever require to estimate the effect of the former,) are of that order which may be regarded as infinitesimal in geometry, and treated as such without fear of error. The whole question, then, is reduced to this:—In a spherical triangle K P X, in which one side K X is constant, and an angle K, and

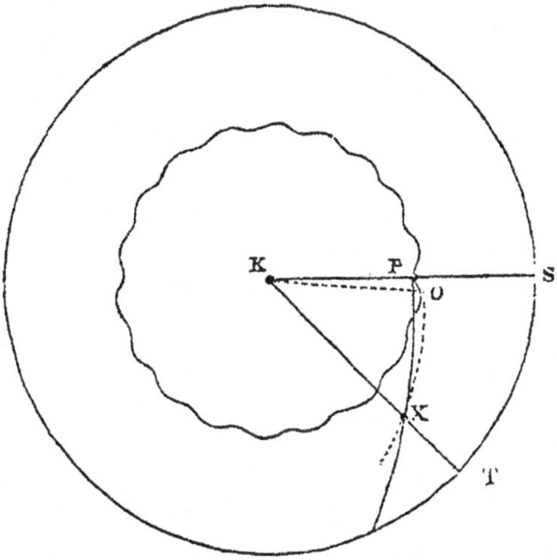

adjacent side K P vary by given infinitesimal changes of the position of P: required the changes thence arising in the other side P X, and the angle K P X? This is a very simple and easy problem of spherical geometry, and being resolved, it gives at once the reductions we are seeking; for P X being the polar distance of the object, and the angle K P X its right ascension *plus* 90°, their variations are the very quantities we seek. It only remains, then, to express in proper form the amount of the precession and nutation in *longitude* and *latitude,* when their amount in right ascension and declination will immediately be obtained.

(273.) The precession in *latitude* is zero, since the latitudes of objects are not changed by it: that in longitude is a quantity proportional to the time at the rate of 50″·10 per annum. With regard to the nutation in *longitude* and *latitude,* these are no other than the abscissa and ordinate of the little ellipse in which the pole moves. The law of its motion, however, therein, cannot be understood till the reader has been made acquainted with the principal features of the moon's motion on which it depends. See Chap. XI.

(274.) Another consequence of what has been shown respecting precession and nutation is, that *sidereal time,* as astronomers use it, *i. e.* as reckoned from the transit of the equinoctial point, is, *not a mean or uniformly flowing quantity,* being affected by nutation; and, moreover, that *so* reckoned, even when cleared of the periodical fluctuation of nutation, it does not *strictly* correspond to the earth's diurnal rotation. As the sun *loses* one day in the year on the stars, by its *direct* motion in longitude; so the equinox *gains* one day in 25,868 years on them by its *retrogradation.* We ought, therefore, as carefully to distinguish between mean and apparent sidereal as between mean and apparent solar time.

(275.) Neither precession nor nutation change the apparent places of celestial objects *inter se.* We see them, so far as these causes go, as they *are,* though from a station more or less *unstable,* as we see distant land objects correctly formed, though appearing to rise and fall when viewed from the heaving deck of a ship in the act of pitching and rolling. But there is an optical cause, independent of refraction or of perspective, which displaces them *one among the other,* and causes us to view the heavens under an aspect always to a certain slight extent false; and whose influence must be estimated and allowed for before we can obtain a precise knowledge of the place of any object. This cause is what is called the aberration of light; a singular and surprising effect arising from this, that we occupy a station not at rest but in rapid motion; and that the apparent directions of the rays of light are not the same to a spectator in motion as to one at rest. As the estimation of its effect belongs to uranography, we must explain it here, though, in so doing, we must anticipate some of the results to be detailed in subsequent chapters.

(276.) Suppose a shower of rain to fall perpendicularly in a dead calm; a person exposed to the shower, who should stand quite still and upright, would receive the drops on his hat, which would thus shelter him, but if he ran forward in any direction they would strike him in the face. The effect would be the same as if he remained still, and a wind should arise of the same velocity, and drift them against him. Suppose a ball let fall from a point A above a horizontal line E F, and that at B were placed to receive it the open mouth of

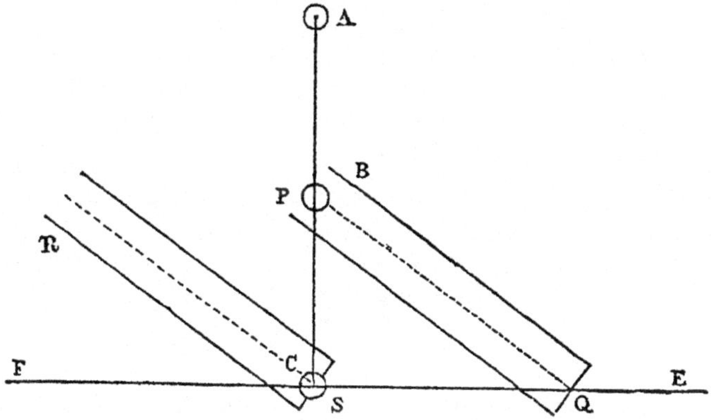

an inclined hollow tube P Q; if the tube were held immoveable the ball would strike on its lower side, but if the tube were carried forward in the direction E F, with a velocity properly adjusted at every instant to that of the ball, while *preserving its inclination* to the horizon, so that when the ball in its natural descent reached C, the tube should have been carried into the position R S, it is evident that the ball would, throughout its whole descent, be found in the axis of the tube; and a spectator, referring to the tube the motion of the ball, and carried along with the former, unconscious of its motion, would fancy that the ball had been moving in the inclined direction R S of the tube's axis.

(277.) Our eyes and telescopes are such tubes. In whatever manner we consider light, whether as an advancing wave in a motionless ether, or a shower of atoms traversing space, if in the interval between the rays traversing the object glass of the one or the cornea of the other (*at which moment* they acquire that convergence which directs them to a certain point *in fixed space*), the cross wires of the one or the retina of the other be *slipped aside*, the point of convergence (which remains unchanged) will no longer correspond to the intersection of the wires or the central point of our visual area. The object then will *appear* displaced; and the amount of this displacement is *aberration*.

(278.) The earth is moving through space with a velocity of about 19 miles per second, in an elliptic path round the sun, and is therefore changing the direction of its motion at every instant. Light travels with a velocity of 192,000 miles per second, which, although much greater than that of the earth, is yet not *infinitely so*. Time is occupied by it in traversing any space, and in that time the earth describes a space which is to the former as 19 to 192,000, or as the tangent of 20″·5 to radius. Suppose now A P S to represent a ray of light from a star at A, and let the tube P Q be that of a telescope so inclined forward that the focus *formed* by its object glass shall be *received* upon its cross wire, it is evident from what has been said, that the inclination of the tube must be such as to make P S : S Q :: velocity of light : velocity of the earth, :: tan. 20″·5 : 1; and, therefore, the angle S P Q, or P S R, by which the axis of the telescope must deviate from the true direction of the star, must be 20″·5.

(279.) A similar reasoning will hold good when the direction of the earth's motion is not perpendicular to the visual ray. If S B be the true direction of the visual ray, and A C the position in which the telescope requires to be held in the apparent direction, we must still have the proportion B C : B A :: velocity of light : velocity of the earth :: rad. : sine of $20''\cdot 5$ (for in such small angles it matters not whether we use the sines or tangents). But we have, also, by trigonometry, B C : B A :: sine of B A C : sine of A C B or C B D, which last is the apparent displacement caused by aberration. Thus it appears that the sine of the aberration, or (since the angle is extremely small) the aberration itself, is proportional to the sine of the angle made by the earth's motion in space with the visual ray, and is therefore a maximum when the line of sight is perpendicular to the direction of the earth's motion.

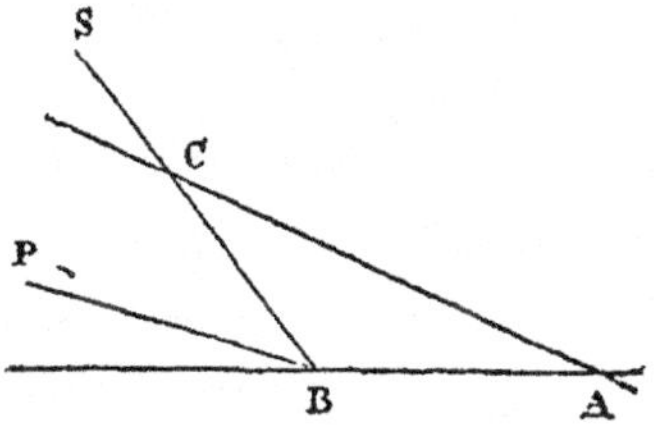

(280.) The uranographical effect of aberration, then, is to distort the aspect of the heavens, causing all the stars to crowd as it were directly towards that point in the heavens which is the vanishing point of all lines parallel to that in which the earth is for the moment moving. As the earth moves round the sun in the plane of the ecliptic, this point must lie in that plane, 90° in advance of the earth's longitude, or 90° *behind* the sun's, and shifts of course continually, describing the circumference of the ecliptic in a year. It is easy to demonstrate that the effect on each particular star will be to make it apparently describe a small ellipse in the heavens, having for its centre the point in which the star would be seen if the earth were at rest.

(281.) Aberration then affects the apparent right ascensions and declinations of all the stars, and that by quantities easily calculable. The formulæ most convenient for that purpose, and which, systematically embracing at the same time the corrections for precession and nutation, enable the observer, with the utmost readiness, to disencumber his observations of right ascension and declination of their influence, have been constructed by Prof. Bessel, and tabulated in the appendix to the first volume of the Transactions of the Astronomical Society, where they will be found accompanied with an extensive catalogue of the places, for 1830, of the principal fixed stars, one of the most useful and best arranged works of the kind which has ever appeared.

(282.) When the body from which the visual ray emanates is, itself, in motion, the best way of conceiving the effect of aberration (independently of theoretical views respecting the nature of light)* is as follows. The ray by which we see any object is not that which it emits at the moment we look at it, but that which it *did* emit some time before, *viz.* the time occupied by light in traversing the interval which separates it from us. The aberration of such a body then arising from the earth's velocity must be applied as a correction, not to the line joining the earth's place at the moment of observation with that occupied by the body *at the same moment*, but at that antecedent instant when the ray quitted it. Hence it is easy to derive the rule given by astronomical writers for the case of a moving object. *From the known laws of its motion and the earth's, calculate its apparent or relative angular motion in the time taken by light to traverse its distance from the earth. This is its aberration, and its effect is* to displace it in a direction contrary to its apparent relative motion among the stars.

We shall conclude this chapter with a few urano-

* The results of the undulatory and corpuscular theories of light, in the matter of aberration are, in the main, the same. We say *in the main.* There is, however, a minute difference even of numerical results. In the undulatory doctrine, the propagation of light takes place with equal velocity in all directions whether the luminary be at rest or in motion. In the corpuscular, with an excess of velocity in the direction of the motion over that in the contrary equal to twice the velocity of the body's motion. In the cases, then, of a body moving with equal velocity directly to and directly from the earth, the aberrations will be *alike* on the undulatory, but different on the corpuscular hypothesis. The utmost difference which can arise from this cause *in our system* cannot amount to above six thousandths of a second.

graphical problems of frequent practical occurrence, which may be resolved by the rules of spherical trigonometry.

(283.) Of the following five quantities, given any three, to find one or both the others.

1st, The latitude of the place; 2d, the declination of an object; 3d, its hour angle east or west from the meridian; 4th, its altitude; 5th, its azimuth.

In the figure of art. 94. P is the pole, Z the zenith, and S the star; and the five quantities above mentioned, or their complements, constitute the sides and angles of the spherical triangle P Z S; P Z being the co-latitude, P S the co-declination or polar distance; S P Z the hour angle; P S the co-altitude or zenith distance; and P Z S the azimuth. By the solution of this spherical triangle, then, all problems involving the relations between these quantities may be resolved.

(284.) For example, suppose the time of rising or setting of the sun or of a star were required, having given its right ascension and polar distance. The star rises when *apparently* on the horizon, or *really* about 34′ below it (owing to refraction), so that, at the moment of its apparent rising, its zenith distance is $90° 34' = ZS$. Its polar distance P S being also given, and the co-latitude Z P of the place, we have given the three sides of the triangle, to find the hour angle Z P S, which, being known, is to be added to or subtracted from the star's right ascension, to give the sidereal time of setting or rising, which, if we please, may be converted into solar time by the proper rules and tables.

(285.) As another example of the same triangle, we may propose to find the local sidereal time, and the latitude of the place of observation, by observing equal altitudes of the same star east and west of the meridian, and noting the interval of the observations in sidereal time.

The hour angles corresponding to equal altitudes of a fixed star being equal, the hour angle east or west will be measured by half the observed interval of the observations. In our triangle, then, we have given this hour angle Z P S, the polar distance P S of the star, and Z S, its co-altitude at the moment of observation. Hence we may find P Z, the co-latitude of the place. Moreover, the hour angle of the star being known, and also its right ascension, the point of the equinoctial is known, which is on the meridian at the moment of observation; and, therefore, the local sidereal time at that moment. This is a very useful observation for determining the latitude and time at an unknown station.

(286.) It is often of use to know the situation of the ecliptic in the visible heavens at any instant; that is to say, the points where it cuts the horizon, and the altitude of its highest point, or, as it is sometimes called, the *nonagesimal* point of the ecliptic, as well as the longitude of this point on the ecliptic itself from the equinox. These, and all questions referable to the same data and quæsita, are resolved by the spherical triangle Z P E, formed by the zenith Z (considered as the pole of the horizon), the pole of the equinoctial P, and the pole of the ecliptic E. The sidereal time being given, and also

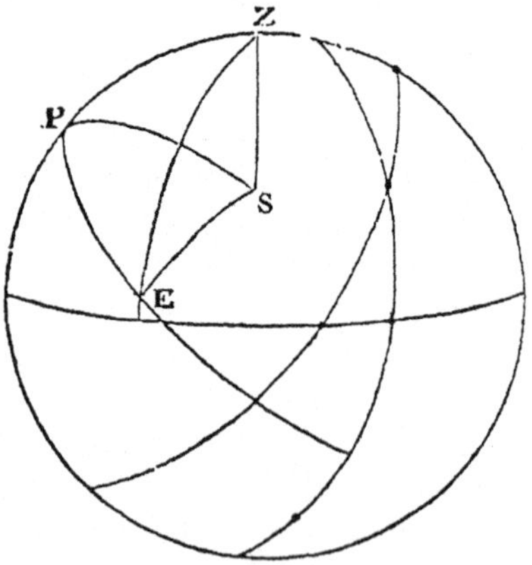

the right ascension of the pole of the ecliptic (which is always the same, viz. $18^h\ 0^m\ 0^s$), the hour angle Z P E of that point is known. Then, in this triangle we have given P Z, the co-latitude; P E, the polar distance of the pole of the ecliptic, 23° 28′, and the angle Z P E; from which we may find, 1st, the side Z E, which is easily seen to be equal to the altitude of the nonagesimal point sought; and, 2dly, the angle P Z E, which is the azimuth of the pole of the ecliptic, and which, therefore, being added to and subtracted from 90°, gives the azimuths of the eastern and western intersections of the ecliptic with the horizon. Lastly, the longitude of the nonagesimal point may be had, by calculating in the same triangle the angle P E Z, which is its complement.

(287.) The *angle of situation* of a star is the angle included at the star between circles of latitude and of declination passing through it. To determine it in any proposed case, we must resolve the triangle P S E, in which are given P S, P E, and the angle S P E, which is the difference between the star's right ascension and 18 hours; from which it is easy to find the angle P S E required. This angle is of use in many enquiries in physical astronomy. It is called in most books on astronomy the angle of *position*; but the latter expression has become otherwise, and more conveniently, appropriated.

(288.) From these instances, the manner of treating such questions in uranography as depend on spherical trigonometry will be evident, and will, for the most part, offer little difficulty, if the student will bear in mind, as a practical maxim, *rather to consider the poles of the great circles which his question refers to, than the circles themselves.*

CHAP. V.

OF THE SUN'S MOTION.

APPARENT MOTION OF THE SUN NOT UNIFORM. — ITS APPARENT DIAMETER ALSO VARIABLE. — VARIATION OF ITS DISTANCE CONCLUDED. — ITS APPARENT ORBIT AN ELLIPSE ABOUT THE FOCUS. — LAW OF THE ANGULAR VELOCITY. — EQUABLE DESCRIPTION OF AREAS. — PARALLAX OF THE SUN. — ITS DISTANCE AND MAGNITUDE. — COPERNICAN EXPLANATION OF THE SUN'S APPARENT MOTION. — PARALLELISM OF THE EARTH'S AXIS. — THE SEASONS. — HEAT RECEIVED FROM THE SUN IN DIFFERENT PARTS OF THE ORBIT.

(289.) In the foregoing chapters, it has been shown that the apparent path of the sun is a great circle of the sphere, which it performs in a period of one sidereal year. From this it follows, that the line joining the earth and sun lies constantly *in one plane;* and that, therefore, whatever be the real motion from which this apparent motion arises, it must be confined to one plane, which is called the *plane of the ecliptic.*

(290.) We have already seen (art. 118.) that the sun's motion in right ascension among the stars is not uniform. This is partly accounted for by the obliquity of the ecliptic, in consequence of which equal variations in longitude do not correspond to equal changes of right ascension. But if we observe the place of the sun daily throughout the year, by the transit and circle, and from these calculate the longitude for each day, it will still be found that, even in its own proper path, its apparent angular motion is far from uniform. The change of longitude in twenty-four mean solar hours *averages* $0^\circ\ 59'\ 8''\cdot33$; but about the 31st of December it amounts to $1^\circ\ 1'\ 9''\cdot9$, and about the 1st of July is only $0^\circ\ 57'\ 11''\cdot5$. Such are the extreme limits, and such the mean value of the sun's apparent angular velocity in its annual orbit.

(291.) This variation of its angular velocity is accompanied with a corresponding change of its distance from us. The change of distance is recognized by a variation observed to take place in its apparent diameter, when measured at different seasons of the year, with an instrument adapted for that purpose, called the *heliometer* *, or, by calculating from the time which its disk takes to traverse the meridian in the transit instrument. The greatest apparent diameter corresponds to the 31st of December, or to the greatest angular velocity, and measures 32′ 35″·6; the least is 31′ 31″·0, and corresponds to the 1st of July; at which epochs, as we have seen, the angular motion is also at its extreme limit either way. Now, as we cannot suppose the sun to alter its real size periodically, the observed change of its apparent size can only arise from an actual change of distance. And the sines or tangents of such small arcs being proportional to the arcs themselves, its distances from us, at the above-named epoch, must be in the inverse proportion of the apparent diameters. It appears, therefore, that the greatest, the mean, and the least distances of the sun from us are in the respective proportions of the numbers 1·01679, 1·00000, and 0·98321; and that its apparent angular velocity diminishes as the distance increases, and *vice versâ*.

(292.) It follows from this, that the real orbit of the sun, as referred to the earth supposed at rest, is not a circle with the earth in the centre. The situation of the earth within it is *excentric*, the *excentricity* amount-

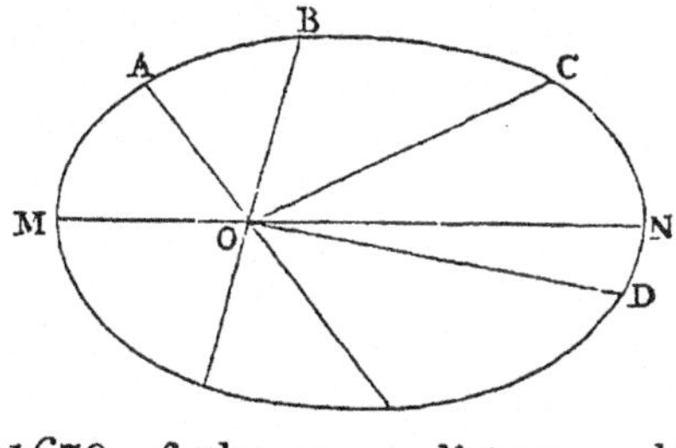

ing to 0·01679 of the mean distance, which may be regarded as our unit of measure in this enquiry. But besides this, the *form* of the orbit is not circular, but

* Ἥλιος, the sun, and μετρεῖν, to measure.

elliptic. If from any point O, taken to represent the earth, we draw a line, O A, in some fixed direction, from which we then set off a series of angles, A O B, A O G, &c. equal to the observed longitudes of the sun throughout the year, and in these respective directions measure off from O the distances O A, O B, O C, &c. representing the distances deduced from the observed diameter, and then connect all the extremities A, B, C, &c. of these lines by a continuous curve, it is evident this will be a correct representation of the relative orbit of the sun about the earth. Now, when this is done, a deviation from the circular figure in the resulting curve becomes apparent; it is found to be evidently longer than it is broad—that is to say, elliptic, and the point O to occupy not the *centre*, but one of the foci of the ellipse. The graphical process here described is sufficient to point out the general figure of the curve in question; but for the purposes of exact verification, it is necessary to recur to the properties of the ellipse*, and to express the distance of any one of its points in terms of the angular situation of that point with respect to the longer axis, or diameter of the ellipse. This, however, is readily done; and when numerically calculated, on the supposition of the excentricity being such as above stated, a perfect coincidence is found to subsist between the distances thus computed, and those derived from the measurement of the apparent diameter.

(293.) The mean distance of the earth and sun being taken for unity, the extremes are 1·01679 and 0·98321. But if we compare, in like manner, the mean or average angular velocity with the extremes, greatest and least, we shall find these to be in the proportions of 1·03386, 1·00000, and 0·96614. The variation of the sun's *angular velocity*, then, is much greater in proportion than that of its distance — fully twice as great; and if we examine its numerical expressions at different periods, comparing them with the mean value, and also with the corresponding distances, it will be found, that, by what-

* See Conic Sections, by the Rev. H. P. Hamilton.

ever fraction of its mean value the distance exceeds the mean, the angular velocity will fall short of *its* mean or average quantity by very nearly *twice* as great a fraction of the latter, and *vice versâ*. Hence we are led to conclude that the *angular velocity* is in the inverse proportion, not of the distance simply, but of its *square;* so that, to compare the daily motion in longitude of the sun, at one point, A, of its path, with that at B, we must state the proportion thus:—

$OB^2 : OA^2 ::$ daily motion at A : daily motion at B. And this is found to be exactly verified in every part of the orbit.

(294.) Hence we deduce another remarkable conclusion—viz. that if the sun be supposed really to move around the circumference of this ellipse, its actual speed cannot be uniform, but must be greatest at its least distance, and less at its greatest. For, were it uniform, the apparent angular velocity would be, of course, inversely proportional to the distance; simply because the same linear change of place, being produced in the same time at different distances from the eye, must, by the laws of perspective, correspond to apparent angular displacements inversely as those distances. Since, then, observation indicates a more rapid law of variation in the angular velocities, it is evident that mere change of distance, unaccompanied with a change of actual speed, is insufficient to account for it; and that the increased proximity of the sun to the earth must be accompanied with an actual increase of its real velocity of motion along its path.

(295.) This elliptic form of the sun's path, the excentric position of the earth within it, and the unequal speed with which it is actually traversed by the sun itself, all tend to render the calculation of its longitude from theory (*i.e.* from a knowledge of the causes and nature of its motion) difficult; and indeed impossible, so long as the *law* of its actual velocity continues unknown. This *law*, however, is not immediately apparent. It does not come forward, as it were, and present itself at once, like the elliptic form of the orbit, by a direct comparison of angles and distances, but requires an attentive consideration of the whole series of observations registered during an entire period. It was not, therefore, without much painful and laborious calculation, that it was discovered by Kepler (who was also the first to ascertain the elliptic form of the orbit), and announced in the following terms:—Let a line be always supposed to connect the sun, supposed in motion, with the earth, supposed at rest; then, as the sun moves along its ellipse, this line (which is called in astronomy the *radius vector*) will *describe* or *sweep over* that portion of the whole *area* or *surface* of the ellipse which is included between its consecutive positions: and the motion of the sun will be such that *equal areas are* thus *swept over* by the revolving radius vector *in equal times,* in whatever part of the circumference of the ellipse the sun may be moving.

(296.) From this it necessarily follows, that in *un*equal times, the areas described must be proportional to the times. Thus, in the figure of art. 292. the time in which the sun moves from A to B, is the time in which it moves from C to D, as the area of the elliptic sector A O B is to the area of the sector D O C.

(297.) The circumstances of the sun's apparent annual motion may, therefore, be summed up as follows:—It is performed in an orbit lying in one plane passing through the earth's centre, called the plane of the ecliptic, and whose projection on the heavens is the great circle so called. In this plane, however, the actual path is not circular, but elliptical; having the earth, not in its centre, but in one focus. The excentricity of this ellipse is 0·01679, in parts of a unit equal to the *mean distance,* or *half the longer diameter of the ellipse;* and the motion of the sun in its circumference is so regulated, that equal areas of the ellipse are passed over by the radius vector in equal times.

(298.) What we have here stated supposes no knowledge of the sun's actual distance from the earth, nor, consequently, of the actual dimensions of its orbit, nor of the body of the sun itself. To come to any conclusions on these points, we must first consider by what means we can arrive at any knowledge of the distance of an object to which we have no access. Now, it is obvious, that its *parallax* alone can afford us any information on this subject. Parallax may be generally defined to be the change of apparent situation of an object arising from a change of real situation of the observer. Suppose, then, P A B Q to represent the earth, C its

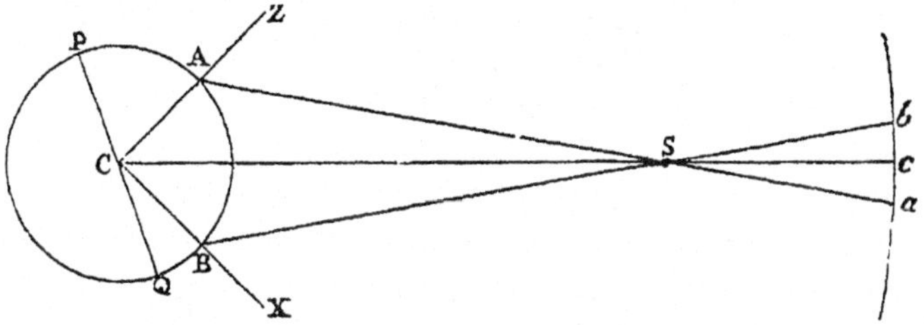

centre, and S the sun, and A, B two situations of a spectator, or, which comes to the same thing, the stations of two spectators, both observing the sun S at the same instant. The spectator A will see it in the direction A S *a*, and will refer it to a point *a* in the infinitely distant sphere of the fixed stars, while the spectator B will see it in the direction B S *b*, and refer it to *b*. The angle included between these directions, or the measure of the celestial arc *a b*, by which it is *displaced*, is equal to the angle A S B; and if this angle be known, and the local situations of A and B, with the part of the earth's surface A B included between them, it is evident that the distance C S may be calculated.

(299.) Parallax, however, in the astronomical acceptation of the word, has a more technical meaning. It is restricted to the difference of apparent positions of any celestial object when viewed from a station on the *surface* of the earth, and from its *centre*. The *centre* of *the earth* is the general station to which all astronomical observations are referred: but, as we observe from the surface, a *reduction to the centre* is needed; and the amount of this reduction is called parallax. Thus, the sun being seen from the earth's centre, in the direction C S, and from A on the surface in the direction A S, the angle A S C, included between these two directions, is the parallax at A, and similarly B S C is that at B.

Parallax, in this sense, may be distinguished by the epithet *diurnal*, or *geocentric*, to discriminate it from the *annual*, or *heliocentric;* of which more hereafter.

(300.) The reduction for parallax, then, in any proposed case, is obtained from the consideration of the triangle A C S, formed by the spectator, the centre of the earth, and the object observed; and since the side C A prolonged passes through the observer's zenith, it is evident that *the effect of parallax*, in this its technical acceptation, *is always to depress the object observed in a vertical circle.* To estimate the amount of this depression, we have, by plane trigonometry,

CS : CA : : sine of CAS = sine of ZAS : sine of ASC.

(301.) The parallax, then, for objects equidistant from the earth, is proportional to the sines of their zenith distances. It is, therefore, at its maximum when the body observed is in the horizon. In this situation it is called the *horizontal parallax;* and when this is known, since small arcs are proportional to their sines, the parallax at any given altitude is easily had by the following rule:—

Parallax = (horizontal parallax) × sine of zenith distance.

The horizontal parallax is given by this proportion:—

Distance of object : earth's radius : : rad. : sine of horizontal parallax.

It is, therefore, known, when the proportion of the object's distance to the radius of the earth is known: and *vice versâ*—if by any method of observation we can come at a knowledge of the horizontal parallax of an object, its distance, expressed in units equal to the earth's radius, becomes known.

(302.) To apply this general reasoning to the case of the sun. Suppose two observers — one in the northern, the other in the southern hemisphere — at stations on the same meridian, to observe on the same day the meridian altitudes of the sun's centre. Having thence derived the apparent zenith distances, and cleared them of the effects of refraction, if the distance of the sun were equal to that of the fixed stars, the sum of the zenith distances thus found would be precisely equal to the sum of the latitudes north and south of the places of observation. For the sum in question would then be equal to the angle Z C X, which is the meridional distance of the stations across the equator. But the effect of parallax being in both cases to increase the apparent zenith distances, their observed sum will be greater than the sum of the latitudes, by the whole amount of the two parallaxes, or by the angle A S B. This angle, then, is obtained by subducting the sum of the latitudes from that of the zenith distances; and this once determined, the horizontal parallax is easily found, by dividing the angle so determined by the sum of the sines of the two latitudes.

(303.) If the two stations be not exactly on the same meridian (a condition very difficult to fulfil), the same process will apply, if we take care to allow for the change of the sun's actual zenith distance in the interval of time elapsing between its arrival on the meridians of the stations. This change is readily ascertained, either from tables of the sun's motion, grounded on the experience of a long course of observations, or by actual observation of its meridional altitude on several days before and after that on which the observations for parallax are taken. Of course, the nearer the stations are to each other in longitude, the less is this interval of time; and, consequently, the smaller the amount of this correction; and, therefore, the less injurious to the accuracy of the final result is any uncertainty in the daily change of zenith distance which may arise from imperfection in the solar tables, or in the observations made to determine it.

(304.) The horizontal parallax of the sun has been concluded from observations of the nature above described, performed in stations the most remote from each other in latitude, at which observatories have been instituted. It has also been deduced from other methods of a more refined nature, and susceptible of much greater exactness, to be hereafter described. Its amount, so obtained, is about 8″·6. Minute as this quantity is, there can be no doubt that it is a tolerably correct approximation to the truth; and in conformity with it, we must admit the sun to be situated at a mean distance from us, of no less than 23,984 times the length of the earth's radius, or about 95,000,000 miles.

(305.) That at so vast a distance the sun should appear to us of the size it does, and should so powerfully influence our condition by its heat and light, requires us to form a very grand conception of its actual magnitude, and of the scale on which those important processes are carried on within it, by which it is enabled to keep up its liberal and unceasing supply of these elements. As to its actual magnitude we can be at no loss, knowing its distance, and the angles under which its diameter appears to us. An object, placed at the distance of 95,000,000 miles, and subtending an angle of 32′ 3″, must have a real diameter of 882,000 miles. Such, then, is the diameter of this stupendous globe. If we compare it with what we have already ascertained of the dimensions of our own, we shall find that in linear magnitude it exceeds the earth in the proportion of 111½ to 1, and in bulk in that of 1,384,472 to 1.

(306.) It is hardly possible to avoid associating our conception of an object of definite globular figure, and of such enormous dimensions, with some corresponding attribute of massiveness and material solidity. That the sun is not a mere phantom, but a body having its own peculiar structure and economy, our telescopes distinctly inform us. They show us dark spots on its surface, which slowly change their places and forms, and by attending to whose situation, at different times, astronomers have ascertained that the sun revolves about an axis inclined at a constant angle of 82° 40′ to the plane of the ecliptic, performing one rotation in a period of 25 days and in the same direction with the diurnal rotation of the earth, *i. e.* from west to east. Here, then, we have an analogy with our own globe; the slower and more majestic movement only corresponding with the greater dimensions of the machinery, and impressing us with the prevalence of similar mechanical laws, and of, at least, such a community of nature as the existence of inertia and obedience to force may argue. Now, in the exact proportion in which we invest our idea of this immense bulk with the attribute of inertia, or weight, it becomes difficult to conceive its circulation round so comparatively small a body as the earth, without, on the one hand, dragging it along, and displacing it, if bound to it by some invisible tie; or, on the other hand, if not so held to it, pursuing its course alone in space, and leaving the earth behind. If we tie two stones together by a string, and fling them aloft, we see them circulate about a point between them, which is their common centre of gravity; but if one of them be greatly more ponderous than the other, this common centre will be proportionally nearer to that one, and even within its surface, so that the smaller one will circulate, in fact, about the larger, which will be comparatively but little disturbed from its place.

(307.) Whether the earth move round the sun, the sun round the earth, or both round their common centre of gravity, will make no difference, so far as appearances are concerned, provided the stars be supposed sufficiently distant to undergo no sensible apparent *parallactic* displacement by the motion so attributed to the earth. Whether they are so or not must still be a matter of enquiry; and from the absence of any measureable amount of such displacement, we can conclude nothing but this, that the scale of the sidereal universe is so great, that the mutual orbit of the earth and sun may be regarded as an imperceptible point in its comparison. Admitting, then, in conformity with the laws of dynamics, that two bodies connected with and revolving about each other in free space do, in fact, revolve about their common centre of gravity, which remains immoveable by their mutual action, it becomes a matter of further enquiry, *whereabouts* between them this centre is situated. Mechanics teaches us that its place will divide their mutual distance in the inverse ratio of their *weights* or *masses* *; and calculations grounded on phenomena, of which an account will be given further on, inform us that this ratio, in the case of the sun and earth, is actually that of 354,936 to 1, — the sun being, in that proportion, more ponderous than the earth. From this it will follow that the common point about which they both circulate is only 267 miles from the sun's centre, or about $\frac{1}{3300}$th part of its own diameter.

(308.) Henceforward, then, in conformity with the above statements, and with the Copernican view of our system, we must learn to look upon the sun as the comparatively motionless centre about which the earth performs an annual elliptic orbit of the dimensions and excentricity, and with a velocity regulated according to the law above assigned; the sun occupying one of the foci of the ellipse, and from that station quietly disseminating on all sides its light and heat; while the earth, travelling round it, and presenting itself differently to it at different times of the year and day, passes through the varieties of day and night, summer and winter, which we enjoy.

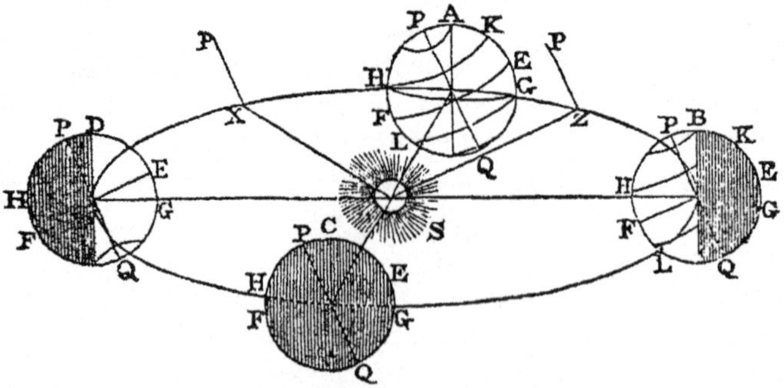

* See Cab. Cyc. MECHANICS, Centre of Gravity.

(309.) In this annual motion of the earth, its axis preserves, at all times, the same direction as if the orbitual movement had no existence; and is carried round parallel to itself, and pointing always to the same vanishing point in the sphere of the fixed stars. This it is which gives rise to the variety of seasons, as we shall now explain. In so doing, we shall neglect (for a reason which will be presently explained) the ellipticity of the orbit, and suppose it a circle, with the sun in the centre.

(310.) Let, then, S represent the sun, and A, B, C, D, four positions of the earth in its orbit 90° apart, viz. A that which it has on the 21st of March, or at the time of the vernal equinox; B that of the 21st of June, or the summer solstice; C that of the 21st of September, or the autumnal equinox; and D that of the 21st of December, or the winter solstice. In each of these positions let P Q represent the axis of the earth, about which its diurnal rotation is performed without interfering with its annual motion in its orbit. Then, since the sun can only enlighten one half of the surface at once, viz. that turned towards it, the shaded portions of the globe in its several positions will represent the dark, and the bright, the enlightened halves of the earth's surface in these positions. Now, 1st, in the position A, the sun is vertically over the intersection of the equinoctial F E and the ecliptic H G. It is, therefore, in the equinox; and in this position the poles P, Q, both fall on the extreme confines of the enlightened side. In this position, therefore, it is day over half the northern and half the southern hemisphere at once; and as the earth revolves on its axis, every point of its surface describes half its diurnal course in light, and half in darkness; in other words, the duration of day and night is here equal over the whole globe hence the term *equinox*. The same holds good at the autumnal equinox on the position C.

(311.) B is the position of the earth at the time of

the *northern, summer* solstice. Here the north pole P, and a considerable portion of the earth's surface in its neighbourhood, as far as B, are situated *within* the enlightened half. As the earth turns on its axis in this position, therefore, the whole of that part remains constantly enlightened; therefore, at this point of its orbit, or at this season of the year, it is continual day at the north pole, and in all that region of the earth which encircles this pole as far as B — that is, to the distance of 23° 28′ from the pole, or within what is called, in geography, the *arctic circle*. On the other hand, the opposite or south pole Q, with all the region comprised within the *antarctic* circle, as far as 23° 28′ from the south pole, are immersed at this season in darkness, during the entire diurnal rotation, so that it is here continual night.

(312.) With regard to that portion of the surface comprehended between the arctic and antarctic circles, it is no less evident that the nearer any point is to the north pole, the larger will be the portion of its diurnal course comprised within the bright, and the smaller within the dark hemisphere; that is to say, the longer will be its day, and the shorter its night. Every station north of the equator will have a day of more and a night of less than twelve hours' duration, and *vice versâ*. All these phenomena are exactly inverted when the earth comes to the opposite point D of its orbit.

(313.) Now, the temperature of any part of the earth's surface depends mainly, if not entirely, on its exposure to the sun's rays. Whenever the sun is above the horizon of any place, that place is receiving heat; when below, parting with it, by the process called radiation; and the whole quantities received and parted with in the year must balance each other at every station, or the equilibrium of temperature would not be supported. Whenever, then, the sun remains more than twelve hours above the horizon of any place, and less beneath, the general temperature of that place will be above the average; when the reverse, below. As the earth, then, moves from A to B, the days growing longer, and the nights shorter, in the northern hemisphere, the temperature of every part of that hemisphere increases, and we pass from spring to summer; while, at the same time, the reverse obtains in the southern hemisphere. As the earth passes from B to C, the days and nights again approach to equality—the excess of temperature in the northern hemisphere above the mean state grows less, as well as its defect in the southern; and at the autumnal equinox C, the mean state is once more attained. From thence to D, and, finally, round again to A, all the same phenomena, it is obvious, must again occur, but reversed,—it being now winter in the northern, and summer in the southern hemisphere.

(314.) All this is exactly consonant to observed fact. The continual day within the polar circles in summer, and night in winter, the general increase of temperature and length of day as the sun approaches the elevated pole, and the reversal of the seasons in the northern and southern hemispheres, are all facts too well known to require further comment. The positions A, C of the earth correspond, as we have said, to the equinoxes; those at B, D to the *solstices*. This term must be explained. If, at any point, X, of the orbit, we draw X P the earth's axis, and X S to the sun, it is evident that the angle P X S will be the sun's *polar distance*. Now, this angle is at its maximum in the position D, and at its minimum at B; being in the former case $=90^\circ+23^\circ\ 28'=103^\circ\ 28'$, and in the latter $90^\circ-23^\circ\ 28'=66^\circ\ 32'$. At these points the sun ceases to approach to or to recede from the pole, and hence the name solstice.

(315.) The elliptic form of the earth's orbit has but a very trifling share in producing the variation of temperature corresponding to the difference of seasons. This assertion may at first sight seem incompatible with what we know of the laws of the communication of heat from a luminary placed at a variable distance. Heat, like light, being equally dispersed from the sun in all directions, and being spread over the surface of a sphere continually enlarging as we recede from the centre, must, of course, diminish in intensity according to the inverse proportion of the surface of the sphere over which it is spread; that is, in the inverse proportion of the square of the distance. But we have seen (art. 293.) that this is also the proportion in which the *angular velocity* of the earth about the sun varies. Hence it appears, that the *momentary supply of heat* received by the earth from the sun varies in the exact proportion of the angular velocity, *i. e.* of the *momentary increase of longitude:* and from this it follows, that equal amounts of heat are received from the sun in passing over equal angles round it, in whatever part of the ellipse those angles may be situated. Let, then, S represent the sun;

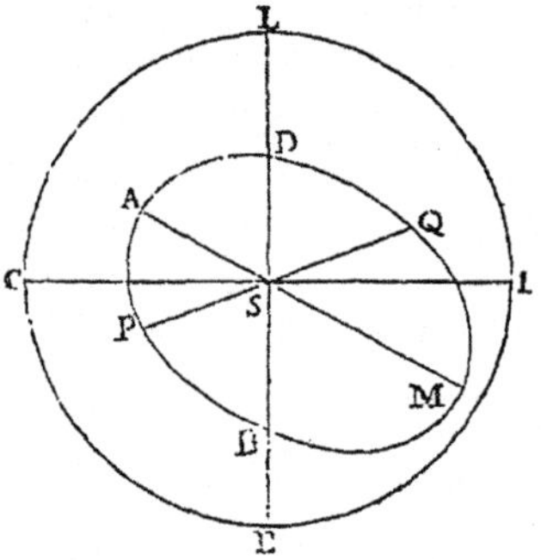

A Q M P the earth's orbit; A its nearest point to the sun, or, as it is called, the *perihelion* of its orbit; M the farthest, or the *aphelion;* and therefore A S M the *axis* of the ellipse. Now, suppose the orbit divided into two segments by a straight line P S Q, drawn through the sun, and any how situated as to direction; then, if we suppose the earth to circulate in the direction P A Q M P, it will have passed over 180° of longitude in moving from P to Q, and as many in moving from Q to P. It appears, therefore, from what has been shown, that the supplies of heat received from the sun will be equal in the two segments, in whatever direction the line P S Q be drawn. They will, indeed, be described in unequal times; that in which the perihelion A lies in a shorter, and the other in a longer, in proportion to their unequal area: but the greater proximity of the sun in the smaller segment compensates exactly for its more rapid description, and thus an equilibrium of heat is, as it were, maintained. Were it not for this, the excentricity of the orbit would materially influence the transition of seasons. The fluctuation of distance amounts to nearly $\frac{1}{30}$th of its mean quantity, and, consequently, the fluctuation in the sun's direct heating power to double this, or $\frac{1}{15}$th of the whole. Now, the perihelion of the orbit is situated nearly at the place of the northern winter solstice; so that, were it not for the compensation we have just described, the effect would be to exaggerate the difference of summer and winter in the southern hemisphere, and to moderate it in the northern; thus producing a more violent alternation of climate in the one hemisphere, and an approach to perpetual spring in the other. As it is, however, no such inequality subsists, but an equal and impartial distribution of heat and light is accorded to both.*

(316.) The great key to simplicity of conception in astronomy, and, indeed, in all sciences where motion is concerned, consists in contemplating every movement as referred to points which are either permanently fixed, or so nearly so, as that their motions shall be too small to interfere materially with and confuse our notions. In the choice of these primary points of reference, too, we must endeavour, as far as possible, to select such as have simple and symmetrical geometrical relations of situation with respect to the curves described by the moving parts of the system, and which are thereby fitted to perform the office of natural centres—advantageous stations for the eye of reason and theory. Having learned to attribute an orbitual motion to the earth, it loses this advantage, which is transferred to the sun, as the fixed centre about which its orbit is performed. Precisely as,

* See Geological Transactions, 1832, "On the Astronomical Causes which may influence Geological Phenomena,"—by the author of this work.

when embarrassed by the earth's diurnal motion, we have learned to transfer, in imagination, our station of observation from its surface to its centre, by the application of the diurnal parallax; so, when we come to enquire into the movements of the planets, we shall find ourselves continually embarrassed by the orbitual motion of our point of view, unless, by the consideration of the *annual or heliocentric parallax*, as it may be termed, we consent to refer all our observations on them to the centre of the sun, or rather to the common centre of gravity of the sun, and the other bodies which are connected with it in our system. Hence arises the distinction between the *geocentric* and *heliocentric* place of an object. The former refers its situation in space to an imaginary sphere of infinite radius, having the centre of the earth for its centre — the latter to one concentric with the sun. Thus, when we speak of the *heliocentric longitudes* and *latitudes* of objects, we suppose the spectator situated in the sun, and referring them, by circles perpendicular to the plane of the ecliptic, to the great circle marked out in the heavens by the infinite prolongation of that plane.

(317.) The point in the imaginary concave of an infinite heaven, to which a spectator in the sun refers the earth, must, of course, be diametrically opposite to that to which a spectator on the earth refers the sun's centre; consequently, the heliocentric *latitude* of the earth is always nothing, and its heliocentric longitude always equal to the sun's geocentric longitude + 180°. The heliocentric equinoxes and solstices are, therefore, the same as the geocentric; and to conceive them, we have only to imagine a plane passing through the sun's centre, parallel to the earth's equator, and prolonged infinitely on all sides. The line of intersection of this plane and the plane of the ecliptic is the line of equinoxes, and the solstices are 90° distant from it.

(318.) The position of the longer axis of the earth's orbit is a point of great importance. In the figure (art. 315.) let E C L I be the ecliptic, E the vernal equinox, L the autumnal (*i. e.* the points *to which the earth is referred from the sun when its heliocentric longitudes are* 0° *and* 180° *respectively*). Supposing the earth's motion to be performed in the direction E C L I, the angle E S A, or the longitude of the perihelion, in the year 1800 was 99° 30′ 5″: we say in the year 1800, because, in point of fact, by the operation of causes hereafter to be explained, its position is subject to an extremely slow variation of about 12″ per annum to the eastward, and which, in the progress of an immensely long period — of no less than 20,984 years — carries the axis A S M of the orbit completely round the whole circumference of the ecliptic. But this motion must be disregarded for the present, as well as many other minute deviations, to be brought into view when they can be better understood.

(319.) Were the earth's orbit a circle, described with a uniform velocity about the sun placed in its centre, nothing could be easier than to calculate its position at any time, with respect to the line of equinoxes, or its longitude, for we should only have to reduce to numbers the proportion following; viz. One year : the time elapsed :: 360° : the arc of longitude passed over. The longitude so calculated is called in astronomy the *mean* longitude of the earth. But since the earth's orbit is neither circular, nor uniformly described, this rule will not give us the true place in the orbit at any proposed moment. Nevertheless, as the excentricity and deviation from a circle are small, the *true place* will never deviate very far from that so determined (which, for distinction's sake, is called the *mean place*), and the former may at all times be calculated from the latter, by applying to it a *correction* or *equation* (as it is termed), whose amount is never very great, and whose computation is a question of pure geometry, depending on the equable description of areas by the earth about the sun. For since, in the elliptic motion, according to Kepler's law above stated, *areas* not *angles* are described uniformly, the proportion must now be stated thus; One year : the time elapsed : : the whole *area* of the ellipse : the *area* of the sector swept over by the radius vector in that time. This area, therefore, becomes known, and it is then, as above observed, a problem of pure geometry to ascertain the *angle* about the sun (A S P, *fig.* art. 315.), which corresponds to any proposed fractional area of the whole ellipse supposed to be contained in the sector A P S. Suppose we set out from A the perihelion, then will the angle A S P at first increase more rapidly than the *mean longitude*, and will, therefore, during the whole semi-revolution from A to M, exceed it in amount; or, in other words, the *true place* will be in advance of the *mean*: at M, one half the year will have elapsed, and one half the orbit have been described, whether it be circular or elliptic. Here, then, the *mean* and *true* places coincide; but in all the other half of the orbit, from M to A, the true place will fall short of the mean, since at M the angular motion is slowest, and the true place from this point begins to lag behind the mean — to make up with it, however, as it approaches A, where it once more overtakes it.

(320.) The quantity by which the *true* longitude of the earth differs from the *mean* longitude is called the equation of the centre, and is *additive* during all the half-year in which the earth passes from A to M, beginning at $0° 0' 0''$, increasing to a maximum, and again diminishing to zero at M; after which it becomes subtractive, attains a maximum of subtractive magnitude between M and A, and again diminishes to 0 at A. Its maximum, both additive and subtractive, is $1° 55' 33''\cdot 3$.

(321.) By applying, then, to the earth's mean longitude, the equation of the centre corresponding to any given time at which we would ascertain its place, the true longitude becomes known; and since the sun is always seen from the earth in 180° more longitude than the earth from the sun, in this way also the sun's true place in the ecliptic becomes known. The calculation of the equation of the centre is performed by a table constructed for that purpose, to be found in all "Solar Tables."

(322.) The maximum value of the equation of the centre depends only on the ellipticity of the orbit, and may be expressed in terms of the excentricity. *Vice versâ*, therefore, if the former quantity can be ascertained by observation, the latter may be derived from it; because, whenever the law, or numerical connection, between two quantities is known, the one can always be determined from the other. Now, by assiduous observation of the sun's transits over the meridian, we can ascertain, for every day, its exact right ascension, and thence conclude its longitude (art. 260.). After this, it is easy to assign the angle by which this *observed* longitude exceeds or falls short of the mean; and the greatest amount of this excess or defect which occurs in the whole year, is the maximum equation of the centre. This, as a means of ascertaining the excentricity of the orbit, is a far more easy and accurate method than that of concluding its distance by measuring its apparent diameter. The results of the two methods coincide, however, perfectly.

(323.) If the ecliptic coincided with the equinoctial, the effect of the equation of the centre, by disturbing the uniformity of the sun's apparent motion in longitude, would cause an inequality in its time of coming on the meridian on successive days. When the sun's centre comes to the meridian, it is *apparent noon*, and if its motion in longitude were uniform, and the ecliptic coincident with the equinoctial, this would always coincide with *mean noon*, or the stroke of 12 on a well-regulated solar clock. But, independent of the want of uniformity in its motion, the obliquity of the ecliptic gives rise to another inequality in this respect; in consequence of which, the sun, even supposing its motion in the ecliptic uniform, would yet alternately, in its time of attaining the meridian, anticipate and fall short of the mean noon as shown by the clock. For the right ascension of a celestial object, forming a side of a right-angled spherical triangle, of which its longitude is the hypothenuse, it is clear that the uniform increase of the latter must necessitate a deviation from uniformity in the increase of the former.

(324.) These two causes, then, acting conjointly, produce, in fact, a very considerable fluctuation in the time as shown per clock, when the sun really attains the meridian. It amounts, in fact, to upwards of half an hour; apparent noon sometimes taking place as much as $16\frac{1}{4}$ min. *before mean* noon, and at others as much as $14\frac{1}{2}$ min. after. This difference between apparent and mean noon is called the *equation of time*, and is calculated and inserted in ephemerides for every day of the year, under that title; or else, which comes to the same thing, the moment, *in mean time*, of the sun's culmination for each day, is set down as an astronomical phenomenon to be observed.

(325.) As the sun, in its apparent annual course, is carried along the ecliptic, its declination is continually varying between the extreme limits of 23° 28′ 40″ north, and as much south, which it attains at the solstices. It is consequently always vertical over some part or other of that zone or belt of the earth's surface which lies between the north and south parallels of 23° 28′ 40″. These parallels are called in geography the *tropics;* the northern one that of *Cancer*, and the southern of *Capricorn;* because the sun, at the respective solstices, is situated in the division, or signs of the ecliptic so denominated. Of these signs there are twelve, each occupying 30° of its circumference. They commence at the vernal equinox, and are named in order — Aries, Taurus, Gemini, Cancer, Leo, Virgo, Libra, Scorpio, Sagittarius, Capricornus, Aquarius, Pisces. They are denoted also by the following symbols:— ♈, ♉, ♊, ♋, ♌, ♍, ♎, ♏, ♐, ♑, ♒, ♓. The ecliptic itself is also divided into signs, degrees, and minutes, &c. thus, 5^s 27° 0′ corresponds to 177° 0′; but this is beginning to be disused.

(326.) When the sun is in either tropic, it enlightens, as we have seen, the pole on that side the equator, and shines over or beyond it to the extent of 23° 28′ 40″. The parallels of latitude, at this distance from either pole, are called the polar circles, and are distinguished from each other by the names *arctic* and *antarctic*. The regions within these circles are sometimes termed frigid zones, while the belt between the tropics is called the torrid zone, and the immediate belts temperate zones. These last, however, are merely names given for the sake naming; as, in fact, owing to the different distribution of land and sea in the two hemispheres, zones of *climate* are not co-terminal with zones of *latitude*.

(327.) Our seasons are determined by the apparent passages of the sun across the equinoctial, and its alternate arrival in the northern and southern hemisphere. Were the equinox invariable, this would happen at intervals precisely equal to the duration of the sidereal year; but, in fact, owing to the slow conical motion of the earth's axis described in art. 264, the equinox retreats on the ecliptic, and *meets* the advancing sun somewhat *before* the whole sidereal circuit is completed. The annual retreat of the equinox is 50″·1, and this arc is described by the sun in the ecliptic in 20′ 19″·9. By so much *shorter*, then, is the periodical return of our seasons than the true sidereal revolution of the earth round the sun. As the latter period, or sidereal year, is equal to $365^d\ 6^h\ 9^m\ 9^s \cdot 6$, it follows, then, that the former must be only $365^d\ 5^h\ 48^m\ 49^s \cdot 7$; and this is what is meant by the *tropical* year.

(328.) We have already mentioned that the longer axis of the ellipse described by the earth has a slow motion of 11″·8 per annum in advance. From this it results, that when the earth, setting out from the perihelion, has completed one sidereal period, the perihelion will have moved forward by 11″·8, which arc must be described before it can again reach the perihelion. In so doing, it occupies 4′ 39″·7 and this must therefore be added to the sidereal period, to give the interval between two consecutive returns to the perihelion. This interval, then, is $365^d\ 6^h\ 13^m\ 49^s\cdot 3$*, and is what is called the *anomalistic year*. All these periods have their uses in astronomy; but that in which mankind in general are most interested is *the tropical year*, on which the return of the seasons depends, and which we thus perceive to be a compound phenomenon, depending chiefly and directly on the annual revolution of the earth round the sun, but subordinately also, and indirectly, on its rotation round its own axis, which is what occasions the precession of the equinoxes; thus affording an instructive example of the way in which a motion, once admitted in any part of our system, may be traced in its influence on others with which at first sight it could not possibly be supposed to have any thing to do.

(329.) As a rough consideration of the appearance of the earth points out the general roundness of its form, and more exact enquiry has led us first to the discovery of its elliptic figure, and, in the further progress of refinement, to the perception of minuter local deviations from that figure; so, in investigating the solar motions, the first notion we obtain is that of an orbit, generally speaking, round, and not far from a circle, which, on more careful and exact examination, proves to be an ellipse of small excentricity, and described in conformity with certain laws, as above stated. Still minuter enquiry, however, detects yet smaller deviations again from this form and from these laws, of which we have a specimen in the slow motion of the axis of the orbit spoken of in art. 318.; and which are generally comprehended under the name of perturbations and secular inequalities. Of these deviations, and their causes, we shall speak hereafter at length. It is the triumph of physical astronomy to have rendered a complete account of them all, and to have left nothing unexplained, either in the motions of the sun or in those of any other of the bodies of our system. But the nature of this explanation cannot be

* These numbers, as well as all the other numerical data of our system, are taken from Mr. Baily's Astronomical Tables and Formulæ, unless the contrary is expressed.

understood till we have developed the law of gravitation, and carried it into its more direct consequences. This will be the object of our three following chapters; in which we shall take advantage of the proximity of the moon, and its immediate connection with and dependence on the earth, to render it, as it were, a stepping-stone to the general explanation of the planeta. movements.

(330.) We shall conclude this by describing what is known of the physical constitution of the sun.

When viewed through powerful telescopes, provided with coloured glasses, to take off the heat, which would otherwise injure our eyes, it is observed to have frequently large and perfectly black spots upon it, surrounded with a kind of border, less completely dark, called a penumbra. Some of these are represented at *a*, *b*, *c*, plate iii. fig. 1, in the plate at the end of this volume. They are, however, not permanent. When watched from day to day, or even from hour to hour, they appear to enlarge or contract, to change their forms, and at length to disappear altogether, or to break out anew in parts of the surface where none were before. In such cases of disappearance, the central dark spot always contracts into a point, and vanishes before the border. Occasionally they break up, or divide into two or more, and in those offer every evidence of that extreme mobility which belongs only to the fluid state, and of that excessively violent agitation which seems only compatible with the atmospheric or gaseous state of matter. The scale on which their movements take place is immense. A single second of angular measure, as seen from the earth, corresponds on the sun's disc to 465 miles; and a circle of this diameter (containing therefore nearly 220,000 square miles) is the least space which can be distinctly discerned on the sun as a *visible area*. Spots have been observed, however, whose linear diameter has been upwards of 45,000 miles*;

* Mayer, Obs. Mar. 15. 1758. "Ingens macula in sole conspiciebatur, cujus diameter $= \frac{1}{20}$ diam. solis."

and even, if some records are to be trusted, of very much greater extent. That such a spot should close up in six weeks' time (for they hardly ever last longer), its borders must approach at the rate of more than 1000 miles a day.

Many other circumstances tend to corroborate this view of the subject. The part of the sun's disc not occupied by spots is far from uniformly bright. Its *ground* is finely mottled with an appearance of minute, dark dots, or *pores*, which, when attentively watched, are found to be in a constant state of change. There is nothing which represents so faithfully this appearance as the slow subsidence of some flocculent chemical precipitates in a transparent fluid, when viewed perpendicularly from above: so faithfully, indeed, that it is hardly possible not to be impressed with the idea of a luminous medium intermixed, but not confounded, with a transparent and non-luminous atmosphere, either floating as clouds in our air, or pervading it in vast sheets and columns like flame, or the streamers of our northern lights.

(331.) Lastly, in the neighbourhood of great spots, or extensive groups of them, large spaces of the surface are often observed to be covered with strongly marked curved, or branching streaks, more luminous than the rest, called *faculæ*, and among these, if not already existing, spots frequently break out. They may, perhaps, be regarded with most probability, as the ridges of immense waves in the luminous regions of the sun's atmosphere, indicative of violent agitation in their neighbourhood.

(332.) But what *are* the spots? Many fanciful notions have been broached on this subject, but only one seems to have any degree of physical probability, viz. that they are the dark, or at least comparatively dark, solid body of the sun itself, laid bare to our view by those immense fluctuations in the luminous regions of its atmosphere, to which it appears to be subject. Respecting the manner in which this disclosure takes place, different ideas again have been advocated.

Lalande (art. 3240.) suggests, that eminences in the nature of mountains are actually laid bare, and project above the luminous ocean, appearing black above it, while their shoaling declivities produce the penumbræ, where the luminous fluid is less deep. A fatal objection to this theory is the perfectly uniform shade of the penumbra and its sharp termination, both inwards, where it joins the spot, and outwards, where it borders on the bright surface. A more probable view has been taken by Sir William Herschel*, who considers the luminous strata of the atmosphere to be sustained far above the level of the solid body by a transparent elastic medium, carrying on its upper surface (*or rather*, to avoid the former objection, *at some considerably lower level within its depth*,) a cloudy stratum which, being strongly illuminated from above, reflects a considerable portion of the light to our eyes, and forms a penumbra, while the solid body, shaded by the clouds, reflects none. The temporary removal of both the strata, but more of the upper than the lower, he supposes effected by powerful upward currents of the atmosphere, arising, perhaps, from spiracles in the body, or from local agitations. See fig. 1. *d*, Plate III.

(333.) The region of the spots is confined within about 30° of the sun's equator, and, from their motion on the surface, carefully measured with micrometers, is ascertained the position of the equator, which is a plane inclined 7° 20′ to the ecliptic, and intersecting it in a line whose direction makes an angle of 80° 21′ with that of the equinoxes. It has been also noticed, (not, we think, without great need of further confirmation,) that extinct spots have again broken out, after long intervals of time, on the same identical points of the sun's globe. Our knowledge of the period of its rotation (which, according to Delambre's calculations, is $25^{d}\cdot01154$, but, according to others, materially different,) can hardly be regarded as sufficiently precise to establish a point of so much nicety.

* Phil. Trans. 1801.

(334.) That the temperature at the visible surface of the sun cannot be otherwise than very elevated, much more so than any artificial heat produced in our furnaces, or by chemical or galvanic processes, we have indications of several distinct kinds: 1st, From the law of decrease of radiant heat and light, which, being inversely as the squares of the distances, it follows, that the heat received on a given area exposed at the distance of the earth, and on an equal area at the visible surface of the sun, must be in the proportion of the area of the sky occupied by the sun's apparent disc to the whole hemisphere, or as 1 to about 300000. A far less intensity of solar radiation, collected in the focus of a burning glass, suffices to dissipate gold and platina in vapour. 2dly, From the facility with which the calorific rays of the sun traverse glass, a property which is found to belong to the heat of artificial fires in the direct proportion of their intensity.* 3dly, From the fact, that the most vivid flames disappear, and the most intensely ignited solids appear only as black spots on the disk of the sun when held between it and the eye.† From this last remark it follows, that the body of the sun, however dark it may appear when seen through its spots, *may*, nevertheless, be in a state of most intense ignition. It does not, however, follow of necessity that it *must* be so. The contrary is at least physically possible. A *perfectly reflective* canopy would effectually defend it from the radiation of the luminous regions above its atmosphere, and no heat would be conducted downwards through a gaseous medium increasing rapidly in density. That the penumbral clouds *are* highly reflective, the fact of their visibility in such a situation can leave no doubt.

* By direct measurement with the *actinometer*, an instrument I have long employed in such enquiries, and whose indications are liable to none of those sources of fallacy which beset the usual modes of estimation, I find that out of 1000 calorific solar rays, 816 penetrate a sheet of plate glass 0·12 inch thick; and that of 1000 rays which have passed through one such plate, 859 are capable of passing through another. — *Author.*

† The ball of ignited quicklime, in lieutenant Drummond's oxy-hydrogen lamp, gives the nearest imitation of the solar splendour which has yet been produced. The appearance of this against the sun was however as described in an imperfect trial at which I was present. The experiment ought to be repeated under favourable circumstances. — *Author.*

(335.) This immense escape of heat by radiation, we may also remark, will fully explain the constant state of tumultuous agitation in which the fluids composing the visible surface are maintained, and the continual generation and filling in of *the pores*, without having recourse to internal causes. The mode of action here alluded to is perfectly represented to the eye in the disturbed subsidence of a precipitate, as described in art. 330., when the fluid from which it subsides is warm, and losing heat from its surface.

(336.) The sun's rays are the ultimate source of almost every motion which takes place on the surface of the earth. By its heat are produced all winds, and those disturbances in the electric equilibrium of the atmosphere which give rise to the phenomena of terrestrial magnetism. By their vivifying action vegetables are elaborated from inorganic matter, and become, in their turn, the support of animals and of man, and the sources of those great deposits of dynamical efficiency which are laid up for human use in our coal strata. By them the waters of the sea are made to circulate in vapour through the air, and irrigate the land, producing springs and rivers. By them are produced all disturbances of the chemical equilibrium of the elements of nature, which, by a series of compositions and decompositions, give rise to new products, and originate a transfer of materials. Even the slow degradation of the solid constituents of the surface, in which its chief geological changes consist, and their diffusion among the waters of the ocean, are entirely due to the abrasion of the wind and rain, and the alternate action of the seasons; and when we consider the immense transfer of matter so produced, the increase of pressure over large spaces in the bed of the ocean, and diminution over corresponding portions of the land, we are not at a loss to perceive how the elastic power of subterranean fires, thus repressed on the one hand and relieved on the other, may break forth in points when the resistance is barely adequate to their retention, and thus bring the phenomena of even volcanic activity under the general law of solar influence.

(337.) The great mystery, however, is to conceive how so enormous a conflagration (if such it be) can be kept up. Every discovery in chemical science here leaves us completely at a loss, or rather, seems to remove farther the prospect of probable explanation. If conjecture might be hazarded, we should look rather to the known possibility of an indefinite generation of heat by friction, or to its excitement by the electric discharge, than to any actual combustion of ponderable fuel, whether solid or gaseous, for the origin of the solar radiation.*

* Electricity traversing excessively rarefied air or vapours, gives out light, and, doubtless, also heat. May not a continual current of electric matter be constantly circulating in the sun's immediate neighbourhood, or traversing the planetary spaces, and exciting, in the upper regions of its atmosphere, those phenomena of which, on however diminutive a scale, we have yet an unequivocal manifestation in our aurora borealis. The possible analogy of the solar light to that of the aurora has been distinctly insisted on by my Father, in his paper already cited. It would be a highly curious subject of experimental enquiry, how far a mere reduplication of sheets of flame, at a distance one behind the other (by which their *light* might be brought to any required intensity), would communicate to the *heat* of the resulting compound ray the *penetrating* character which distinguishes the solar calorific rays. We may also observe, that the tranquillity of the sun's polar, as compared with its equatorial regions (if its spots be really atmospheric), cannot be accounted for by its rotation on its axis only, but *must* arise from some cause external to the sun, as we see the belts of Jupiter and Saturn, and our trade-winds, arise from a cause, external to these planets, combining itself with their rotation, which *alone* can produce no motions when once the form of equilibrium is attained.

The prismatic analysis of the solar beam exhibits in the spectrum a series of "fixed lines," totally unlike those which belong to the light of any known terrestrial flame. This may hereafter lead us to a clearer insight into its origin. But, before we can draw any conclusions from such an indication, we must recollect, that previous to reaching us it has undergone the whole absorptive action of our atmosphere, as well as of the sun's. Of the latter we know nothing, and may conjecture every thing; but of the blue colour of the former we are sure; and if this be an inherent (*i. e.* an absorptive) colour, the air must be expected to act on the spectrum after the analogy of other coloured media, which often (and *especially light blue* media) leave unabsorbed portions separated by dark intervals. It deserves enquiry, therefore, whether some or all the fixed lines observed by Wollaston and Fraunhofer may not have their origin in our own atmosphere. Experiments made on lofty mountains, or the cars of balloons, on the one hand, and on the other with reflected beams which have been made to traverse several miles of additional air near the surface, would decide this point. The absorptive effect of the sun's atmosphere, and possibly also of the medium surrounding it (whatever it be), which resists the motions of comets, cannot be thus eliminated.—*Author.*

CHAP. VI.

OF THE MOON. — ITS SIDEREAL PERIOD. — ITS APPARENT DIAMETER. — ITS PARALLAX, DISTANCE, AND REAL DIAMETER. — FIRST APPROXIMATION TO ITS ORBIT. — AN ELLIPSE ABOUT THE EARTH IN THE FOCUS. — ITS ECCENTRICITY AND INCLINATION. — MOTION OF THE NODES OF ITS ORBIT. — OCCULTATIONS. — SOLAR ECLIPSES. — PHASES OF THE MOON. — ITS SYNODICAL PERIOD. — LUNAR ECLIPSES. — MOTION OF THE APSIDES OF ITS ORBIT. — PHYSICAL CONSTITUTION OF THE MOON. — ITS MOUNTAINS. — ATMOSPHERE. — ROTATION ON AXIS. — LIBRATION. — APPEARANCE OF THE EARTH FROM IT.

(338.) THE moon, like the sun, appears to advance among the stars with a movement contrary to the general diurnal motion of the heavens, but much more rapid, so as to be very readily perceived (as we have before observed) by a few hours' cursory attention on any moonlight night. By this continual advance, which, though sometimes quicker, sometimes slower, is never intermitted or reversed, it makes the tour of the heavens in a mean or average period of $27^d\ 7^h\ 43^m\ 11^s{\cdot}5$, returning, in that time, to a position among the stars nearly coincident with that it had before, and which would be exactly so, but for causes presently to be stated.

(339.) The moon, then, like the sun, apparently describes an orbit round the earth, and this orbit cannot be *very* different from a circle, because the apparent angular diameter of the full moon is not liable to any great extent of variation.

(340.) The distance of the moon from the earth is concluded from its horizontal parallax, which may be found either directly, by observations at remote geographical stations, exactly similar to those described in art. 302., in the case of the sun, or by means of the phenomena called occultations (art. 346.), from which also its apparent diameter is most readily and correctly found.

From such observations it results that the mean or average distance of the center of the moon from that of the earth is 59·9643 of the earth's equatorial radii, or about 237000 miles. This distance, great as it is, is little more than one fourth of the diameter of the sun's body, so that the globe of the sun would nearly twice include the whole orbit of the moon; a consideration wonderfully calculated to raise our ideas of that stupendous luminary!

(341.) The distance of the moon's center from an observer at any station on the earth's surface, compared with its apparent angular diameter as measured from that station, will give its real or linear diameter. Now, the former distance is easily calculated when the distance from the earth's center is known, and the apparent zenith distance of the moon also determined by observation; for if we turn to the figure of art. 298., and suppose S the moon, A the station, and C the earth's center, the distance S C, and the earth's radius C A, two sides of the triangle A C S are given, and the angle C A S, which is the supplement of Z A S, the observed zenith distance, whence it is easy to find A S, the moon's distance from A. From such observations and calculations it results, that the real diameter of the moon is 2160 miles, or about 0·2729 of that of the earth, whence it follows that, the bulk of the latter being considered as 1, that of the former will be 0·0204, or about $\frac{1}{49}$.

(342.) By a series of observations, such as described in art. 340., if continued during one or more revolutions of the moon, its real distance may be ascertained at every point of its orbit; and if at the same time its apparent places in the heavens be observed, and reduced by means of its parallax to the earth's center, their angular intervals will become known, so that the path of the moon may then be laid down on a chart supposed to represent the plane in which its orbit lies, just as was explained in the case of the solar ellipse (art. 292.). Now, when this is done, it is found that, neglecting certain small (though very perceptible) deviations of which a satisfactory account will hereafter be rendered), the form of the apparent orbit, like that of the sun, is elliptic, but considerably more eccentric, the eccentricity amounting to 0·05484 of the mean distance, or the major semi-axis of the ellipse, and the earth's center being situated in its focus.

(343.) The plane in which this orbit lies is not the ecliptic, however, but is inclined to it at an angle of 5° 8′ 48″, which is called the inclination of the lunar orbit, and intersects it in two opposite points, which are called its nodes—the *ascending node* being that in which the moon passes from the southern side of the ecliptic to the northern, and the *descending* the reverse. The points of the orbit at which the moon is nearest to, and farthest from, the earth, are called respectively its *perigee* and *apogee*, and the line joining them and the earth the line of *apsides*.

(344.) There are, however, several remarkable circumstances which interrupt the closeness of the analogy, which cannot fail to strike the reader, between the motion of the moon around the earth, and of the earth round the sun. In the latter case, the ellipse described remains, during a great many revolutions, unaltered in its position and dimensions; or, at least, the changes which it undergoes are not perceptible but in a course of very nice observations, which have disclosed, it is true, the existence of "perturbations," but of so minute an order, that, in ordinary parlance, and for common purposes, we may leave them unconsidered. But this cannot be done in the case of the moon. Even in a single revolution, its deviation from a perfect ellipse is very sensible. It does not return to the same exact position among the stars from which it set out, thereby indicating a continual change in the *plane* of its orbit. And, in effect, if we trace by observation, from month to month, the point where it traverses the ecliptic, we shall find that the *nodes* of its orbit are in a continual state of *retreat* upon the ecliptic. Suppose O to be the earth, and A *b a d* that portion of the plane of the ecliptic which is intersected by the moon, in its alternate passages through it, from south to north, and *vice versâ*; and let A B C D E F be a portion of the moon's orbit, embracing a complete sidereal revolution. Sup-

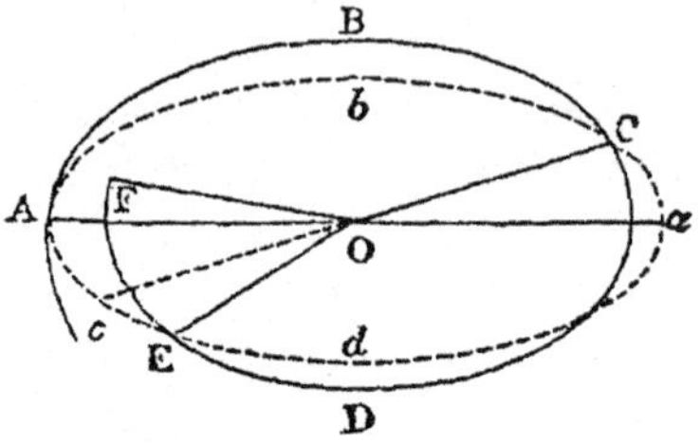

pose it to set out from the ascending node, A; then, if the orbit lay all in one plane, passing through O, it would have *a*, the opposite point in the ecliptic, for its descending node; after passing which, it would again ascend at A. But, in fact, its real path carries it not to *a*, but along a certain curve, A B C, to C, a point in the ecliptic less than 180° distant from A; so that the angle A O C, or the arc of longitude described between the ascending and the descending node, is somewhat less than 180°. It then pursues its course below the ecliptic, along the curve C D E, and rises again above it, not at the point *c*, diametrically opposite to C, but at a point E, less advanced in longitude. On the whole, then, the arc described in longitude between two consecutive passages from south to north, through the plane of the ecliptic, falls short of 360° by the angle A O E; or, in other words, the ascending node appears to have retreated in one lunation, on the plane of the ecliptic by that amount. To complete a sidereal revolution, then, it must still go on to describe an arc, A F, on its orbit, which will no longer, however, bring it exactly back to A, but to a point somewhat above it, or *having north latitude.*

(345.) The actual amount of this retreat of the moon's node is about 3′ 10″·64 *per diem,* on an average, and in a period of 6793·39 mean solar days, or about 18·6 years, the ascending node is carried round in a direction contrary to the moon's motion in its orbit (or from east to west) over a whole circumference of the ecliptic. Of course, in the middle of this period the position of the orbit must have been precisely reversed from what it was at the beginning. Its apparent path, then, will lie among totally different stars and constellations at different parts of this period; and, this kind of spiral revolution being continually kept up, it will, at one time or other, cover with its disc every point of the heavens within that limit of latitude or distance from the ecliptic which its inclination permits; that is to say, a belt or zone of the heavens, of 10° 18′ in breadth, having the ecliptic for its middle line. Nevertheless, it still remains true that the *actual place* of the moon in consequence of this motion, deviates in a single revolution *very little* from what it would be were the nodes at rest. Supposing the moon to set out from its node A, its latitude, when it comes to F, having completed a revolution in longitude, will not exceed 8′; and it must be borne in mind that it is to account for, and represent geometrically, a deviation of this small order, that the *motion of the nodes* is devised.

(346.) Now, as the moon is at a very moderate distance from us (astronomically speaking), and is in fact our nearest neighbour, while the sun and stars are in comparison immensely beyond it, it must of necessity happen, that at one time or other it must *pass over* and *occult* or *eclipse* every star and planet within the zone above described (and, as seen from the *surface* of earth, even somewhat beyond it, by reason of parallax, which may throw it apparently nearly a degree either way from its place as seen from the center, according to the observer's station). Nor is the sun itself exempt from being thus hidden, whenever any part of the moon's disc, in this her tortuous course, comes to *overlap* any part of the space occupied in the heavens by that luminary. On these occasions is exhibited the most striking and impressive of all the occasional phenomena of astronomy, an *eclipse of the sun*, in which a greater or less portion, or even in some rare conjunctures the whole, of its disc is obscured, and, as it were, obliterated, by the superposition of that of the moon, which appears upon it as a circularly-terminated black spot, producing a temporary diminution of daylight, or even nocturnal darkness, so that the stars appear as if at midnight. In other cases, when, at the moment that the moon is centrally superposed on the sun, it so happens that her distance from the earth is such as to render her angular diameter less than the sun's, the very singular phenomenon of an *annular solar eclipse* takes place, when the edge of the sun appears for a few minutes as a narrow ring of light, projecting on all sides beyond the dark circle occupied by the moon in its center.

(347.) A solar eclipse can only happen when the sun and moon are *in conjunction*, that is to say, have the *same*, or nearly the same, position in the heavens, or the same longitude. It will presently be seen that this condition can only be fulfilled at the time of a *new moon*, though it by no means follows, that at *every* conjunction there *must* be an eclipse of the sun. If the lunar orbit coincided with the ecliptic, this would be the case, but as it is inclined to it at an angle of upwards of 5°, it is evident that the conjunction, or equality of longitudes, may take place when the moon is in the part of her orbit too remote from the ecliptic to permit the discs to meet and overlap. It is easy, however, to assign the limits within which an eclipse is possible. To this end we must consider, that, by the effect of parallax, the moon's *apparent* edge may be thrown in *any* direction, according to a spectator's geographical station, by *any* amount not exceeding the horizontal parallax. Now, this comes to the same (so far as the possibility of an eclipse is concerned) as if the apparent diameter of the moon, seen from the earth's center, were dilated by twice its horizontal parallax; for if, when so dilated, it can touch or overlap the sun, there *must* be an eclipse at *some* part or other of the earth's surface. If, then, at the moment of the nearest conjunction, the geocentric distance of the centers of the two luminaries do not exceed the sum of their semidiameters and of the moon's horizontal parallax, there will be an eclipse. This sum is, at its maximum, about 1° 34′ 27″. In the spherical triangle S N M, then, in which S is the sun's center, M the moon's, S N the ecliptic, M N the moon's orbit, and N the node, we may suppose the angle N S M a right angle, S M = 1° 34′ 27″, and the angle M N S = 5° 8′ 48″, the inclination of the orbit.

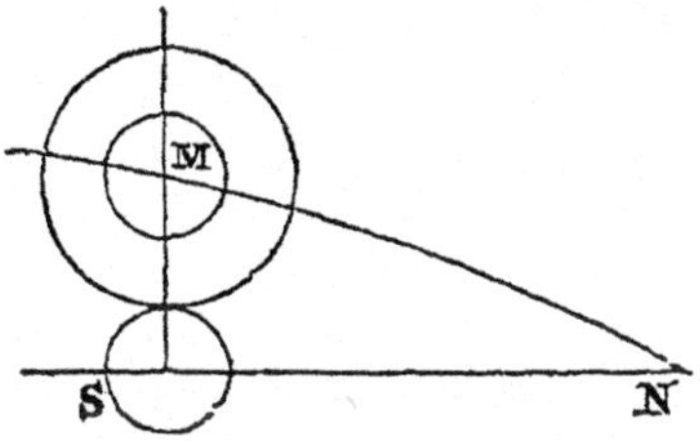

Hence we calculate S N, which comes out 16° 58′. If, then, at the moment of the new moon, the moon's node is farther from the sun in longitude than this limit, there can be no eclipse; if within, there may, and probably will, at some part or other of the earth. To ascertain precisely whether there will or not, and, if there be, how great will be the part eclipsed, the solar and lunar tables must be consulted, the place of the node and the semidiameters exactly ascertained, and the local parallax, and apparent augmentation of the moon's diameter due to the difference of her distance from the observer and from the center of the earth (which may amount to a sixtieth part of her horizontal diameter), determined; after which it is easy, from the above considerations, to calculate the amount overlapped of the two discs, and their moment of contact.

(348.) The calculation of the occultation of a star depends on similar considerations. An occultation is *possible*, when the moon's course, *as seen* from the earth's center, carries her within a distance from the star equal to the sum of her semidiameter and horizontal parallax; and it *will happen at any particular spot*, when her apparent path, as seen from that spot, carries her center within a distance equal to the sum of her *augmented* semidiameter and *actual* parallax. The details of these calculations, which are somewhat troublesome, must be sought elsewhere.*

(349.) The phenomenon of a solar eclipse and of an occultation are highly interesting and instructive in a physical point of view. They teach us that the moon is an opaque body, terminated by a real and sharply defined surface intercepting light like a solid. They prove to us, also, that at those times when we cannot *see* the moon, she really exists, and pursues her course, and that when we see her only as a crescent, however narrow, the whole globular body *is there*, filling up the deficient outline, though unseen. For occultations take place indifferently at the dark and bright, the visible and invisible outline, whichever happens to be towards the direction in which the moon is moving; with this only difference, that a star *occulted* by the bright limb, if the phenomenon be watched with a telescope, gives notice, by its gradual approach to the visible edge, when to expect its disappearance, while, if occulted at the dark limb, if the moon, at least, be more than a few days old, it is, as it were, extinguished in mid-air, without notice or visible cause for its disappearance, which, as it happens *instantaneously*, and without the slightest previous diminution of its light, is always surprising; and, if the star be a large and bright one, even startling from its suddenness. The re-appearance of the star, too, when the moon has passed over it, takes place in those cases when the bright side of the moon is foremost, not at the concave outline of the crescent, but at the invisible outline of the complete circle, and is scarcely less surprising, from its suddenness, than its disappearance in the other case.†

* Woodhouse's Astronomy, vol. i. See also Trans. Ast. Soc. vol. i. p. 325.

† There is an optical illusion of a very strange and unaccountable nature which has often been remarked in occultations. The star appears to advance actually *upon* and *within* the edge of the disc before it disappears, and that sometimes to a considerable depth. I have never myself witnessed this singular effect, but it rests on most unequivocal testimony. I have called it an optical illusion; but it is *barely possible* that a star may shine on such occasions through deep fissures in the substance of the moon. The

(350.) The existence of the complete circle of the disc, even when the moon is not full, does not, however, rest only on the evidence of occultations and eclipses. It may be *seen*, when the moon is crescent or waning, a few days before and after the *new moon*, with the naked eye, as a pale round body, to which the crescent seems attached, and somewhat projecting beyond its outline (which is an optical illusion arising from the greater intensity of its light). The cause of this appearance will presently be explained. Meanwhile the fact is sufficient to show that the moon is not *inherently* luminous like the sun, but that her light is of an adventitious nature. And its crescent form, increasing regularly from a narrow semicircular line to a complete circular disc, corresponds to the appearance a globe would present, one hemisphere of which was black, the other white, when differently turned towards the eye, so as to present a greater or less portion of each. The obvious conclusion from this is, that the moon is such a globe, one half of which is brightened by the rays of some luminary sufficiently distant to enlighten the complete hemisphere, and sufficiently intense to give it the degree of splendour we see. Now, the sun alone is competent to such an effect. Its distance and light suffice; and, moreover, it is invariably observed that, when a crescent, the bright edge is *towards the sun*, and that in proportion as the moon in her monthly course becomes more and more distant from the sun, the breadth of the crescent increases, and *vice versâ*.

(351.) The sun's distance being 23984 radii of the earth, and the moon's only 60, the former is nearly 400 times the latter. Lines, therefore, drawn from the sun to every part of the moon's orbit may be re-

---

occultations of close double stars ought to be narrowly watched, to see whether *both* individuals are thus *projected*, as well as for other purposes connected with their theory. I will only hint at one, viz. that a double star, *too close* to be seen divided with any telescope, may yet be detected to be double by the mode of its disappearance. Should a considerable star, for instance, instead of undergoing instantaneous and complete extinction, go out by two distinct steps, following close upon each other; first losing a portion, then the whole remainder of its light, we may be sure it is a double star, though we cannot see the individuals separately. — *Author.*

garded as parallel. Suppose, now, O to be the earth, A B C D, &c. various positions of the moon in its orbit, and S the sun, at the vast distance above stated; as is shown, then, in the figure, the hemisphere of the lunar globe turned towards it (on the right) will be bright, the opposite dark, wherever it may stand in its orbit. Now, in the position A, when in conjunction with the sun, the dark part is entirely turned towards O, and the bright from it. In this case, then, the moon is not seen, it is *new* moon. When the moon has come to C, half the bright and half

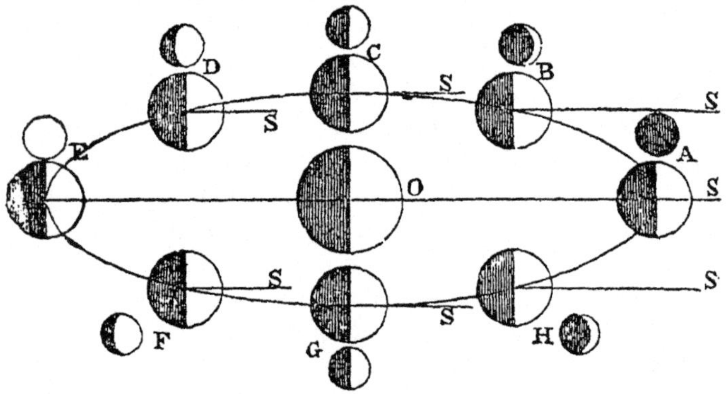

the dark hemisphere are presented to O, and the same in the opposite situation G: these are the first and third quarters of the moon. Lastly, when at E, the whole bright face is towards the earth, the whole dark side from it, and it is then seen wholly bright or *full moon*. In the intermediate positions B D F H, the portions of the bright face presented to O will be at first less than half the visible surface, then greater, and finally less again, till it vanishes altogether, as it comes round again to A.

(352.) These monthly changes of appearance, or *phases*, as they are called, arise, then, from the moon, an opaque body, being illuminated on one side by the sun, and reflecting from it, in all directions, a portion of the light so received. Nor let it be thought surprising that a solid substance thus illuminated should appear to *shine* and again illuminate the earth. It is no more than a white cloud does standing off upon the clear blue sky. By day, the moon can hardly be distinguished in brightness from such a cloud; and, in the dusk of evening, clouds catching the last rays of the sun appear with a dazzling splendour, not inferior to the seeming brightness of the moon at night. That the earth sends also such a light to the moon, only probably more powerful by reason of its greater apparent size*, is agreeable to optical principles, and explains the appearance of the dark portion of the young moon completing its crescent (art. 350.). For, when the moon is nearly new to the earth, the latter (so to speak) is nearly full to the former; it then illuminates its dark half by strong *earth-light;* and it is a portion of this, reflected back again, which makes it visible to us in the twilight sky. As the moon gains age, the earth offers it a less portion of its bright side, and the phenomenon in question dies away.

(353.) The lunar month is determined by the recurrence of its phases: it reckons from new moon to new moon; that is, from leaving its conjunction with the sun to its return to conjunction. If the sun stood still, like a fixed star, the interval between two conjunctions would be the same as the period of the moon's sidereal revolution (art. 338.); but, as the sun apparently advances in the heavens in the same direction with the moon, only slower, the latter has more than a complete sidereal period to perform to come up with the sun again, and will require for it a longer time, which is the lunar month, or, as it is generally termed in astronomy, a *synodical* period. The difference is easily calculated by considering that the superfluous arc (whatever it be) is described by the sun with his velocity of $0^\circ{\cdot}98565$ *per diem, in the same time* that the moon describes that arc *plus* a complete revolution, with her velocity of

* The apparent diameter of the moon is 32′ from the earth; that of the earth seen from the moon is twice her horizontal parallax, or $1^\circ\ 54'$. The apparent surfaces, therefore, are as $(114)^2 : (32)^2$, or as 13 : 1 nearly.

$13^\circ{\cdot}17640$ *per diem;* and, the times of description being identical, the spaces are to each other in the proportion of the velocities.* From these data a slight knowledge of arithmetic will suffice to derive the arc in question, and the time of its description by the moon; which, being the excess of the synodic over the sidereal period, the former will be had, and will appear to be $29^d\ 12^h\ 44^m\ 2^s{\cdot}87$.

(354.) Supposing the position of the nodes of the moon's orbit to permit it, when the moon stands at A (or at the new moon), it will intercept a part or the whole of the sun's rays, and cause a solar eclipse. On the other hand, when at E (or at the full moon), the earth O will intercept the rays of the sun, and *cast a shadow* on the moon, thereby causing a lunar eclipse. And this is perfectly consonant to fact, such eclipses never happening but at the exact time of the full moon. But, what is still more remarkable, as confirmatory of the position of the earth's sphericity, this shadow, which we plainly see to enter upon and, as it were, eat away the disc of the moon, is always terminated by a circular outline, though, from the greater *size* of the circle, it is only partially seen at any one time. Now, a body which always casts a circular shadow must itself be spherical.

(355.) Eclipses of the sun are best understood by regarding the sun and moon as two independent luminaries, each moving according to known laws, and viewed from the earth; but it is also instructive to consider eclipses generally as arising from the shadow of one body thrown on another by a luminary *much larger than either*. Suppose, then, A B to represent the sun, and C D a spherical body, whether earth or moon, illuminated by it. If we join and prolong A C, B D; since B is greater than C D, these lines will meet in a point

* Let V and v be the mean angular velocities, x the superfluous arc; then $V : v :: 1 + x : x$; and $V - v : v :: 1 : x$, whence x is found, and $\frac{x}{v}$ = the time of describing x, or the difference of the sidereal and synodical periods. We shall have occasion for this again.

E, more or less distant from the body C D, according to its size, and within the space C E D (which represents

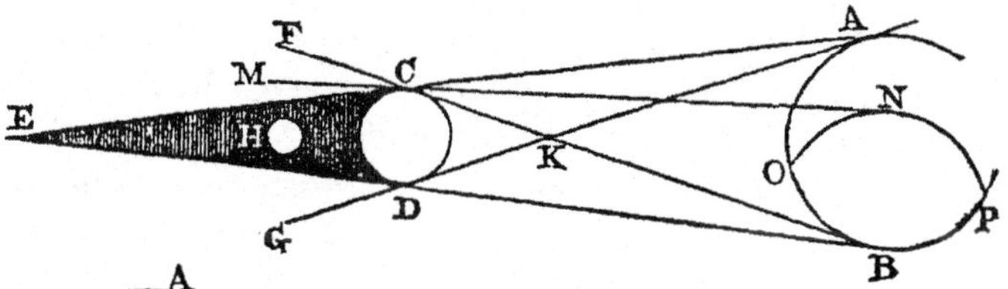

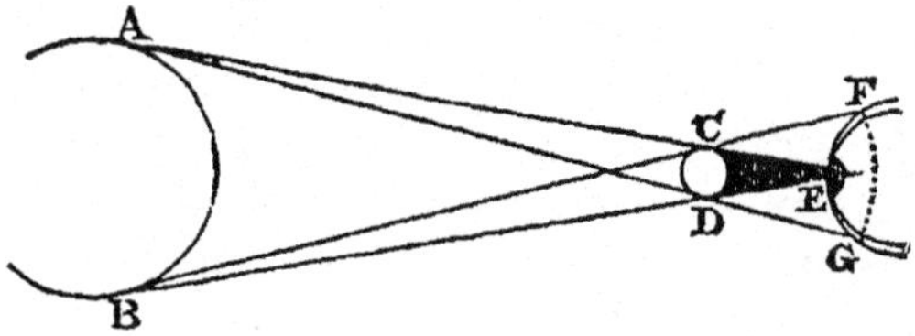

a cone, since C D and A B are spheres), there will be a total shadow. This shadow is called the *umbra*, and a spectator situated within it can see no part of the sun's disc. Beyond the umbra are two diverging spaces (or rather, a portion of a single conical space, having K for its vertex), where if a spectator be situated, as at M, he will see a portion only (A O N P) of the sun's surface, the rest (B O N P) being obscured by the earth. He will, therefore, receive only partial sunshine; and the more, the nearer he is to the exterior borders of that cone which is called the *penumbra*. Beyond this he will see the whole sun, and be in full illumination. All these circumstances may be perfectly well shown by holding a small globe up in the sun, and receiving its shadow at different distances on a sheet of paper.

(356.) In a lunar eclipse (represented in the upper figure), the moon is seen to enter the *penumbra* first, and, by degrees, get involved in the *umbra*, the former surrounding the latter like a haze. Owing to the great size of the earth, the cone of its *umbra* always projects far beyond the moon; so that, if, at the time of the eclipse, the moon's path be properly directed, it is sure to pass through the *umbra*. This is not, however, the case in solar eclipses. It so happens, from the adjustment of the size and distance of the moon, that the extremity of her *umbra* always falls *near* the earth, but sometimes attains and sometimes falls short of its surface. In the former case (represented in the lower figure), a black spot, surrounded by a fainter shadow, is formed, beyond which there is no eclipse on any part of the earth, but within which there may be either a total or partial one, as the spectator is within the *umbra* or *penumbra*. When the apex of the umbra falls *on* the surface, the moon at that point will appear, for an instant, to *just* cover the sun; but, when it falls short, there will be no total eclipse on any part of the earth; but a spectator, situated in or near the prolongation of the axis of the cone, will see the whole of the moon on the sun, although not large enough to cover it, *i. e.* he will witness an annular eclipse.

(357.) Owing to a remarkable enough adjustment of the periods in which the moon's *synodical* revolution, and that of her nodes, are performed, eclipses return after a certain period, very nearly in the same order and of the same magnitude. For 223 of the moon's mean synodical revolutions, or *lunations*, as they are called, will be found to occupy 6585·32 days, and nineteen complete synodical revolutions of the node to occupy 6585·78. The difference in the mean position of the node, then, at the beginning and end of 223 lunations, is nearly insensible; so that a recurrence of all eclipses within that interval must take place. Accordingly, this period of 223 lunations, or eighteen years and ten days, is a very important one in the calculation of eclipses. It is supposed to have been known to the Chaldeans, under the name of the *saros;* the regular return of eclipses having been known as a physical fact for ages before their exact theory was understood.

(358.) The commencement, duration, and magnitude of a lunar eclipse are much more easily calculated than those of a solar, being independent of the position of the spectator on the earth's surface, and the same as if viewed from its center. The common center of the *umbra* and *penumbra* lies always in the ecliptic, at a point opposite to the sun and the path described by the moon in passing through it is its true orbit, as it stands at the moment of the full moon. In this orbit, its position, at every instant, is known from the lunar tables and ephemeris; and all we have, therefore, to ascertain, is, the *moment when* the distance between the moon's center and the center of the shadow is exactly equal to the sum of the semidiameters of the moon and *penumbra,* or of the moon and *umbra,* to know when it enters upon and leaves them respectively.

(359.) The dimensions of the shadow, at the place where it crosses the moon's path, require us to know the distances of the sun and moon at the time. These are variable; but are calculated and set down, as well as their semidiameters, for every day, in the ephemeris, so that none of the data are wanting. The sun's distance is easily calculated from its elliptic orbit; but the moon's is a matter of more difficulty, for a reason we will now explain.

(360.) The moon's orbit, as we have before hinted, is not, strictly speaking, an ellipse returning into itself, by reason of the variation of the plane in which it lies, and the motion of its nodes. But even laying aside this consideration, the axis of the ellipse is itself constantly changing its direction in space, as has been already stated of the solar ellipse, but much more rapidly; making a complete revolution, in the same direction with the moon's own motion, in 3232·5753 mean solar days, or about nine years, being about 3° of angular motion in a whole revolution of the moon. This is the phenomenon known by the name of the revolution of the moon's *apsides.* Its cause will be hereafter explained. Its immediate effect is to produce a variation in the moon's distance from the earth, which is not included in the laws of exact elliptic motion. In a single revolution of the moon, this variation of distance is trifling; but in the course of many it becomes considerable, as is easily seen, if we consider that in four years and a half the position of the axis will be completely reversed, and the apogee of the moon will occur where the perigee occurred before.

(361.) The best way to form a distinct conception of the moon's motion is to regard it as describing an ellipse about the earth in the focus, and, at the same time, to regard this ellipse itself to be in a twofold state of revolution; 1st, in its own plane, by a continual advance of its axis in that plane; and 2dly, by a continual *tilting* motion of the plane itself, exactly similar to, but much more rapid than, that of the earth's equator produced by the conical motion of its axis described in art. 266.

(362.) The physical constitution of the moon is better known to us than that of any other heavenly body. By the aid of telescopes, we discern inequalities in its surface which can be no other than mountains and valleys,—for this plain reason, that we see the shadows cast by the former in the exact proportion as to length which they ought to have, when we take into account the inclination of the sun's rays to that part of the moon's surface on which they stand. The convex outline of the limb turned towards the sun is always circular, and very nearly smooth; but the opposite border of the enlightened part, which (were the moon a perfect sphere) ought to be an exact and sharply defined ellipse, is always observed to be extremely ragged, and indented with deep recesses and prominent points. The mountains near this edge cast long black shadows, as they should evidently do, when we consider that the sun is in the act of rising or setting to the parts of the moon so circumstanced. But as the enlightened edge advances beyond them, *i. e.* as the sun to them gains altitude, their shadows shorten; and at the full moon, when all the light falls in our line of sight, no shadows are seen on any part of her surface. From micrometrical measures of the lengths of the shadows of many of the more conspicuous mountains, taken under the most favourable circumstances, the heights of many of them have been calculated,—the highest being about $1\frac{3}{4}$ English miles in perpendicular altitude. The existence of such mountains is corroborated by their appearance as small points or islands of light beyond the extreme edge of the enlightened part, which are their tops catching the sun-beams before the intermediate plain, and which, as the light advances, at length connect themselves with it, and appear as prominences from the general edge.

(363.) The generality of the lunar mountains present a striking uniformity and singularity of aspect. They are wonderfully numerous, occupying by far the larger portion of the surface, and almost universally of an exactly circular or cup-shaped form, foreshortened, however, into ellipses towards the limb; but the larger have for the most part flat bottoms within, from which rises centrally a small, steep, conical hill. They offer, in short, in its highest perfection, the true *volcanic* character, as it may be seen in the crater of Vesuvius, and in a map of the volcanic districts of the Campi Phlegræi* or the Puy de Dôme. And in some of the principal ones, decisive marks of volcanic stratification, arising from successive deposits of ejected matter, may be clearly traced with powerful telescopes. † What is, moreover, extremely singular in the geology of the moon is, that although nothing having the character of seas can be traced, (for the dusky spots which are commonly called seas, when closely examined, present appearances incompatible with the supposition of deep water,) yet there are large regions perfectly level, and apparently of a decided alluvial character.

(364.) The moon has no clouds, nor any other indications of an atmosphere. Were there any, it could not fail to be perceived in the occultations of stars and the phænomena of solar eclipses. Hence its climate must be very extraordinary; the alternation being that of unmitigated and burning sunshine fiercer than an equatorial noon, continued for a whole fortnight, and the

* See Breislak's map of the environs of Naples, and Desmarest's of Auvergne.

† From my own observations.—*Author.*

keenest severity of frost, far exceeding that of our polar winters, for an equal time. Such a disposition of things must produce a constant transfer of whatever moisture may exist on its surface, from the point beneath the sun to that opposite, by distillation *in vacuo* after the manner of the little instrument called a *cryophorus*. The consequence must be absolute aridity below the vertical sun, constant accretion of hoar frost in the opposite region, and, perhaps, a narrow zone of running water at the borders of the enlightened hemisphere. It is possible, then, that evaporation on the one hand, and condensation on the other, may to a certain extent preserve an equilibrium of temperature, and mitigate the extreme severity of both climates.

(365.) A circle of one second in diameter, as seen from the earth, on the surface of the moon, contains about a square mile. Telescopes, therefore, must yet be greatly improved, before we could expect to see signs of inhabitants, as manifested by edifices or by changes on the surface of the soil. It should, however, be observed, that, owing to the small density of the materials of the moon, and the comparatively feeble gravitation of bodies on her surface, muscular force would there go six times as far in overcoming the weight of materials as on the earth. Owing to the want of air, however, it seems impossible that any form of life analogous to those on earth can subsist there. No appearance indicating vegetation, or the slightest variation of surface which can fairly be ascribed to change of season, can any where be discerned.

(366.) The lunar summer and winter arise, in fact, from the rotation of the moon on its own axis, the period of which rotation is *exactly* equal to its sidereal revolution about the earth, and is performed in a plane $1^\circ 30' 11''$ inclined to the ecliptic, and therefore nearly coincident with her own orbit. This is the cause why we always see the same face of the moon, and have no knowledge of the other side. This remarkable coincidence of two periods, which at first sight would seem perfectly distinct, is said to be a consequence of the general laws to be explained hereafter.

(367.) The moon's rotation on her axis is uniform; but since her motion in her orbit (like that of the sun) is not so, we are enabled to look a few degrees round the equatorial parts of her visible border, on the eastern or western side, according to circumstances; or, in other words, the line joining the centers of the earth and moon fluctuates a little in its position, from its mean or average intersection with her surface, to the east or westward. And, moreover, since the axis about which she revolves is not exactly perpendicular to her orbit, her poles come alternately into view for a small space at the edges of her disc. These phenomena are known by the name of *librations*. In consequence of these two distinct kinds of libration, the same identical point of the moon's surface is not always the center of her disc, and we therefore get sight of a zone of a few degrees in breadth on all sides of the border, beyond an exact hemisphere.

(368.) If there be inhabitants in the moon, the earth must present to them the extraordinary appearance of a moon of nearly 2° in diameter, exhibiting the same phases as we see the moon to do, but *immoveably fixed in their sky*, (or, at least, changing its apparent place only by the small amount of the libration,) while the stars must seem to pass slowly beside and behind it. It will appear clouded with variable spots, and belted with equatorial and tropical zones corresponding to our trade-winds; and it may be doubted whether, in their perpetual change, the outlines of our continents and seas can ever be clearly discerned.

CHAP. VII.

OF TERRESTRIAL GRAVITY.—OF THE LAW OF UNIVERSAL GRAVITATION.—PATHS OF PROJECTILES; APPARENT—REAL.—THE MOON RETAINED IN HER ORBIT BY GRAVITY.—ITS LAW OF DIMINUTION.—LAWS OF ELLIPTIC MOTION.—ORBIT OF THE EARTH ROUND THE SUN IN ACCORDANCE WITH THESE LAWS.—MASSES OF THE EARTH AND SUN COMPARED.—DENSITY OF THE SUN.—FORCE OF GRAVITY AT ITS SURFACE.—DISTURBING EFFECT OF THE SUN ON THE MOON'S MOTION.

(369.) The reader has now been made acquainted with the chief phenomena of the motions of the earth in its orbit round the sun, and of the moon about the earth. —We come next to speak of the physical cause which maintains and perpetuates these motions, and causes the massive bodies so revolving to deviate continually from the directions they would naturally seek to follow, in pursuance of the first law of motion *, and bend their courses into curves concave to their centers.

(370.) Whatever attempts may have been made by metaphysical writers to reason away the connection of cause and effect, and fritter it down into the unsatisfactory relation of habitual sequence †, it is certain that the conception of some more real and intimate connection is quite as strongly impressed upon the human mind as that of the existence of an external world,—the vindication of whose reality has (strange to say) been regarded as an achievement of no common merit in the annals of this branch of philosophy. It is our own immediate consciousness *of effort*, when we exert

* See Cab. Cyc. Mechanics, chap. iii.

† See Brown "On Cause and Effect,"—a work of great acuteness and subtlety of reasoning on some points, but in which the whole train of argument is vitiated by one enormous oversight; the omission, namely, of a *distinct and immediate personal consciousness of causation* in his enumeration of that *sequence of events*, by which the volition of the mind is made to terminate in the motion of material objects. I mean the consciousness of *effort*, as a thing entirely distinct from mere *desire* or *volition* on the one hand, and from mere spasmodic contraction of muscles on the other. Brown, 3d edit. Edin. 1818, p. 47.—*Author.*

force to put matter in motion, or to oppose and neutralize force, which gives us this internal conviction of *power* and *causation* so far as it refers to the material world, and compels us to believe that whenever we see material objects put in motion from a state of rest, or deflected from their rectilinear paths, and changed in their velocities if already in motion, it is in consequence of such an EFFORT *somehow* exerted, though not accompanied with *our* consciousness. That such an effort should be exerted with success through an interposed space, is no more difficult to conceive, than that our hand should communicate motion to a stone, with which it is *demonstrably not in contact.*

(371.) All bodies with which we are acquainted, when raised into the air and quietly abandoned, descend to the earth's surface in lines perpendicular to it. They are therefore urged thereto by a force or effort, the direct or indirect result of a *consciousness* and a *will* existing *somewhere,* though beyond our power to trace, which force we term *gravity;* and whose tendency or direction, as universal experience teaches, is towards the earth's center; or rather, to speak strictly, with reference to its spheroidal figure, perpendicular to the surface of still water. But if we cast a body obliquely into the air, this tendency, though not extinguished or diminished, is materially modified in its ultimate effect. The upward impetus we give the stone is, it is true, after a time destroyed, and a downward one communicated to it, which ultimately brings it to the surface, where it is opposed in its further progress, and brought to rest. But all the while it has been continually deflected or bent aside from its rectilinear progress, and made to describe a curved line concave to the earth's center; and having a *highest point, vertex,* or *apogee,* just as the moon has in its orbit, where the direction of its motion is perpendicular to the radius.

(372.) When the stone which we fling obliquely upwards meets and is stopped in its descent by the earth's surface, its motion is not *towards the center,* but inclined to the earth's radius at the same angle as when it quitted our hand. As we are sure that, if not stopped by the resistance of the earth, it would continue to descend, and that *obliquely*, what presumption, we may ask, is there that it would ever reach the center, to which its motion, in no part of its visible course, was ever directed ? What reason have we to believe that it might not rather circulate round it, as the moon does round the earth, returning again to the point it set out from, after completing an elliptic orbit of which the center occupies the lower focus ? And if so, is it not reasonable to imagine that the same force of gravity *may* (since we know that it is exerted at all accessible heights above the surface, and even in the highest regions of the atmosphere) extend as far as 60 radii of the earth, or to the moon ? and may not this be the power,—for *some* power there *must* be,—which deflects *her* at every instant from the tangent of her orbit, and keeps her in the elliptic path which experience teaches us she actually pursues ?

(373.) If a stone be whirled round at the end of a string, it will stretch the string by a *centrifugal* force*, which, if the speed of rotation be sufficiently increased, will at length break the string, and let the stone escape. However strong the string, it may, by a sufficient rotatory velocity of the stone, be brought to the utmost tension it will bear without breaking; and if we know what weight it is capable of carrying, the velocity necessary for this purpose is easily calculated. Suppose, now, a string to connect the earth's center, with a weight at its surface, whose strength should be just sufficient to sustain that weight suspended from it. Let us, however, for a moment imagine gravity to have no existence, and that the weight is made to revolve with the *limiting velocity* which that string can barely counteract: then will its tension be just equal to the weight of the revolving body; and any power which should continually urge the body towards the center with a force equal to its weight would perform the office, and might supply the place of

* See Cab. Cyc. MECHANICS, chap. viii.

the string, if divided. Divide it then, and in its place let gravity act, and the body will circulate as before; its tendency to the center, or *its weight*, being just balanced by its centrifugal force. Knowing the radius of the earth, we can calculate the periodical time in which a body so balanced must circulate to keep it up; and this appears to be $1^h\ 23^m\ 22^s$.

(374.) If we make the same calculation for a body at the distance of the moon, *supposing its weight* or *gravity the same as at the earth's surface*, we shall find the period required to be $10^h\ 45^m\ 30^s$. The actual period of the moon's revolution, however, is $27^d\ 7^h\ 43^m$; and hence it is clear that the moon's velocity is not nearly sufficient to sustain it against *such* a power, supposing it to revolve in a circle, or neglecting (for the present) the slight ellipticity of its orbit. In order that a body at the distance of the moon (or the moon itself) should be capable of *keeping its distance* from the earth by the outward effort of its centrifugal force, while yet its time of revolution should be what the moon's actually is, it will appear (on executing the calculation from the principles laid down in Cab. Cyc. MECHANICS) that *gravity*, instead of being as intense as at the surface, would require to be very nearly 3600 times less energetic; or, in other words, that its intensity is so enfeebled by the remoteness of the body on which it acts, as to be capable of producing in it, in the same time, only $\frac{1}{3600}$th part of the motion which it would impart to the same mass of matter at the earth's surface.

(375.) The distance of the moon from the earth's center is somewhat less than sixty times the distance from the center to the surface, and $3600 : 1 :: 60^2 : 1^2$; so that the proportion in which we must admit the earth's gravity to be enfeebled at the moon's distance, if it be really the force which retains the moon in her orbit, must be (at least in this particular instance) that of the squares of the distances at which it is compared. Now, in such a diminution of energy with increase of distance, there is nothing *primâ facie* inadmissible. Emanations from a center, such as light and heat, do really diminish in intensity by increase of distance, and in this identical proportion; and though we cannot certainly argue much from this analogy, yet we do see that the power of magnetic and electric attractions and repulsions is actually enfeebled by distance, and much more rapidly than in the simple proportion of the increased distances. The argument, therefore, stands thus: — On the one hand, *Gravity* is a real power, of whose agency we have daily experience. We know that it extends to the greatest accessible heights, and far beyond; and we see no reason for drawing a line at any particular height, and there asserting that it must cease entirely; though we have analogies to lead us to suppose its energy may diminish rapidly as we ascend to great heights from the surface, such as that of the moon. On the other hand, we are sure the moon *is* urged towards the earth by *some* power which retains her in her orbit, and that the intensity of this power is such as would correspond to a diminished gravity, in the proportion, — otherwise not improbable, — of the squares of the distances. If gravity be *not* that power, there must exist some other; and, besides this, gravity must cease at some inferior level, or the nature of the moon must be different from that of ponderable matter; — for if not, it would be urged by *both* powers, and therefore *too much* urged, and forced inwards from her path.

(376.) It is on such an argument that Newton is understood to have rested, in the first instance, and provisionally, his law of universal gravitation, which may be thus abstractly stated: — "Every particle of matter in the universe attracts every other particle, with a force directly proportioned to the mass of the attracting particle, and inversely to the square of the distance between them." In this abstract and general form, however, the proposition is not applicable to the case before us. The earth and moon are not mere *particles*, but great spherical bodies, and to such the general law does not immediately apply; and, before we can make it applicable, it becomes necessary to enquire *what* will be the force with which a congeries of particles, constituting a solid mass of any assigned figure, will attract another such collection of material atoms. This problem is one purely dynamical, and, in its general form, is of extreme difficulty. Fortunately, however, for human knowledge, when the attracting and attracted bodies are spheres, it admits of an easy and direct solution. Newton himself has shown (*Princip.* b. i. prop. 75.) that, in that case, the attraction is precisely the same as if the whole matter of each sphere were collected into its center, and the spheres were single particles there placed; so that, in this case, the general law applies in its strict wording. The effect of the trifling deviation of the earth from a spherical form is of too minute an order to need attention at present. It is, however, perceptible, and may be hereafter noticed.

(377.) The next step in the Newtonian argument is one which divests the law of gravitation of its provisional character, as derived from a loose and superficial consideration of the lunar orbit as a circle described with an average or mean velocity, and elevates it to the rank of a general and primordial relation, by proving its applicability to the state of existing nature in all its detail of circumstances. This step consists in demonstrating, as he has done * (*Princip.* i.17., i. 75.), that, under the influence of such an attractive force mutually urging two spherical gravitating bodies towards each other, they will each, when moving in each other's neighbourhood, be deflected into an orbit concave towards the other, and describe, one about the other regarded as fixed, or both round their common center of gravity, curves whose forms are limited to those figures known in geometry by the general name of conic sections. It

* We refer for these fundamental propositions, as a point of duty, to the immortal work in which they were first propounded. It is impossible for us in this volume to go into these investigations: even did our limits permit, it would be utterly inconsistent with our plan; a general idea, however, of their conduct will be given in the next chapter.

will depend, he shows, in any assigned case, upon the particular circumstances of velocity, distance, and direction, *which* of these curves shall be described,—whether an ellipse, a circle, a parabola, or an hyperbola; but one or other it *must* be; and any one of any degree of eccentricity it *may* be, according to the circumstances of the case; and, in all cases. the point to which the motion is referred, whether it be the center of one of the spheres, or their common center of gravity, will of necessity be the *focus* of the conic section described. He shows, furthermore (*Princip.* i. 1.), that, in every case, the *angular velocity* with which the line joining their centers moves, must be inversely proportional to the square of their mutual distance, and that equal areas of the curves described will be swept over by their line of junction in equal times.

(378.) All this is in conformity with what we have stated of the solar and lunar movements. Their orbits are ellipses, but of different degrees of eccentricity; and this circumstance already indicates the general applicability of the principles in question.

(379.) But here we have already, by a natural and ready implication (such is always the progress of generalisation), taken a further and most important step, almost unperceived. We have extended the action of gravity to the case of the earth and sun, to a distance immensely greater than that of the moon, and to a body apparently quite of a different nature from either. Are we justified in this? or, at all events, are there no modifications introduced by the change of data, if not into the general expression, at least into the particular interpretation, of the law of gravitation? Now, the moment we come to numbers, an obvious incongruity strikes us. When we calculate, as above, from the known distance of the sun (art. 304.), and from the period in which the earth circulates about it (art. 327.), what must be the centrifugal force of the latter by which the sun's attraction is balanced, (and which, therefore, becomes an exact measure of the sun's attractive energy as exerted on the earth,) we find it to be immensely greater than would suffice to counteract the *earth's* attraction on an equal body at that distance—greater in the high proportion of 354936 to 1. It is clear, then, that if the earth be retained in its orbit about the sun by *solar attraction,* conformable in its rate of diminution with the general law, this force must be no less than 354936 times more intense than what the earth would be capable of exerting, *cæteris paribus,* at an equal distance.

(380.) What, then, are we to understand from this result? Simply this,—that the sun attracts as a collection of 354936 earths occupying its place would do, or, in other words, that the sun contains 354936 times the mass or quantity of ponderable matter that the earth consists of. Nor let this conclusion startle us. We have only to recall what has been already shown in art. 305. of the gigantic dimensions of this magnificent body, to perceive that, in assigning to it so vast a mass, we are not outstepping a reasonable proportion. In fact, when we come to compare its *mass* with its *bulk,* we find its density* to be less than that of the earth, being no more than 0·2543. So that it must consist, in reality, of far *lighter* materials, especially when we consider the force under which its central parts must be condensed. This consideration renders it highly probable that an intense heat prevails in its interior, by which its elasticity is reinforced, and rendered capable of resisting this almost inconceivable pressure without collapsing into smaller dimensions.

(381.) This will be more distinctly appreciated, if we estimate, as we are now prepared to do, the intensity of gravity at the sun's surface.

The attraction of a sphere being the same (art. 376.) as if its whole mass were collected in its center, will, of course, be proportional to the mass directly, and the

* The density of a material body is as the *mass* directly, and the volume inversely: hence density of ☉ : density of ⊕ :: $\frac{354936}{1384472}$: 1 : 0·2543 : 1

square of the distance inversely; and, in this case, the distance is the radius of the sphere. Hence we conclude*, that the intensities of solar and terrestrial gravity at the surfaces of the two globes are in the proportions of 27·9 to 1. A pound of terrestrial matter at the sun's surface, then, would exert a pressure equal to what 27·9 such pounds would do at the earth's. An ordinary man, for example, would not only be unable to sustain his own weight on the sun, but would literally be crushed to atoms under the load.†

(382.) Henceforward, then, we must consent to dismiss all idea of the earth's immobility, and transfer that attribute to the sun, whose ponderous mass is calculated to exhaust the feeble attractions of such comparative atoms as the earth and moon, without being perceptibly dragged from its place. Their center of gravity lies, as we have already hinted, almost close to the center of the solar globe, at an interval quite imperceptible from our distance; and whether we regard the earth's orbit as being performed about the one or the other center makes no appreciable difference in any one phenomenon of astronomy.

(383.) It is in consequence of the *mutual* gravitation of all the several parts of matter, which the Newtonian law supposes, that the earth and moon, while in the act of revolving, monthly, in their mutual orbits about their common center of gravity, yet continue to circulate, without parting company, in a greater annual orbit round the sun. We may conceive this motion by connecting two unequal balls by a stick, which, at their center of gravity, is tied by a long string, and whirled round. Their joint *systems* will circulate as one body about the common center to which the string is attached, while yet they may go on circulating round each other in subordinate gyrations, as if the stick were quite free

* Solar gravity : terrestrial :: $\frac{354936}{(440000)^2} : \frac{1}{(4000)^2}$ :: 27·9 : 1; the respective radii of the sun and earth being 440000, and 4000 miles.

† A mass weighing 12 stone or 170 lbs. on the earth, would produce a pressure of 4600 lbs. on the sun.

from any such tie, and merely hurled through the air. If the earth alone, and not the moon, gravitated to the sun, it would be dragged away, and leave the moon behind — and *vice versâ*; but, acting on both, they continue together under its attraction, just as the loose parts of the earth's surface continue to rest upon it. It is, then, in strictness, not the earth or the moon which describes an ellipse around the sun, but their common center of gravity. The effect is to produce a small, but very perceptible, monthly *equation* in the sun's apparent motion as seen from the earth, which is always taken into account in calculating the sun's place.

(384.) And here, *i. e.* in the attraction of the sun, we have the key to all those differences from an exact elliptic movement of the moon in her monthly orbit, which we have already noticed (arts. 344. 360.), viz. to the retrograde revolution of her nodes; to the direct circulation of the axis of her ellipse; and to all the other deviations from the laws of elliptic motion at which we have further hinted. If the moon simply revolved about the earth under the influence of its gravity, none of these phenomena would take place. Its orbit would be a perfect ellipse, returning into itself, and always lying in one and the same plane: that it *is not so*, is a proof that some cause *disturbs* it, and interferes with the earth's attraction; and this cause is no other than the sun's attraction — or rather, that part of it which is not *equally* exerted on the earth.

(385.) Suppose two stones, side by side, or otherwise situated with respect to each other, to be let fall together; then, as gravity accelerates them equally, they will retain their relative positions, and fall together as if they formed one mass. But suppose gravity to be rather more intensely exerted on one than the other; then would that one be rather more accelerated in its fall, and would gradually leave the other: and thus a relative motion between them would arise from the difference of action, however slight.

(386.) The sun is about 400 times more remote than the moon; and, in consequence, while the moon describes her monthly orbit round the earth, her distance from the sun is alternately $\frac{1}{400}$th part greater and as much less than the earth's. Small as this is, it is yet sufficient to produce a perceptible excess of attractive tendency of the moon towards the sun, above that of

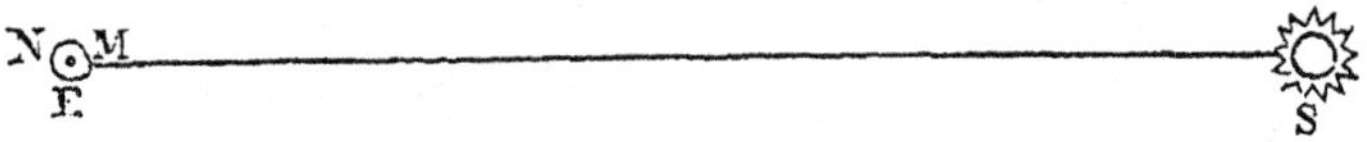

the earth when in the nearer point of her orbit, M, and a corresponding defect on the opposite part, N; and, in the intermediate positions, not only will a difference of *forces* subsist, but a difference of *directions* also; since, however small the lunar orbit M N, it is not a *point*, and, therefore, the lines drawn from the sun S to its several parts cannot be regarded as strictly parallel. If, as we have already seen, the force of the sun were equally exerted, and in parallel directions on both, no disturbance of their relative situations would take place; but from the non-verification of these conditions arises a *disturbing force*, oblique to the line joining the moon and earth, which in some situations acts to *accelerate*, in others to *retard*, her elliptic orbitual motion; in some to draw the earth from the moon, in others the moon from the earth. Again, the lunar orbit, though very nearly, is yet not quite coincident with the plane of the ecliptic; and hence the action of the sun, which is very nearly parallel to the last-mentioned plane, tends to draw her somewhat *out of the plane* of her orbit, and does actually do so — producing the revolution of her nodes, and other phenomena less striking. We are not yet prepared to go into the subject of these *perturbations*, as they are called; but they are introduced to the reader's notice as early as possible, for the purpose of re-assuring his mind, should doubts have arisen as to the logical correctness of our argument, in consequence of our temporary neglect of them while working our way upward to the law of gravity from a general consideration of the moon's orbit.

CHAP. VIII.

OF THE SOLAR SYSTEM.

APPARENT MOTIONS OF THE PLANETS. — THEIR STATIONS AND RETROGRADATIONS. — THE SUN THEIR NATURAL CENTER OF MOTION. — INFERIOR PLANETS. — THEIR PHASES, PERIODS, ETC. — DIMENSIONS AND FORM OF THEIR ORBITS. — TRANSITS ACROSS THE SUN. — SUPERIOR PLANETS. — THEIR DISTANCES, PERIODS, ETC. — KEPLER'S LAWS AND THEIR INTERPRETATION. — ELLIPTIC ELEMENTS OF A PLANET'S ORBIT. — ITS HELIOCENTRIC AND GEOCENTRIC PLACE. — BODE'S LAW OF PLANETARY DISTANCES. — THE FOUR ULTRA-ZODIACAL PLANETS. — PHYSICAL PECULIARITIES OBSERVABLE IN EACH OF THE PLANETS.

(387.) THE sun and moon are not the only celestial objects which appear to have a motion independent of that by which the great constellation of the heavens is daily carried round the earth. Among the stars there are several, — and those among the brightest and most conspicuous, — which, when attentively watched from night to night, are found to change their relative situations among the rest; some rapidly, others much more slowly. These are called *planets*. Four of them, — Venus, Mars, Jupiter, and Saturn, — are remarkably large and brilliant; another, Mercury, is also visible to the naked eye as a large star, but, for a reason which will presently appear, is seldom conspicuous; a fifth, Uranus, is barely discernible without a telescope; and four others, — Ceres, Pallas, Vesta, and Juno, — are never visible to the naked eye. Besides these ten, others yet undiscovered may exist; and it is extremely probable that such is the case, — the multitude of telescopic stars being so great that only a small fraction of their number has been sufficiently noticed to ascertain whether they retain the same places or not, and the five last-mentioned planets having all been discovered within half a century from the present time.

(388.) The apparent motions of the planets are much more irregular than those of the sun or moon. Generally speaking, and comparing their places at distant times, they all advance, though with very different *average* or *mean* velocities, in the same direction as those luminaries, *i. e.* in opposition to the apparent diurnal motion, or from west to east: all of them make the entire tour of the heavens, though under very different circumstances; and all of them, with the exception of the four telescopic planets,—Ceres, Pallas, Juno, and Vesta (which may therefore be termed *ultra-zodiacal*),—are confined in their visible paths within very narrow limits on either side the ecliptic, and perform their movements within that zone of the heavens we have called, above, the Zodiac (art. 254.).

(389.) The obvious conclusion from this is, that whatever be, otherwise, the nature and law of their motions, they are all performed *nearly in the plane of the ecliptic*,—that plane, namely, in which our own motion about the sun is performed. Hence it follows, that we see their evolutions, not in *plan*, but in *section*; their real angular movements and linear distances being all *foreshortened* and confounded undistinguishably, while only their deviations from the ecliptic appear of their natural magnitude, undiminished by the effect of perspective.

(390.) The apparent motions of the sun and moon, though not uniform, do not deviate very greatly from uniformity; a moderate acceleration and retardation, accountable for by the ellipticity of their orbits, being all that is remarked. But the case is widely different with the planets: sometimes they advance rapidly; then relax in their apparent speed—come to a momentary stop; and then actually reverse their motion, and run back upon their former course, with a rapidity at first increasing, then diminishing, till the reversed or retrograde motion ceases altogether. Another *station*, or moment of apparent rest or indecision, now takes place; after which the movement is again reversed, and resumes its original direct character. On the whole, however, the amount of direct motion more than compensates the retrograde; and by the excess of the former over the latter, the gradual advance of the planet from west to east is maintained. Thus, supposing the Zodiac to be unfolded into a plane surface, (or represented as in Mercator's projection, art. 234., taking the ecliptic E C for its ground line,) the track of a planet, when mapped down by observation from day to day, will offer

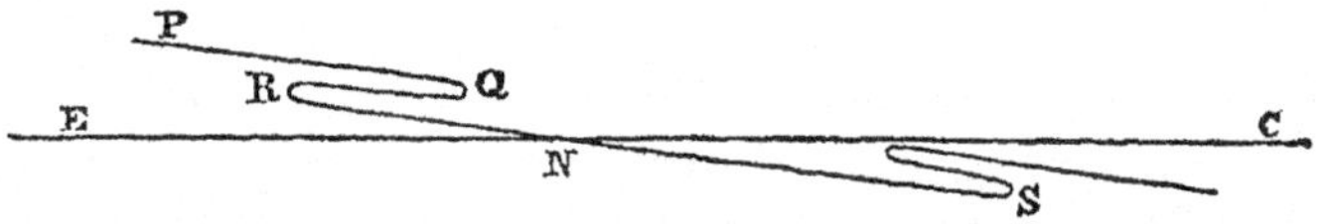

the appearance P Q R S, &c.; the motion from P to Q being direct, at Q stationary, from Q to R retrograde, at R again stationary, from R to S direct, and so on.

(391.) In the midst of the irregularity and fluctuation of this motion, one remarkable feature of uniformity is observed. Whenever the planet crosses the ecliptic, as at N in the figure, it is said (like the moon) to be in its node; and as the earth necessarily lies in the plane of the ecliptic, the planet cannot be *apparently* or *uranographically* situated in the celestial circle so called, without being *really* and *locally* situated *in that plane*. The visible passage of a planet through its node, then, is a phenomenon indicative of a circumstance in its real motion quite independent of the station from which we view it. Now, it is easy to ascertain, by observation, when a planet passes from the north to the south side of the ecliptic: we have only to convert its right ascensions and declinations into longitudes and latitudes, and the change from north to south latitude on two successive days will advertise us on what *day* the transition took place; while a simple proportion, grounded on the observed state of its motion *in latitude* in the interval, will suffice to fix the precise hour and minute of its arrival on the ecliptic. Now, this being done for several transitions from side to side of the ecliptic, and their dates thereby fixed, we find, universally, that the interval of time elapsing between the successive passages of each planet through *the same node* (whether it be the ascending or the descending) is always alike, whether the planet at the moment of such passage be direct or retrograde, swift or slow, in its apparent movement.

(392.) Here, then, we have a circumstance which, while it shows that the motions of the planets are in fact subject to certain laws and fixed periods, may lead us very naturally to suspect that the apparent irregularities and complexities of their movements may be owing to our not seeing them from their natural center (art. 316.), and from our mixing up with their own proper motions movements of a parallactic kind, due to our own change of place, in virtue of the orbitual motion of the earth about the sun.

(393.) If we abandon the earth as a center of the planetary motions, it cannot admit of a moment's hesitation where we should place that center with the greatest probability of truth. It must surely be the sun which is entitled to the first trial, as a station to which to refer them. If it be not connected with them by any physical relation, it at least possesses the advantage, which the earth does not, of comparative immobility. But after what has been shown in art. 380., of the immense mass of that luminary, and of the office it performs to us as a quiescent center of our orbitual motion, nothing can be more natural than to suppose it may perform the same to other globes which, like the earth, may be revolving round it; and these globes may be visible to us by its light reflected from them, as the moon is. Now there are many facts which give a strong support to the idea that the planets are in this predicament.

(394.) In the first place, the planets really are great globes, of a size commensurate with the earth, and several of them much greater. When examined through powerful telescopes, they are seen to be round bodies, of sensible and even of considerable apparent diameter, and offering distinct and characteristic peculiarities, which show them to be solid masses, each possessing its individual structure and mechanism; and that, in one instance at least, an exceedingly artificial and complex one. (See the representations of Jupiter, Saturn, and Mars, in Plate I.) That their distances from us are great, much greater than that of the moon, and some or them even greater than that of the sun, we infer from the smallness of their diurnal parallax, which, even for the nearest of them, when most favourably situated, does not exceed a few seconds, and for the more remote ones is almost imperceptible. From the comparison of the diurnal parallax of a celestial body, with its apparent semidiameter, we can at once estimate its real size. For the parallax is, in fact, nothing else than the apparent semidiameter of the earth as seen from the body in question (art. 298. et seq.); and, the intervening distance being the same, the real diameters must be to each other in the proportion of the apparent ones. Without going into particulars, it will suffice to state it as a general result of that comparison, that the planets are all of them incomparably smaller than the sun, but some of them as large as the earth, and others much greater.

(395.) The next fact respecting them is, that their distances from us, as estimated from the measurement of their angular diameters, are in a continual state of change, periodically increasing and decreasing within certain limits, but by no means corresponding with the supposition of regular circular or elliptic orbits described by them about the earth as a center or focus, but maintaining a constant and obvious relation to their apparent angular distances or *elongations* from the sun. For example; the apparent diameter of Mars is greater when in opposition (as it is called) to the sun, *i. e.* when in the opposite part of the ecliptic, or when it comes on the meridian at midnight,—being then about 18″,—but diminishes rapidly from that amount to about

4″, which is its apparent diameter when in *conjunction*, or when seen in nearly the same direction as that luminary. This, and facts of a similar character, observed with respect to the apparent diameters of the other planets, clearly point out the sun as having more than an accidental relation to their movements.

(396.) Lastly, certain of the planets, when viewed through telescopes, exhibit the appearance of phases like those of the moon. This proves that they are opaque bodies, shining only by reflected light, which can be no other than the sun's; not only because there is no other source of light external to them sufficiently powerful, but because the appearance and succession of the phases themselves are (like their visible diameters) intimately connected with their elongations from the sun, as will presently be shown.

(397.) Accordingly, it is found, that, when we refer the planetary movements to the sun as a center, all that apparent irregularity which they offer when viewed from the earth disappears at once, and resolves itself into one simple and general law, of which the earth's motion, as explained in a former chapter, is only a particular case. In order to show how this happens, let us take the case of a single planet, which we will suppose to revolve round the sun, in a plane nearly, but not quite, coincident with the ecliptic, but passing through the sun, and of course intersecting the ecliptic in a fixed line, which is the line of the planet's nodes. This line must of course divide its orbit into two segments; and it is evident that, so long as the circumstances of the planet's motion remain otherwise unchanged, the times of describing these segments must remain the same. The interval, then, between the planet's quitting either node, and returning to *the same* node again, must be that in which it describes one complete revolution round the sun, or its periodic time; and thus we are furnished with a direct method of ascertaining the periodic time of each planet.

(398.) We have said (art. 388.) that the planets make

the entire tour of the heavens under very different circumstances. This must be explained. Two of them —Mercury and Venus—perform this circuit evidently as attendants upon the sun, from whose vicinity they never depart beyond a certain limit. They are seen sometimes to the east, sometimes to the west of it. In the former case they appear conspicuous over the western horizon, just after sunset, and are called evening stars: Venus, especially, appears occasionally in this situation with a dazzling lustre; and in favourable circumstances may be observed to cast a pretty strong shadow.* When they happen to be to the west of the sun, they rise before that luminary in the morning and appear over the eastern horizon as morning stars: they do not, however, attain the same *elongation* from the sun. Mercury never attains a greater angular distance from it than about 29°, while Venus extends her excursions on either side to about 47°. When they have receded from the sun, *eastward,* to their respective distances, they remain for a time, as it were, immovable *with respect to it,* and are carried along with it in the ecliptic with a motion equal to its own; but presently they begin to approach it, or, which comes to the same, their motion in longitude diminishes, and the sun gains upon them. As this approach goes on, their continuance above the horizon after sunset becomes daily shorter, till at length they set before the darkness has become sufficient to allow of their being seen. For a time, then, they are not seen at all, unless on very rare occasions, when they are to be observed *passing across the sun's disc as small, round, well-defined black spots* totally different in appearance from the solar spots (art. 330.). These phenomena are emphatically called *transits* of the respective planets across the sun, and take place when the earth happens to be passing the line of their nodes while they are in that part of their

* It must be thrown upon a white ground. An open window in a white-washed room is the best exposure. In this situation, I have observed no only the shadow, but the diffracted fringes edging its outline.—*Author.*

orbits, just as in the account we have given (art. 355.) of a solar eclipse. After having thus continued invisible for a time, however, they begin to appear on the other side of the sun, at first showing themselves only for a few minutes before sunrise, and gradually longer and longer as they recede from him. At this time their motion in longitude is rapidly retrograde. Before they attain their greatest elongation, however, they become stationary in the heavens; but their recess from the sun is still maintained by the advance of that luminary along the ecliptic, which continues to leave them behind, until, having reversed their motion, and become again *direct*, they acquire sufficient speed to commence overtaking him — at which moment they have their greatest *western* elongation; and thus is a kind of oscillatory movement kept up, while the general advance along the ecliptic goes on.

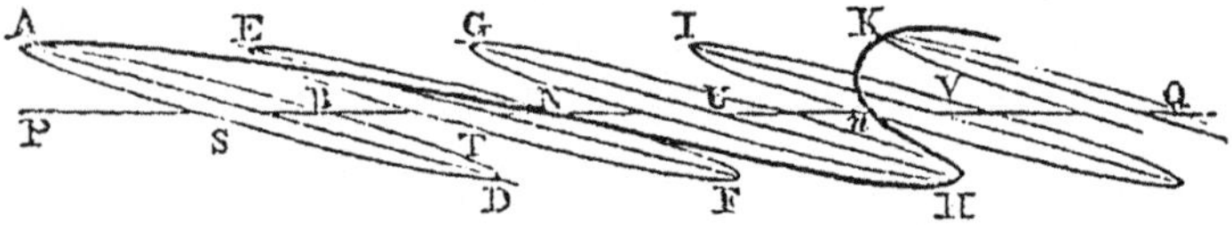

(399.) Suppose P Q to be the ecliptic, and A B D the orbit of one of these planets, (for instance, Mercury,) seen almost edgewise by an eye situated very nearly in its plane; S, the sun, its center; and A, B, D, S successive positions of the planet, of which B and S are in the nodes. If, then, the sun S stood apparently still in the ecliptic, the planets would simply appear to oscillate backwards and forwards from A to D, alternately passing before and behind the sun; and, if the eye happened to lie exactly *in* the plane of the orbit, *transiting* his disc in the former case, and being covered by it in the latter. But as the sun is not so stationary, but apparently carried along the ecliptic P Q, let it be supposed to move over the spaces S T, T U, U V, while the planet in each case executes one quarter of its period. Then will its orbit be apparently carried along with the sun, into the successive positions represented in the figure; and while its real motion round the sun brings it into the respective points B, D, S, A, its apparent movement in the heavens will seem to have been along the wavy or zigzag line A N H K. In this, its motion in longitude will have been direct in the parts A N, N H, and retrograde in the parts H *n* K; while at the turns of the zigzag, at H, K, it will have been stationary.

(400.) The only two planets—Mercury and Venus—whose evolutions are such as above described, are called inferior planets; their points of farthest recess from the sun are called (as above) their *greatest* eastern and western *elongations;* and their points of nearest approach to it, their *inferior* and *superior* conjunctions,—the former when the planet passes between the earth and the sun, the latter when behind the sun.

(401.) In art. 398. we have traced the apparent path of an inferior planet, by considering its orbit in section, or as viewed from a point in the plane of the ecliptic. Let us now contemplate it *in plan,* or as viewed from a station above that plane, and projected on it. Suppose, then, S to represent the sun, *a b c d* the orbit of Mercury, and A B C D a part of that of the earth—the direction of the circulation being the same in both, viz. that of the arrow. When the planet stands at *a,* let the earth be situated at A, in the direction of a tangent, *a* A, to its orbit; then it is evident that it will appear at its *greatest elongation* from the sun,—the angle *a* A S, which measures their apparent interval as seen from A, being then greater than in any other situation of *a* upon its own circle.

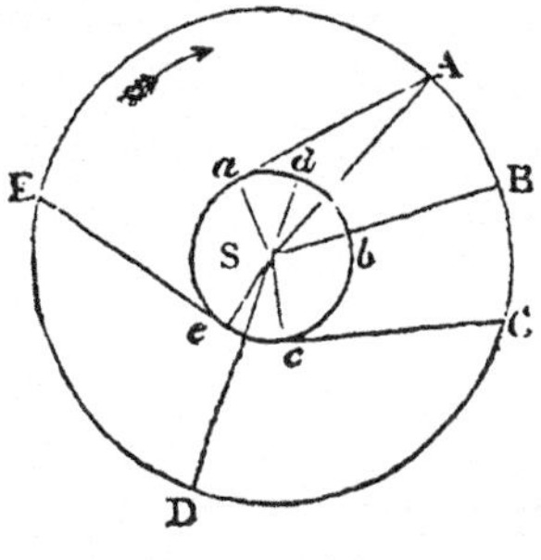

(402.) Now, this angle being known by observation, we are hereby furnished with a ready means of ascertaining, at least approximately, the distance of the planet from the sun, or the radius of its orbit, supposed a circle. For the triangle S A *a* is right-angled at *a*, and consequently we have S *a* : S A :: sin. S A *a* : radius, by which proportion the radii S *a*, S A of the two orbits are directly compared. If the orbits were both exact circles, this would of course be a perfectly rigorous mode of proceeding: but (as is proved by the inequality of the resulting values of S *a* obtained at different times) this is not the case; and it becomes necessary to admit an excentricity of position, and a deviation from the exact circular form in *both* orbits, to account for this difference. Neglecting, however, at present this inequality, a mean or average value of S *a* may, at least, be obtained from the frequent repetition of this process in all varieties of situation of the two bodies. The calculations being performed, it is concluded that the mean distance of Mercury from the sun is about 36000000 miles; and that of Venus, similarly derived, about 68000000; the radius of the earth's orbit being 95000000.

(403.) The siderealperiods of the planets may be obtained (as before observed), with a considerable approach to accuracy, by observing their passages through the nodes of their orbits; and, indeed, when a certain very minute motion of these nodes (similar to that of the moon's nodes, but incomparably slower,) is allowed for, with a precision only limited by the imperfection of the appropriate observations. By such observation, so corrected, it appears that the sidereal period of Mercury is $87^d\ 23^h\ 15^m\ 43{\cdot}9^s$; and that of Venus, $224^d\ 16^h\ 49^m\ 8{\cdot}0^s$. These periods, however, are widely different from the intervals at which the successive appearances of the two planets at their eastern and western elongations from the sun are observed to happen. Mercury is seen at its greatest splendour as an evening star, at average intervals of about 116, and Venus at intervals of about 584 days. The difference between the *sidereal* and *synodical* revolutions (art. 353.) accounts for this. Referring again to the figure of art. 401., if the earth stood still at A, while the planet advanced in its orbit, the lapse of a sidereal period, which should bring it round again to *a*, would also reproduce a similar elongation from the sun. But, meanwhile, the earth has advanced in its orbit in the same direction towards E, and therefore the next greatest elongation on the same side of the sun will happen — not in the position *a* A of the two bodies, but in some more advanced position, *e* E. The determination of this position depends on a calculation exactly similar to what has been explained in the article referred to; and we need, therefore, only here state the resulting synodical revolutions of the two planets, which come out respectively $115 \cdot 877^{d}$, and $583 \cdot 920^{d}$.

(404.) In this interval, the planet will have described a whole revolution *plus* the arc *a e*, and the earth only the arc A C E of its orbit. During its lapse, the *inferior conjunction* will happen when the earth has a certain intermediate situation, B, and the planet has reached *b*, a point between the sun and earth. The greatest elongation on the opposite side of the sun will happen when the earth has come to C, and the planet to *c*, where the line of junction C *c* is a tangent to the interior circle on the opposite side from M. Lastly, the superior conjunction will happen when the earth arrives at D, and the planet at *d* in the same line prolonged on the other side of the sun. The intervals at which these phenomena happen may easily be computed from a knowledge of the synodical periods and the radii of the orbits.

(405.) The circumferences of circles are in the proportion of their radii. If, then, we calculate the circumferences of the orbits of Mercury and Venus, and the earth, and compare them with the times in which their revolutions are performed, we shall find that the actual velocities with which they move in their orbits differ greatly; that of Mercury being about 109400 miles per hour, of Venus 80060 and of the earth 68080.

From this it follows, that at the inferior conjunction, or at *b*, either planet is moving in the *same* direction as the earth, but with a greater velocity; it will, therefore, leave the earth *behind* it; and the apparent motion of the planet viewed from the earth, will be *as if* the planet stood still, and the earth moved in a contrary direction from what it really does. In this situation, then, the apparent motion of the planet must be contrary to the apparent motion of the sun; and, therefore, retrograde. On the other hand, at the superior conjunction, the real motion of the planet being in the opposite direction to that of the earth, the relative motion will be the same as if the planet stood still, and the earth advanced with their united velocities in its own proper direction. In this situation, then, the apparent motion will be direct. Both these results are in accordance with observed fact.

(406.) The stationary points may be determined by the following consideration. At *a* or *c*, the points of greatest elongation, the motion of the planet is directly to or from the earth, or *along* their line of junction, while that of the earth is nearly perpendicular to it. Here, then, the apparent motion must be direct. At *b*, the inferior conjunction, we have seen that it must be retrograde, owing to the planet's motion (which is there, as well as the earth's, *perpendicular* to the line of junction,) surpassing the earth's. Hence, the stationary points ought to lie, as it is found by observation they do, between *a* and *b*, or *c* and *b*, viz. in such a position that the obliquity of the planet's motion with respect to the line of junction shall just compensate for the excess of its velocity, and cause an equal advance of each extremity of that line, by the motion of the planet at one end, and of the earth at the other: so that, for an instant of time, the whole line shall move parallel to itself. The question thus proposed is purely geometrical, and its solution on the supposition of circular orbits is easy; but when we regard them as otherwise than circles (which they really are), it becomes somewhat complex — too much so to be here entered upon. It will suffice to state the results which experience verifies, and which assigns the stationary points of Mercury at from 15° to 20° of elongation from the sun, according to circumstances; and of Venus, at an elongation never varying much from 29°. The former continues to retrograde during about 22 days; the latter, about 42.

(407.) We have said that some of the planets exhibit phases like the moon. This is the case with both Mercury and Venus; and is readily explained by a consideration of their orbits, such as we have above supposed them. In fact, it requires little more than

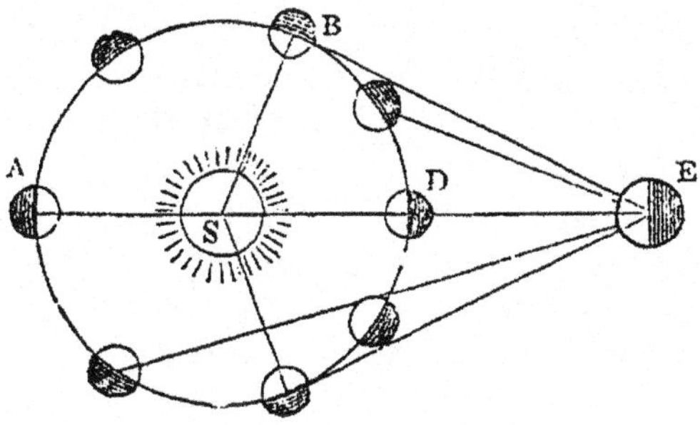

mere inspection of the figure annexed, to show, that to a spectator situated on the earth E, an inferior planet, illuminated by the sun, and therefore bright on the side next to him, and dark on that turned from him, will appear *full* at the superior conjunction A; *gibbous* (*i. e.* more than half full, like the moon between the first and second quarter,) between that point and the points B C of its greatest elongation; half-mooned at these points; and crescent-shaped, or horned, between these and the inferior conjunction D. As it approaches this point, the crescent ought to thin off till it vanishes altogether, rendering the planet invisible, unless in those cases where it *transits* the sun's disc, and appears on it as a black spot. All these phenomena are exactly conformable to observation; and, what is not a little satisfactory, they were predicted as necessary consequences of the Copernican theory before the invention of the telescope.*

(408.) The variation in brightness of Venus in different parts of its apparent orbit is very remarkable. This arises from two causes: 1st, the varying proportion of its visible illuminated area to its whole disc; and, 2dly, the varying angular diameter, or whole apparent magnitude of the disc itself. As it approaches its inferior conjunction from its greater elongation, the half-moon becomes a crescent, which *thins off*; but this is more than compensated, for some time, by the increasing apparent magnitude, in consequence of its diminishing distance. Thus the total light received from it goes on increasing, till at length it attains a maximum, which takes place when the planet's elongation is about 40°.

(409.) The transits of Venus are of very rare occurrence, taking place alternately at intervals of 8 and 113 years, or thereabouts. As astronomical phenomena, they are, however, extremely important; since they afford the best and most exact means we possess of ascertaining the sun's distance, or its parallax. Without going into the niceties of calculation of this problem, which, owing to the great multitude of circumstances to be attended to, are extremely intricate, we shall here explain its principle, which, in the abstract, is very simple and obvious. Let E be the earth, V Venus, and S the sun, and C D the portion of Venus's relative orbit which she describes while in the act of transiting the sun's disc. Suppose A B two spectators at opposite extremities of that diameter of the earth which is perpendicular to the ecliptic, and, to avoid complicating the case, let us lay

* See Essay on the Study of Natural Philosophy, Cab. Cyclo. Vol. XIV. p. 269.

out of consideration the earth's rotation, and suppose A, B, to retain that situation during the whole time of

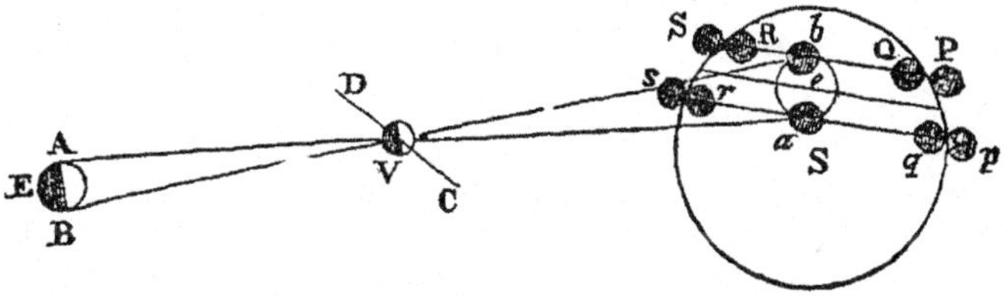

the transit. Then, at any moment when the spectator at A sees the center of Venus projected at *a* on the sun's disc, he at B will see it projected at *b*. If then one or other spectator could suddenly transport himself from A to B, he would see Venus suddenly displaced on the disc from *a* to *b;* and if he had any means of noting accurately the place of the points on the disc, either by micrometrical measures from its edge, or by other means, he might ascertain the angular measure of *a b* as seen from the earth. Now, since A V *a*, B V *b*, are straight lines, and therefore make equal angles on each side V, *a b* will be to A B as the distance of Venus from the sun is to its distance from the earth, or as 68 to 27, or nearly as $2\frac{1}{2}$ to 1: *a b* therefore, occupies on the sun's disc a space $2\frac{1}{2}$ times as great as the earth's diameter; and its angular measure is therefore equal to about $2\frac{1}{2}$ times the earth's apparent diameter at the distance of the sun, or (which is the same thing) to five times the sun's horizontal parallax (art. 298.). Any error, therefore, which may be committed in measuring *a b*, will entail only *one fifth* of that error on the horizontal parallax concluded from it.

(410.) The thing to be ascertained, therefore, is, in fact, neither more nor less than the breadth of the zone P Q R S, *p q r s*, included between the extreme apparent paths of the center of Venus across the sun's disc, from its entry on one side to its quitting it on the other. The whole business of the observers at A, B, therefore, resolves itself into this;—to ascertain, with all possible care and precision, each at his own station, this path,—where it enters, where it quits, and what segment of the sun's disc it cuts off. Now, one of the most exact ways in which (conjoined with careful micrometric measures) this can be done, is by noting the *time* occupied in the whole transit: for the relative angular motion of Venus being, in fact, very precisely known from the tables of her motion, and the apparent path being very nearly a straight line, these times give us a measure (*on a very enlarged scale*) of the lengths of the chords of the segments cut off; and the sun's diameter being known also with great precision, their versed sines, and therefore their difference, or the breadth of the zone required, becomes known. To obtain these times correctly, each observer must ascertain the instants of ingress and egress of the *center* To do this, he must note, 1st, the instant when the first visible impression or notch on the edge of the disc at P is produced, or the *first external contact;* 2dly, when the planet is just wholly immersed, and the broken edge of the disc just closes again at Q, or the first *internal* contact; and, lastly, he must make the same observations at the egress at R, S. The mean of the internal and external contacts gives the entry and egress of the planet's center.

(411.) The modifications introduced into this process by the earth's rotation on its axis, and by other geographical stations of the observers thereon than here supposed, are similar in their principles to those which enter into the calculation of a solar eclipse, or the occultation of a star by the moon, only more refined. Any consideration of them, however, here, would lead us too far; but in the view we have taken of the subject, it affords an admirable example of the way in which minute elements in astronomy may become magnified in their effects, and, by being made subject to measurement on a greatly enlarged scale, or by substituting the measure of time for space, may be ascertained with a degree of precision adequate to every purpose, by only watching favourable opportunities, and taking advantage of nicely adjusted combinations of circumstance. So important has this observation appeared to astronomers, that at the last transit of Venus, in 1769, expeditions were fitted out, on the most efficient scale, by the British, French, Russian, and other governments, to the remotest corners of the globe, for the express purpose of performing it. The celebrated expedition of Captain Cook to Otaheite was one of them. The general result of all the observations made on this most memorable occasion gives 8″·5776 for the sun's horizontal parallax.

(412.) The orbit of Mercury is very elliptical, the excentricity being nearly one fourth of the mean distance. This appears from the inequality of the greatest elongations from the sun, as observed at different times, and which vary between the limits 16° 12′ and 28° 48′, and, from exact measures of such elongations, it is not difficult to show that the orbit of Venus also is slightly excentric, and that both these planets, in fact, describe ellipses, having the sun in their common focus.

(413.) Let us now consider the superior planets, or those whose orbits enclose on all sides that of the earth. That they do so is proved by several circumstances: — 1st, They are not, like the inferior planets, confined to certain limits of elongation from the sun, but appear at all distances from it, even in the opposite quarter of the heavens, or, as it is called, in *opposition;* which could not happen, did not the earth at such times place itself between them and the sun: 2dly, They never appear horned, like Venus or Mercury, nor even semilunar. Those, on the contrary, which, from the minuteness of their parallax, we conclude to be the most distant from us, viz. Jupiter, Saturn, and Uranus, never appear otherwise than round; a sufficient proof, of itself, that we see them always in a direction not very remote from that in which the sun's rays illuminate them; and that, therefore, we occupy a station which is never very widely removed from the center of their orbits, or, in other words, that the earth's orbit is entirely enclosed within theirs, and of comparatively small diameter. One only of them, Mars, exhibits any perceptible *phase,* and in its deficiency from a circular outline, never surpasses a moderately *gibbous* appearance,—the enlightened portion of the disc being never less than seven-eighths of the whole. To understand this, we need only cast our eyes on the annexed figure, in which E is the earth, at its apparent greatest elongation from the sun S, as seen from Mars, M. In this position, the angle S M E, included between the lines S M and E M, is at its maximum; and, therefore, in this state of things, a spectator on the earth is enabled to see a greater portion of the dark hemisphere of Mars than in any other situation. The extent of the phase, then, or greatest observable degree of gibbosity, affords a measure—a sure, although a coarse and rude one—of the angle S M E, and therefore of the proportion of the distance S M, of Mars, to S E, that of the earth from the sun, by which it appears that the diameter of the orbit of Mars cannot be less than $1\frac{1}{2}$ that of the earth's. The phases of Jupiter, Saturn, and Uranus being imperceptible, it follows that their orbits must include not only that of the earth, but of Mars also.

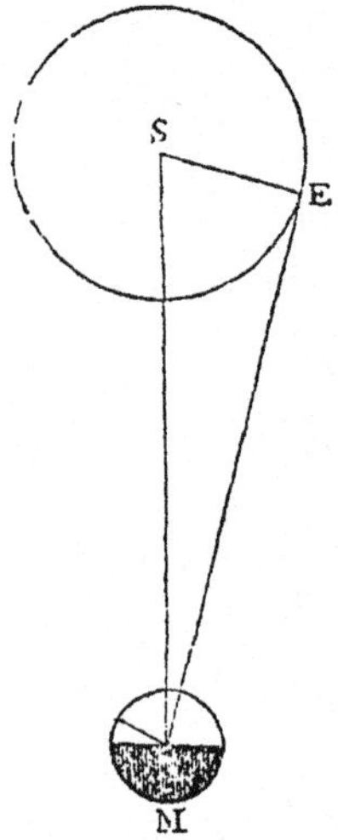

(414.) All the superior planets are retrograde in their apparent motions when in *opposition,* and for some time before and after; but they differ greatly from each other, both in the extent of their arc of retrogradation, in the duration of their retrograde movement, and in its rapidity when swiftest. It is more extensive and rapid in the case of Mars than of Jupiter, of Jupiter than of Saturn, and of that planet than Uranus. The angular velocity with which a planet appears to retrograde is easily ascertained by observing its apparent place in the heavens from day to day; and from such observations, made about the time of opposition, it is easy to conclude the relative magnitudes of their orbits as compared with the earth's, supposing their periodic times known. For, from these, their mean angular velocities are known also, being inversely as the times. Suppose, then, E *e* to be a very small portion of the earth's orbit, and M *m* a corresponding portion of that

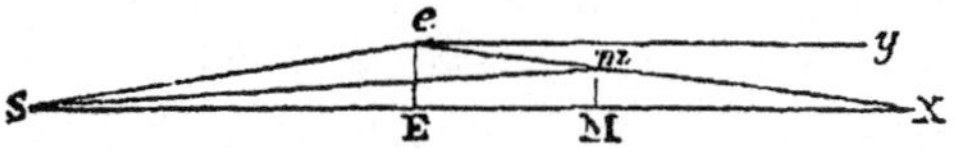

of a superior planet, described on the day of opposition, about the sun S, on which day the three bodies lie in one straight line S E M X. Then the angles E S *e* and M S *m* are given. Now, if *e m* be joined and prolonged to meet S M continued in X, the angle *e* X E, which is equal to the alternate angle X *e y*, is evidently the retrogradation of Mars on that day, and is, therefore, also given. E *e*, therefore, and the angle E X *e*, being given in the right-angled triangle E *e* X, the side E X is easily calculated, and thus S X becomes known. Consequently, in the triangle S *m* X, we have given the side S X and the two angles *m* S X and *m* X S, whence the other sides, S *m*, *m* X, are easily determined. Now, S *m* is no other than the radius of the orbit of the superior planet required, which in this calculation is supposed circular as well as that of the earth; a supposition not exact, but sufficiently so to afford a satisfactory approximation to the dimensions of its orbit, and which, if the process be often repeated, in every variety of situation at which the opposition can occur, will ultimately afford an average or mean value of its diameter fully to be depended upon.

(415.) To apply this principle, however, to practice, it is necessary to know the periodic times of the several planets. These may be obtained directly, as has been already stated, by observing the intervals of their passages through the ecliptic; but, owing to the very small inclination of the orbits of some of them to its plane, they cross it so obliquely that the precise moment of their arrival on it is not ascertainable, unless by very nice observations. A better method consists in determining, from the observations of several successive days, the exact moments of their arriving *in opposition* with the sun, the criterion of which is a difference of longitudes between the sun and planet of exactly 180°. The interval between successive oppositions thus obtained is nearly one *synodical* period; and would be exactly so, were the planet's orbit and that of the earth both circles, and uniformly described; but as that is found not to be the case (and the criterion is, the *inequality* of successive synodical revolutions so observed), the average of a great number, taken in all varieties of situation in which the oppositions occur, will be freed from the elliptic inequality, and may be taken as a *mean synodical* period. From this, by the considerations employed in art. 353., and by the process of calculation indicated in the note to that article, the sidereal periods are readily obtained. The accuracy of this determination will, of course, be greatly increased by embracing a long interval between the extreme observations employed. In point of fact, that interval extends to nearly 2000 years in the cases of the planets known to the ancients, who have recorded their observations of them in a manner sufficiently careful to be made use of. Their periods may, therefore, be regarded as ascertained with the utmost exactness. Their numerical values will be found stated, as well as the mean distances, and all the other elements of the planetary orbits, in the synoptic table at the end of the volume, to which (to avoid repetition) the reader is once for all referred.

(416.) In casting our eyes down the list of the planetary distances, and comparing them with the periodic times, we cannot but be struck with a certain correspondence. The greater the distance, or the larger the orbit, evidently the longer the period. The order of the planets, beginning from the sun, is the same, whether we arrange them according to their distances, or to the time they occupy in completing their revolutions; and is as follows:—Mercury, Venus, Earth, Mars,—the four ultra-zodiacal planets,—Jupiter, Saturn, and Uranus. Nevertheless, when we come to examine the numbers expressing them, we find that the relation between the two series is not that of simple *proportional* increase. The periods increase more than in proportion to the distances. Thus, the period of Mercury is about 88 days, and that of the Earth 365—being in proportion as 1 to 4·15, while their distances are in the less proportion of 1 to 2·56; and a similar remark holds good in every instance. Still, the ratio of increase of the times is not so rapid as that of the *squares* of the distances. The square of 2·56 is 6.5536, which is considerably greater than 4·15. An intermediate rate of increase, between the simple proportion of the distances and that of their squares, is therefore clearly pointed out by the sequence of the numbers; but it required no ordinary penetration in the illustrious Kepler, backed by uncommon perseverance and industry, at a period when the data themselves were involved in obscurity, and when the processes of trigonometry and of numerical calculation were encumbered with difficulties, of which the more recent invention of logarithmic tables has happily left us no conception, to perceive and demonstrate the real law of their connection. This connection is expressed in the following proposition:—"The squares of the periodic times of any two planets are to each other, in the same proportion as the cubes of their mean distances from the sun." Take, for example, the earth and Mars*, whose periods are in the proportion of 3652564 to 6869796, and whose distances from the sun is that of 100000 to

* The expression of this law of Kepler requires a slight modification when we come to the extreme nicety of numerical calculation, for the greater planets, due to the influence of their masses. This correction is imperceptible for the earth and Mars.

152369; and it will be found, by any one who will take the trouble to go through the calculation, that—

$$(3652564)^2 : (6869796)^2 :: (100000)^3 : (152369)^3.$$

(417.) Of all the laws to which induction from pure observation has ever conducted man, this *third law* (as it is called) *of Kepler* may justly be regarded as the most remarkable, and the most pregnant with important consequences. When we contemplate the constituents of the planetary system from the point of view which this relation affords us, it is no longer mere analogy which strikes us—no longer a general resemblance among them, as individuals independent of each other, and circulating about the sun, each according to its own peculiar nature, and connected with it by its own peculiar tie. The resemblance is now perceived to be a true *family* likeness; they are bound up in one chain—interwoven in one web of mutual relation and harmonious agreement—subjected to one pervading influence, which extends from the center to the farthest limits of that great system, of which all of them, the earth included, must henceforth be regarded as members.

(418.) The laws of elliptic motion about the sun as a focus, and of the equable description of areas by lines joining the sun and planets, were originally established by Kepler, from a consideration of the observed motions of Mars; and were by him extended, analogically, to all the other planets. However precarious such an extension might then have appeared, modern astronomy has completely verified it as a matter of fact, by the general coincidence of its results with entire series of observations of the apparent places of the planets. These are found to accord satisfactorily with the assumption of a particular ellipse for each planet, whose magnitude, degree of excentricity, and situation in space, are numerically assigned in the synoptic table before referred to. It is true, that when observations are carried to a high degree of precision, and when each planet is traced through many successive revolutions, and its history carried back, by the aid of calculations founded on these data, for many centuries, we learn to regard the laws of Kepler as only *first approximations* to the much more complicated ones which actually prevail; and that to bring remote observations into rigorous and mathematical accordance with each other, and at the same time to retain the extremely convenient nomenclature and relations of the ELLIPTIC SYSTEM, it becomes necessary to modify, to a certain extent, our verbal expression of the laws, and to regard the numerical data or *elliptic elements* of the planetary orbits as not absolutely permanent, but subject to a series of extremely slow and almost imperceptible changes. These changes may be neglected when we consider only a few revolutions; but going on from century to century, and continually accumulating, they at length produce considerable departures in the orbits from their original state. Their explanation will form the subject of a subsequent chapter; but for the present we must lay them out of consideration, as of an order too minute to affect the general conclusions with which we are now concerned. By what means astronomers are enabled to compare the results of the elliptic theory with observation, and thus satisfy themselves of its accordance with nature, will be explained presently.

(419.) It will first, however, be proper to point out what particular theoretical conclusion is involved in each of the three laws of Kepler, considered as satisfactorily established,—what indication each of them, separately, affords of the mechanical forces prevalent in our system, and the mode in which its parts are connected,—and how, when thus considered, they constitute the basis on which the Newtonian explanation of the mechanism of the heavens is mainly supported. To begin with the first law, that of the equable description of areas.—Since the planets move in curvilinear paths, they *must* (if they be bodies obeying the laws of dynamics) be deflected from their otherwise natural rectilinear progress *by force*. And from this law, taken as a matter of observed fact, it follows, that the *direction* of such force, at every point of the orbit of each planet, always *passes through the sun.* No matter from what ultimate cause the power which is called gravitation originates,—be it a virtue lodged in the sun as its receptacle, or be it pressure from without, or the resultant of many pressures or sollicitations of unknown fluids, magnetic or electric ethers, or impulses,—still, when finally brought under our contemplation, and summed up into a single resultant energy—its *direction* is, *from* every point an all sides, *towards the sun's center.* As an abstract dynamical proposition, the reader will find it demonstrated by Newton, in the 1st proposition of the *Principia,* with an elementary simplicity to which we really could add nothing but obscurity by amplification, that any body, urged towards a certain central point by a force continually directed thereto, and thereby deflected into a curvilinear path, will describe about that center equal areas in equal times; and *vice versâ,* that such equable description of areas is itself the essential criterion of a continual direction of the acting force towards the center to which this character belongs. The first law of Kepler, then, gives us no information as to the nature or intensity of the force urging the planets to the sun; the only conclusion it involves, is that it does so urge them. It is a property of orbitual rotation under the influence of central forces *generally,* and, as such, we daily see it exemplified in a thousand familiar instances. A simple experimental illustration of it is to tie a bullet to a thin string, and, having whirled it round with a moderate velocity in a vertical plane, to draw the end of the string through a small ring, or allow it to coil itself round the finger, or a cylindrical rod held very firmly in a horizontal position. The bullet will then approach the center of motion in a spiral line; and the increase not only of its angular but of its linear velocity, and the rapid diminution of its periodic time when near the center, will express, more clearly than any words, the compensation by which its uniform description of areas is maintained under a constantly diminishing distance. If the motion be reversed, and the thread allowed to uncoil, beginning with a rapid impulse, the velocity will diminish by the same degrees as it before increased. The increasing rapidity of a dancer's *pirouette*, as he draws in his limbs and straightens his whole person, so as to bring every part of his frame as near as possible to the axis of his motion, is another instance where the connection of the observed effect with the central force exerted, though equally real, is much less obvious.

(420.) The second law of Kepler, or that which asserts that the planets describe ellipses about the sun as their focus, involves, as a consequence, the *law* of solar gravitation (so be it allowed to call the force, whatever it be, which urges them towards the sun) as exerted on each individual planet, apart from all connection with the rest. A straight line, dynamically speaking, is the only path which can be pursued by a body *absolutely free*, and under the action of *no* external force. All *deflection* into a curve is evidence of the exertion of a force; and the greater the deflection in equal times, the more intense the force. Deflection from a straight line is only another word for *curvature* of path; and as a circle is characterized by the uniformity of its curvature in all its parts—so is every other curve (as an ellipse) characterized by the particular *law* which regulates the increase and diminution of its curvature as we advance along its circumference. The deflecting force, then, which continually bends a moving body into a curve, may be ascertained, provided its direction, in the first place, and, secondly, the law of curvature of the curve itself, be known. Both these enter as elements into the expression of the force. A body may describe, for instance, an ellipse, under a great variety of dispositions of the acting forces: it may glide along it, for example, as a bead upon a polished wire, bent into an elliptic form; in which case the acting force is always perpendicular to the wire, and the velocity is uniform. In this case the *force* is directed to *no fixed* center, and there is no equable description of areas at all. Or it may describe it as we may see done, if we suspend a ball by a *very* long string, and, drawing it a little aside from the perpendicular, throw it *round* with a gentle impulse. In this case the acting force is directed to the center of the ellipse, about which areas are described equably, and *to* which a force *proportional* to the distance (the decomposed result of terrestrial gravity) perpetually urges it. This is at once a very easy experiment, and a very instructive one, and we shall again refer to it. In the case before us, of an ellipse described by the action of a force directed to the *focus,* the steps of the investigation of the law of force are these: 1st, The law of the areas determines the actual *velocity* of the revolving body at every point, or the space really run over by it in a given minute portion of time; 2dly, The law of curvature of the ellipse determines the linear amount of deflection from the tangent *in the direction of the focus,* which corresponds to that space so run over; 3dly, and lastly, The laws of accelerated motion declare that the intensity of the acting force causing such deflection *in its own direction,* is measured by or proportional to the amount of that deflection, and may therefore be calculated in any particular position, or generally expressed by geometrical or algebraic symbols, *as a law* independent of particular positions, when that deflection is so calculated or expressed. We have here the spirit of the process by which Newton has resolved this interesting problem. For its geometrical detail, we must refer to the 3d section of his *Principia.* We know of no artificial mode of imitating this species of elliptic motion; though a rude approximation to it—enough, however, to give a conception of the alternate approach and recess of the revolving body to and from the focus, and the variation of its velocity—may be had by suspending a small steel bead to a fine and very long silk fibre, and setting it to revolve in a small orbit round the pole of a powerful cylindrical magnet, held upright, and vertically under the point of suspension.

(421.) The third law of Kepler, which connects the distances and periods of the planets by a general rule, bears with it, as its theoretical interpretation, this important consequence, viz. that it is one and the same force, modified only by distance from the sun, which retains *all* the planets in their orbits about it. That the attraction of the sun (if such it be) is exerted upon all the bodies of our system indifferently, without regard to the peculiar materials of which they may consist, in the exact proportion of their inertiæ, or quantities of matter; that it is not, therefore, of the nature of the elective attractions of chemistry, or of magnetic action, which is powerless on other substances than iron and some one or two more, but is of a more universal character, and extends equally to all the material constituents of our system, and (as we shall hereafter see abundant reason to admit) to those of other systems than our own. This law, important and general as it is, results, as the simplest of corollaries, from the relations established by Newton in the section of the *Principia* referred to (Prop. xv.), from which proposition it results, that if the earth were taken from its actual orbit, and launched anew in space at the place, in the direction, and with the velocity of any of the other planets, it would describe the very same orbit, and in the same period, which that planet actually does, a very minute correction of the period only excepted, arising from the difference between the mass of the earth and that of the planet. Small as the planets are compared to the sun, some of them are not, as the earth is, mere atoms in the comparison. The strict wording of Kepler's law, as Newton has proved in his fifty-ninth proposition, is applicable only to the case of planets whose proportion to the central body is absolutely inappreciable. When this is not the case, the periodic time is shortened in the proportion of the square root of the number expressing the sun's mass or inertia, to that of the sum of the numbers expressing the masses of the sun and planet; and in general, whatever be the masses of two bodies revolving round each other under the influence of the Newtonian law of gravity, the square of their periodic time will be expressed by a fraction whose numerator is the cube of their mean distance, *i. e.* the greater semi-axis of their elliptic orbit, and whose denominator is the sum of their masses. When one of the masses is incomparably greater than the other, this resolves itself into Kepler's law; but when this is not the case, the proposition thus generalized stands in lieu of that law. In the system of the sun and planets, however, the numerical correction thus introduced into the results of Kepler's law is too small to be of any importance, the mass of the largest of the planets (Jupiter) being much less than a thousandth part of that of the sun. We shall presently, however, perceive all the importance of this generalization, when we come to speak of the satellites.

(422.) It will first, however, be proper to explain by what process of calculation the expression of a planet's elliptic orbit by its *elements* can be compared with observation, and how we can satisfy ourselves that the numerical data contained in a table of such elements for the whole system does really exhibit a true picture of it, and afford the means of determining its state at every instant of time, by the mere application of Kepler's laws. Now, for each planet, it is necessary for this purpose to know, 1st, the magnitude and form of its ellipse; 2dly, the situation of this ellipse in space, with respect to the ecliptic, and to a fixed line drawn therein; 3dly, the local situation of the planet in its ellipse at some known epoch, and its periodic time or mean angular velocity, or, as it is called, its mean motion.

(423.) The magnitude and form of an ellipse are determined by its greatest length and least breadth, or its two principal axes; but for astronomical uses it is preferable to use the semi-axis major (or half the greatest length), and the excentricity or distance of the focus from the center, which last is usually estimated in parts of the former. Thus, an ellipse, whose length is 10 and breadth 8 parts of any scale, has for its major semi-axis 5, and for its excentricity 3 such parts; but when estimated in parts of the semi-axis, regarded as a unit, the excentricity is expressed by the fraction $\frac{3}{5}$.

(424.) The ecliptic is the plane to which an inhabitant of the earth most naturally refers the rest of the solar system, as a sort of ground-plane; and the axis of its orbit might be taken for a line of departure in that plane or origin of angular reckoning. Were the axis *fixed*, this would be the best possible origin of longitudes; but as it has a motion (though an excessively slow one), there is, in fact, no advantage in reckoning from the axis more than from the line of the equinoxes, and astronomers therefore prefer the latter, taking account of its variation by the effect of precession, and restoring it, by calculation at every instant, to a fixed position. Now, to determine the situation of the ellipse described by a planet with respect to this plane, three *elements* require to be known: — 1st, the *inclination* of the plane of the planet's orbit to the plane of the ecliptic; 2dly, the line in which these two planes intersect each other, which of necessity passes through the sun, and whose position with respect to the line of the equinoxes is therefore given by stating its longitude. This line is called the line of the nodes. When the planet is in this line, in the act of passing from the south to the north side of the ecliptic, it is *in its ascending node*, and its longitude at that moment is the element called the *songitude of the node.* These two data determine the *l*ituation of *the plane* of the orbit; and there only remains, for the complete determination of the situation of the planet's ellipse, to know how it is placed *in* that plane, which (since its focus is necessarily in the sun) is ascertained by stating the *longitude of its perihelion*, or the place which the extremity of the axis nearest the sun occupies, when orthographically projected on the ecliptic.

(425.) The dimensions and situation of the planet's

orbit thus determined, it only remains, for a complete acquaintance with its history, to determine the circumstances of its motion in the orbit so precisely fixed. Now, for this purpose, all that is needed is to know the moment of time when it is either at the perihelion, or at any other precisely determined point of its orbit, and its whole period; for these being known, the law of the areas determines the place at every other instant. This moment is called (when the perihelion is the point chosen) the *perihelion passage*, or, when some point of the orbit is fixed upon, without special reference to the perihelion, the *epoch*.

(426.) Thus, then, we have *seven* particulars or elements, which must be numerically stated, before we can reduce to calculation the state of the system at any given moment. But, these known, it is easy to ascertain the apparent positions of each planet, as it would be seen from the sun, or is seen from the earth at any time. The former is called the *heliocentric*, the latter the *geocentric*, place of the planet.

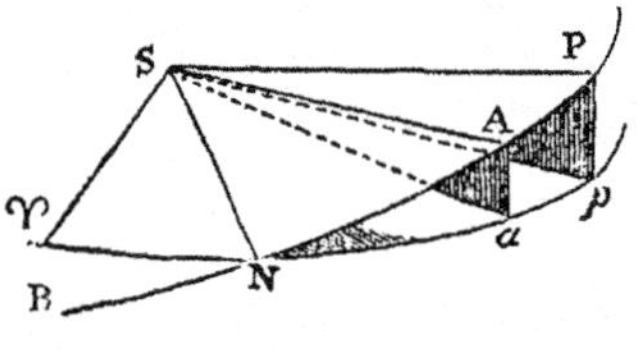

(427.) To commence with the heliocentric places. Let S represent the sun; A P N the orbit of the planet, being an ellipse, having the sun S in its focus, and A for its perihelion; and let *p a* N ♈ represent the projection of the orbit on the plane of the ecliptic, intersecting the line of equinoxes S ♈ in ♈, which, therefore, is the origin of longitudes. Then will S N be the line of nodes; and if we suppose B to lie on the south, and A on the north side of the ecliptic, and the direction of the planet's motion to be from B to A, N will be the ascending node, and the angle ♈ S N the *longitude of the node*. In like manner, if P be the place of the planet at any time, and if it and the perihelion A be projected on the ecliptic, upon the points *p a*, the angles ♈ S *p*, ♈ S *a*, will be the respective heliocentric longitudes of the planet and of the perihelion, the former of which is to be determined, and the latter is one of the given elements. Lastly, the angle *p* S P is the heliocentric latitude of the planet which is also required to be known.

(428.) Now, the time being given, and also the moment of the planet's passing the perihelion, the interval, or the time of describing the portion A P of the orbit, is given, and the periodical time, and the whole area of the ellipse being known, the law of proportionality of areas to the times of their description gives the magnitude of the area A S P. From this it is a problem of pure geometry to determine the corresponding *angle* A S P, which is called the planet's *true anomaly*. This problem is of the kind called transcendental, and has been resolved by a great variety of processes, some more, some less intricate. It offers, however, no peculiar difficulty, and is practically resolved with great facility by the help of tables constructed for the purpose, adapted to the case of each particular planet.*

(429.) The true anomaly thus obtained, the planet's angular distance from the node, or the angle N S P, is to be found. Now, the longitudes of the perihelion and node being respectively ♈ *a* and ♈ N, which are given, their difference *a* N is also given, and the angle N of the spherical right-angled triangle A N *a*, being the inclination of the plane of the orbit to the ecliptic, is known. Hence we calculate the arc N A, or the angle N S A, which, added to A S P, gives the angle N S P required. And from this, regarded as the measure of

* It will readily be understood, that, except in the case of uniform circular motion, an equable description of *areas* about any center is incompatible with an equable description of *angles*. The object of the problem in the text is to pass from *the area*, supposed known, to the *angle*, supposed unknown: in other words, to derive the true amount of angular motion from the perihelion, or the *true anomaly* from what is technically called the mean anomaly, that is, the mean angular motion which would have been performed had the motion *in angle* been uniform instead of the motion *in area*. It happens, fortunately, that this is the simplest of all problems of the transcendental kind, and can be resolved, in the most difficult case, by the rule of "false position," or trial and error, in a very few minutes. Nay, it may even be resolved instantly on inspection by a simple and easily constructed piece of mechanism, of which the reader may see a description in the Cambridge Philosophical Transactions, vol. iv. p. 425., by the author of this work.

the arc N P, forming the hypothenuse of the right-angled spherical triangle P N *p*, whose angle N, as before, is known, it is easy to obtain the other two sides, N *p* and P *p*. The latter, being the measure of the angle *p* S P, expresses the planet's heliocentric latitude; the former measures the angle N S *p*, or the planet's distance in longitude from its node, which, added to the known angle ♈ S N, the longitude of the node, gives the heliocentric longitude. This process, however circuitous it may appear, when once well understood, may be gone through numerically, by the aid of the usual logarithmic and trigonometrical tables, in little more time than it will have taken the reader to peruse its description.

(430.) The geocentric differs from the heliocentric place of a planet by reason of that parallactic change of apparent situation which arises from the earth's motion in its orbit. Were the planets' distance as vast as those of the stars, the earth's orbitual motion would be insensible when viewed from them, and they would always appear to us to hold the same relative situations among the fixed stars, as if viewed from the sun, *i. e.* they would then be seen in their *heliocentric* places. The difference, then, between the heliocentric and geocentric places of a planet is, in fact, the same thing with its *parallax* arising from the earth's removal from the center of the system and its annual motion. It follows from this, that the first step towards a knowledge of its amount, and the consequent determination of the apparent place of each planet, as referred from the earth to the sphere of the fixed stars, must be to ascertain the proportion of its linear distances from the earth and from the sun, as compared with the earth's distance from the sun, and the angular positions of all three with respect to each other.

(431.) Suppose, therefore, S to represent the sun, E the earth, and P the planet; S ♈ the line of equinoxes, ♈ E the earth's orbit, and P *p* a perpendicular let fall from the planet on the ecliptic. Then will the angle

S P E (according to the general notion of parallax conveyed in art. 69.) represent the parallax of the planet arising from the change of station from S to E, E P will be the apparent direction of the planet seen from E; and if S Q be drawn parallel to E *p*, the angle ♈ S Q will be the geocentric longitude of the planet, while ♈ S E represents the heliocentric longitude of the earth, and ♈ S *p* that of the planet. The former of these, ♈ S E, is given by the solar tables; the latter, ♈ S *p* is found by the process above described (art. 429.). Moreover, S P is the radius vector of the planet's orbit, and S E that of the earth's, both of which are determined from the known dimensions of their respective ellipses, and the places of the bodies in them at the assigned time. Lastly, the angle P S *p* is the planet's heliocentric latitude.

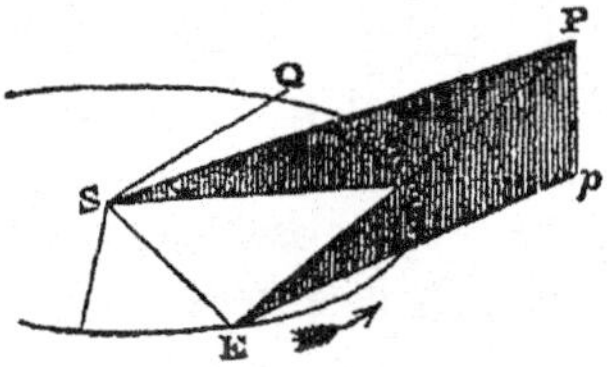

(432.) Our object, then, is, from all these data, to determine the angle ♈ S Q and P E *p*, which is the geocentric latitude. The process, then, will stand as follows:—1st, In the triangle S P *p*, right-angled at P, given S P, and the angle P S *p* (the planet's radius vector and heliocentric latitude), find S *p*, and P *p*; 2dly, In the triangle S E *p*, given S *p* (just found), S E (the earth's radius vector), and the angle E S *p* (the difference of heliocentric longitudes of the earth and planet), find the angle S *p* E, and the side E *p*. The former being equal to the alternate angle *p* S Q, is the parallactic removal of the planet in longitude, which, added to ♈ S *p*, gives its heliocentric longitude. The latter, E *p* (which is called the *curtate distance* of the planet from the earth, gives at once the geocentric latitude, by means of the right-angled triangle P E *p*, of which E *p* and P *p* are known sides, and the angle P E *p* is the longitude sought.

(433.) The calculations required for these purposes are nothing but the most ordinary processes of plane trigonometry; and, though somewhat tedious, are neither intricate nor difficult. When executed, however, they afford us the means of comparing the places of the planets actually observed with the elliptic theory, with the utmost exactness, and thus putting it to the severest trial; and it is upon the testimony of such computations, so brought into comparison with observed facts, that we declare that theory to be a true representation of nature.

(434.) The planets Mercury, Venus, Mars, Jupiter, and Saturn, have been known from the earliest ages in which astronomy has been cultivated. Uranus was discovered by Sir W. Herschel in 1781, March 13., in the course of a review of the heavens, in which every star visible in a telescope of a certain power was brought under close examination, when the new planet was immediately detected by its disc, under a high magnifying power. It has since been ascertained to have been observed on many previous occasions, with telescopes of insufficient power to show its disc, and even entered in catalogues as a star; and some of the observations which have been so recorded have been used to improve and extend our knowledge of its orbit. The discovery of the ultra-zodiacal planets dates from the first day of 1801, when Ceres was discovered by Piazzi, at Palermo; a discovery speedily followed by those of Juno by professor Harding, of Göttingen; and of Pallas and Vesta, by Dr. Olbers, of Bremen. It is extremely remarkable that this important addition to our system had been in some sort surmised as a thing not unlikely, on the ground that the intervals between the planetary orbits go on doubling as we recede from the sun, or nearly so. Thus, the interval between the orbits of the earth and Venus is nearly twice that between those of Venus and Mercury; that between the orbits of Mars and the earth nearly twice that between the earth and Venus; and so on. The interval between the orbits of Jupiter and Mars, however, is too great, and would form an exception to this law, which is, however, again resumed in the case of the three remoter planets. It was, therefore, thrown out, by the late professor Bode of Berlin, as a possible surmise, that a planet might exist between Mars and Jupiter; and it may easily be imagined what was the astonishment of astronomers to find four, revolving in orbits tolerably well corresponding with the law in question. No account, *à priori*, or from theory, can be given of this singular progression, which is not, like Kepler's laws, strictly exact in its numerical verification; but the circumstances we have just mentioned lead to a strong belief that it is something beyond a mere accidental coincidence, and belongs to the essential structure of the system. It has been conjectured that the ultra-zodiacal planets are fragments of some greater planet, which formerly circulated in that interval, but has been blown to atoms by an explosion; and that more such fragments exist, and may be hereafter discovered. This may serve as a specimen of the dreams in which astronomers, like other speculators, occasionally and harmlessly indulge.

(435.) We shall devote the rest of this chapter to an account of the physical peculiarities and probable condition of the several planets, so far as the former are known by observation, or the latter rest on probable grounds of conjecture. In this, three features principally strike us, as necessarily productive of extraordinary diversity in the provisions by which, if they be, like our earth, inhabited, animal life must be supported. There are, first, the difference in their respective supplies of light and heat from the sun; secondly, the difference in the intensities of the gravitating forces which must subsist at their surfaces, or the different ratios which, on their several globes, the *inertiæ* of bodies must bear to their *weights*; and, thirdly, the difference in the nature of the materials of which, from what we know of their mean density, we have every reason to believe they consist. The intensity of solar radiation is nearly seven times greater on Mercury than on the earth, and on Uranus 330 times less; the proportion between the two extremes being that of upwards of 2000 to one. Let any one figure to himself the condition of our globe, were the sun to be septupled, to say nothing of the greater ratio! or were it diminished to a seventh, or to a 300th of its actual power! Again, the intensity of gravity, or its efficacy in counteracting muscular power and repressing animal activity on Jupiter is nearly three times that on the Earth, on Mars not more than one third, on the Moon one sixth, and on the four smaller planets probably not more than one twentieth; giving a scale of which the extremes are in the proportion of sixty to one. Lastly, the density of Saturn hardly exceeds one eighth of the mean density of the earth, so that it must consist of materials not much heavier than cork. Now, under the various combinations of elements so important to life as these, what immense diversity must we not admit in the conditions of that great problem, the maintenance of animal and intellectual existence and happiness, which seems, so far as we can judge by what we see around us in our own planet, and by the way in which every corner of it is crowded with living beings, to form an unceasing and worthy object for the exercise of the Benevolence and Wisdom which presides over all!

(436.) Quitting, however, the region of mere speculation, we will now show what information the telescope affords us of the actual condition of the several planets within its reach. Of Mercury we can see little more than that it is round, and exhibits phases. It is too small, and too much lost in the constant neighbourhood of the Sun, to allow us to make out more of its nature. The real diameter of Mercury is about 3200 miles: its apparent diameter varies from 5″ to 12″. Nor does Venus offer any remarkable peculiarities: although its real diameter is 7800 miles, and although it occasionally attains the considerable apparent diameter of 61″, which is larger than that of any other planet, it is yet the most difficult of them all to define with telescopes. The intense lustre of its illuminated part dazzles the sight, and exaggerates every imperfection of the telescope; yet we see clearly that its surface is not mottled over with permanent spots like the moon; we perceive in it neither mountains nor shadows, but a uniform brightness, in which sometimes we may, indeed, fancy obscurer portions, but can seldom or never rest fully satisfied of the fact. It is from some observations of this kind that both Venus and Mercury have been concluded to revolve on their axes in about the same time as the Earth. The most natural conclusion, from the very rare appearance and want of permanence in the spots, is, that we do not see, as in the Moon, the real surface of these planets, but only their atmospheres, much loaded with clouds, and which may serve to mitigate the otherwise intense glare of their sunshine.

(437.) The case is very different with Mars. In this planet we discern, with perfect distinctness, the outlines of what may be continents and seas. (See Plate I. *fig.* 1., which represents Mars in its gibbous state, as seen on the 16th of August, 1830, in the 20-feet reflector at Slough.) Of these, the former are distinguished by that ruddy colour which characterizes the light of this planet (which always appears red and fiery), and indicates, no doubt, an ochrey tinge in the general soil, like what the red sandstone districts on the Earth may possibly offer to the inhabitants of Mars, only more decided. Contrasted with this (by a general law in optics), the seas, as we may call them, appear greenish.* These spots, however, are not always to be seen equally distinct, though, *when seen*, they offer always the same appearance. This may arise from the planet not being entirely destitute of atmosphere and clouds †; and what adds greatly to the probability of this is the appearance of brilliant white spots at its poles, — one of which appears in our figure, —which have been conjectured with a great deal of probability to be snow; as they disap-

* I have noticed the phænomena described in the text on many occasions, but never more distinct than on the occasion when the drawing was made from which the figure in Plate I. is engraved. — *Author.*

† It has been surmised to have a very extensive atmosphere, but on no sufficient or even plausible grounds.

pear when they have been long exposed to the sun, and are greatest when just emerging from the long night of their polar winter. By watching the spots during a whole night, and on successive nights, it is found that Mars has a rotation on an axis inclined about 30° 18′ to the ecliptic, and in a period of $24^h\ 39^m\ 21^s$ in the same direction as the earth's, or from west to east. The greatest and least apparent diameters of Mars are 4″ and 18″, and its real diameter about 4100 miles.

(438.) We come now to a much more magnificent planet, Jupiter, the largest of them all, being in diameter no less than 87,000 miles, and in bulk exceeding that of the Earth nearly 1300 times. It is, moreover, dignified by the attendance of four *moons, satellites,* or *secondary planets,* as they are called, which constantly accompany and revolve about it, as the moon does round the earth, and in the same direction, forming with their principal, or *primary,* a beautiful miniature system, entirely analogous to that greater one of which their central body is itself a member, obeying the same laws, and exemplifying, in the most striking and instructive manner, the prevalence of the gravitating power as the ruling principle of their motions: of these, however, we shall speak more at large in the next chapter.

(439.) The disc of Jupiter is always observed to be crossed in one certain direction by dark bands or belts, presenting the appearance, in Plate I. *fig.* 2., which represents this planet as seen on the 23d of September, 1832, in the 20-feet reflector at Slough. These belts are, however, by no means alike at all times; they vary in breadth and in situation on the disc (though never in their general direction). They have even been seen broken up, and distributed over the whole face of the planet; but this phænomenon is extremely rare. Branches running out from them, and subdivisions, as represented in the figure, as well as evident dark spots, like strings of clouds, are by no means uncommon; and from these, attentively watched, it is concluded that this planet revolves in the surprisingly short period of

$9^h\ 55^m\ 50^s$ (sid. time), on an axis perpendicular to the direction of the belts. Now, it is very remarkable, and forms a most satisfactory comment on the reasoning by which the spheroidal figure of the earth has been deduced from its diurnal rotation, that the outline of Jupiter's disc is evidently not circular, but elliptic, being considerably flattened in the direction of its axis of rotation. This appearance is no optical illusion, but is authenticated by micrometrical measures, which assign 107 to 100 for the proportion of the equatorial and polar diameters. And to confirm, in the strongest manner, the truth of those principles on which our former conclusions have been founded, and fully to authorize their extension to this remote system, it appears, on calculation, that this is really the degree of oblateness which corresponds, on those principles, to the dimensions of Jupiter, and to the time of his rotation.

(440.) The parallelism of the belts to the equator of Jupiter, their occasional variations, and the appearances of spots seen upon them, render it extremely probable that they subsist in the atmosphere of the planet, forming tracts of comparatively clear sky, determined by currents analogous to our trade-winds, but of a much more steady and decided character, as might indeed be expected from the immense velocity of its rotation. That it is the comparatively darker body of the planet which appears in the belts is evident from this,— that they do not come up in all their strength to the edge of the disc, but fade away gradually before they reach it. (See Plate I. *fig.* 2.) The apparent diameter of Jupiter varies from $30''$ to $46''$.

(441.) A still more wonderful, and, as it may be termed, elaborately artificial mechanism, is displayed in Saturn, the next in order of remoteness to Jupiter, to which it is not much inferior in magnitude, being about 79,000 miles in diameter, nearly 1000 times exceeding the earth in bulk, and subtending an apparent angular diameter at the earth, of about $16''$. This stupendous globe, besides being attended by no less than seven satellites, or moons is surrounded with two broad flat, extremely thin rings, concentric with the planet and with each other; both lying in one plane, and separated by a very narrow interval from each other throughout their whole circumference, as they are from the planet by a much wider. The dimensions of this extraordinary appendage are as follows *:—

		Miles.
Exterior diameter of exterior ring	=	176418.
Interior ditto - -	=	155272.
Exterior diameter of interior ring	=	151690.
Interior ditto - -	=	117339.
Equatorial diameter of the body	=	79160.
Interval between the planet and interior ring	=	19090.
Interval of the rings - -	=	1791.
Thickness of the rings not exceeding	=	100.

The figure (*fig.* 3. Plate I.) represents Saturn surrounded by its rings, and having its body striped with dark belts, somewhat similar, but broader and less strongly marked than those of Jupiter, and owing, doubtless, to a similar cause. That the ring is a solid opake substance is shown by its throwing its shadow on the body of the planet, on the side nearest the sun, and on the other side receiving that of the body, as shown in the figure. From the parallelism of the belts with the plane of the ring, it may be conjectured that the axis of rotation of the planet is perpendicular to that plane; and this conjecture is confirmed by the occasional appearance of extensive dusky spots on its surface, which, when watched, like the spots on Mars or Jupiter, indicate a rotation in $10^h\ 29^m\ 17^s$ about an axis so situated.

(442.) The axis of rotation, like that of the earth, preserves its parallelism to itself during the motion of the planet in its orbit; and the same is also the case with the ring, whose plane is constantly inclined at the

* These dimensions are calculated from Prof. Struve's micrometric measures, Mem. Ast. Soc. iii. 301. with the exception of the thickness of the ring, which is concluded from my own observations, during its gradual extinction now in progress. The interval of the rings here stated is possibly somewhat too small.

same, or very nearly the same, angle to that of the orbit, and, therefore, to the ecliptic, viz. 28° 40′; and intersects the latter plane in a line, which makes an angle with the line of equinoxes of 170°. So that the *nodes of the ring* lie in 170° and 350° of longitude. Whenever, then, the planet happens to be situated in one or other of these longitudes, as at A B, the plane of the ring passes through the sun, which then illuminates only the edge of it; and as, at the same moment, owing to the smallness of the earth's orbit, E, compared with that of Saturn, the earth is necessarily not far out of that plane, and must, at all events, pass through it a little before or after that moment, it only then appears to us as a very fine straight line, drawn across the disc, and projecting out on each side,—indeed, so very thin is the ring, as to be quite invisible, in this situation, to any but telescopes of extraordinary power. This remarkable phænomenon takes place at intervals of 15 years, but the disappearance of the ring is generally double, the earth passing *twice* through its plane before it is carried past our orbit by the slow motion of Saturn. This second disappearance is now in progress*. As the planet, however, recedes from these points of its orbit, the line of sight becomes gradually more and more inclined to the plane of the ring, which, according to the laws of perspective, appears to open out into an ellipse which attains its greatest breadth when the planet is 90° from either node, as at C D. Supposing the upper part of the figure to be north, and the lower south of the ecliptic, the north side only of the ring will be seen when the planet lies in the semicircle A C B, and the southern only when in A D B. At the time of the greatest opening, the longer diameter is almost exactly double the shorter.

(443.) It will naturally be asked how so stupendous an arch, if composed of solid and ponderous materials,

* The disappearance of the rings is complete, when observed with a reflector eighteen inches in aperture, and twenty feet in focal length April 29, 1833.—*Author.*

can be sustained without collapsing and falling in upon the planet? The answer to this is to be found in a swift rotation of the ring in its own plane, which observation has detected, owing to some portions of the ring being a little less bright than others, and assigned its period at $10^h\ 29^m\ 17^s$, which, from what we know of its dimensions, and of the force of gravity in the Saturnian system, is very nearly the periodic time of a satellite revolving at the same distance as the middle of its breadth. It is the centrifugal force, then, arising from this rotation, which sustains it; and, although no observation nice enough to exhibit a difference of periods between the outer and inner rings have hitherto been made, it is more than probable that such a difference does subsist as to place each independently of the other in a similar state of equilibrium.

(444.) Although the rings are, as we have said, very nearly concentric with the body of Saturn, yet recent micrometrical measurements of extreme delicacy have demonstrated that the coincidence is not mathematically exact, but that the center of gravity of the rings oscillates round that of the body describing a very minute orbit, probably under laws of much complexity. Trifling as this remark may appear, it is of the utmost importance to the stability of the system of the rings. Supposing them mathematically perfect in their circular form, and exactly concentric with the planet, it is demonstrable that they would form (in spite of their centrifugal force) a system in a state of *unstable equilibrium*, which the slightest external power would subvert — not by causing a rupture in the substance of the rings — but by precipitating them, *unbroken*, on the surface of the planet. For the attraction of such a ring or rings on a point or sphere excentrically situate within them, is not the same in all directions, but tends to draw the point or sphere towards the nearest part of the ring, or away from the center. Hence, supposing the body to become, from any cause, ever so little excentric to the ring, the tendency of their mutual gravity is, not to correct but to increase this excentricity, and to bring the nearest parts of them together. (See Chap. XI.) Now, external powers, capable of producing such excentricity, exist in the attractions of the satellites, as will be shown in Chap. XI.; and in order that the system may be *stable*, and possess within itself a power of resisting the first inroads of such a tendency, while yet nascent and feeble, and opposing them by an opposite or maintaining power, it has been shown that it is sufficient to admit the rings to be *loaded* in some part of their circumference, either by some minute inequality of thickness, or by some portions being denser than others. Such a load would give to the whole ring to which it was attached somewhat of the character of a heavy and sluggish satellite, maintaining itself in an orbit with a certain energy sufficient to overcome minute causes of disturbance, and establish an average bearing on its center. But even without supposing the existence of any such load,— of which, after all, we have no proof,— and granting, therefore, in its full extent, the general instability of the equilibrium, we think we perceive, in the periodicity of all the causes of disturbance, a sufficient guarantee of its preservation. However homely be the illustration, we can conceive nothing more apt in every way to give a general conception of this maintenance of equilibrium under a constant tendency to subversion, than the mode in which a practised hand will sustain a long pole in a perpendicular position resting on the finger by a continual and almost imperceptible variation of the point of support. Be that, however, as it may, the observed oscillation of the centers of the rings about that of the planet is in itself the evidence of a perpetual contest between conservative and destructive powers — both extremely feeble, but so antagonizing one another as to prevent the latter from ever acquiring an uncontrollable ascendancy, and rushing to a catastrophe.

(445.) This is also the place to observe, that, as the

smallest difference of velocity between the body and rings must infallibly precipitate the latter on the former, never more to separate, (for they would, once in contact, have attained a position of *stable equilibrium*, and be held together ever after by an immense force;) it follows, either that their motions in their common orbit round the sun must have been adjusted to each other by an external power, with the minutest precision, or that the rings must have been formed about the planet while subject to their common orbitual motion, and under the full and free influence of all the acting forces.

(446.) The rings of Saturn must present a magnificent spectacle from those regions of the planet which lie above their enlightened sides, as vast arches spanning the sky from horizon to horizon, and holding an invariable situation among the stars. On the other hand, in the regions beneath the dark side, a solar eclipse of fifteen years in duration, under their shadow, must afford (to our ideas) an inhospitable asylum to animated beings, ill compensated by the faint light of the satellites. But we shall do wrong to judge of the fitness or unfitness of their condition from what we see around us, when, perhaps, the very combinations which convey to our minds only images of horror, may be in reality theatres of the most striking and glorious displays of beneficent contrivance.

(447.) Of Uranus we see nothing but a small round uniformly illuminated disc, without rings, belts, or discernible spots. Its apparent diameter is about 4″, from which it never varies much, owing to the smallness of our orbit in comparison of its own. Its real diameter is about 35,000 miles, and its bulk 80 times that of the earth. It is attended by satellites — two at least, probably five or six — whose orbits (as will be seen in the next chapter) offer remarkable peculiarities.

(448.) If the immense distance of Uranus precludes all hope of coming at much knowledge of its physical state, the minuteness of the four ultra-zodiacal planets is no less a bar to any enquiry into theirs. One of them, Pallas, is said to have somewhat of a nebulous or hazy appearance, indicative of an extensive and vaporous atmosphere, little repressed and condensed by the inadequate gravity of so small a mass. No doubt the most remarkable of their peculiarities must lie in this condition of their state. A man placed on one of them would spring with ease 60 feet high, and sustain no greater shock in his descent than he does on the earth from leaping a yard. On such planets giants might exist; and those enormous animals, which on earth require the buoyant power of water to counteract their weight, might there be denizens of the land. But of such speculation there is no end.

(449.) We shall close this chapter with an illustration calculated to convey to the minds of our readers a general impression of the relative magnitudes and distances of the parts of our system. Choose any well levelled field or bowling green. On it place a globe, two feet in diameter; this will represent the Sun; Mercury will be represented by a grain of mustard seed, on the circumference of a circle 164 feet in diameter for its orbit; Venus a pea, on a circle 284 feet in diameter; the Earth also a pea, on a circle of 430 feet; Mars a rather large pin's head, on a circle of 654 feet; Juno, Ceres, Vesta, and Pallas, grains of sand, in orbits of from 1000 to 1200 feet; Jupiter a moderate-sized orange, in a circle nearly half a mile across; Saturn a small orange, on a circle of four-fifths of a mile; and Uranus a full sized cherry, or small plum, upon the circumference of a circle more than a mile and a half in diameter. As to getting correct notions on this subject by drawing circles on paper, or, still worse, from those very childish toys called orreries, it is out of the question. To imitate the motions of the planets, in the above mentioned orbits, Mercury must describe its own diameter in 41 seconds; Venus, in $4^m\ 14^s$; the Earth, in 7 minutes; Mars, in $4^m\ 48^s$; Jupiter, in $2^h\ 56^m$; Saturn, in $3^h\ 13^m$; and Uranus, in $2^h\ 16^m$.

CHAP. IX.

OF THE SATELLITES.

OF THE MOON, AS A SATELLITE OF THE EARTH. — GENERAL PROXIMITY OF SATELLITES TO THEIR PRIMARIES, AND CONSEQUENT SUBORDINATION OF THEIR MOTIONS. — MASSES OF THE PRIMARIES CONCLUDED FROM THE PERIODS OF THEIR SATELLITES. — MAINTENANCE OF KEPLER'S LAWS IN THE SECONDARY SYSTEMS. — OF JUPITER'S SATELLITES. — THEIR ECLIPSES, ETC. — VELOCITY OF LIGHT DISCOVERED BY THEIR MEANS. — SATELLITES OF SATURN — OF URANUS.

(450.) In the annual circuit of the earth about the sun, it is constantly attended by its satellite, the moon, which revolves round it, or rather both round their common center of gravity; while this center, strictly speaking, and not either of the two bodies thus connected, moves in an elliptic orbit, undisturbed by their mutual action, just as the center of gravity of a large and small stone tied together and flung into the air describes a parabola as if it were a real material substance under the earth's attraction, while the stones circulate round it or round each other, as we choose to conceive the matter.

(451.) If we trace, therefore, the *real* curve actually described by either the moon's or earth's centers, in virtue of this compound motion, it will appear to be, not an exact ellipse, but an undulated curve, like that represented in the figure to article 272., only that the number of undulations in a whole revolution is but 13, and their actual deviation from the general ellipse, which serves them as a central line, is comparatively very much smaller — so much so, indeed, that every part of the curve described by either the earth or moon is *concave* towards the sun. The excursions of the earth on either side of the ellipse, indeed, are so very small as to be hardly appreciable. In fact, the center of gravity of the earth and moon lies always *within* the surface of the earth, so that the monthly orbit described by the earth's center about the common center of gravity is comprehended within a space less than the size of the earth itself. The effect *is*, nevertheless, sensible, in producing an apparent monthly displacement of the sun in longitude, of a parallactic kind, which is called the *menstrual equation;* whose greatest amount is, however, less than the sun's horizontal parallax, or than 8·6″.

(452.) The moon, as we have seen, is about 60 radii of the earth distant from the center of the latter. Its proximity, therefore, to its center of attraction, thus estimated, is much greater than that of the planets to the sun; of which Mercury, the nearest, is 84, and Uranus 2026 solar radii from *its* center. It is owing to this proximity that the moon remains attached to the earth as a satellite. Were it much farther, the feebleness of its gravity towards the earth would be inadequate to produce that alternate acceleration and retardation in its motion about the sun, which divests it of the character of an independent planet, and keeps its movements subordinate to those of the earth. The one would outrun, or be left behind the other, in their revolutions round the sun (by reason of Kepler's third law), according to the relative dimensions of their heliocentric orbits, after which the whole influence of the earth would be confined to producing some considerable periodical disturbance in the moon's motion, as it passed or was passed by it in each synodical revolution.

(453.) At the distance at which the moon really is from us, its gravity towards the earth is actually less than towards the sun. That this is the case, appears sufficiently from what we have already stated, that the moon's *real* path, even when between the earth and sun, is *concave towards the latter*. But it will appear still more clearly if, from the known periodic times * in which

* R and *r* radii of two orbits (supposed circular), P and *p* the periodic

the earth completes its annual and the moon its monthly orbit, and from the dimensions of those orbits, we calculate the amount of deflection, in either, from their tangents, in equal very minute portions of time, as one second. These are the versed sines of the arcs described in that time in the two orbits, and these are the measures of the acting forces which produce those deflections. If we execute the numerical calculation in the case before us, we shall find 2·209 : 1 for the proportion in which the intensity of the force which retains the earth in its orbit round the sun actually exceeds that by which the moon is retained in *its* orbit about the earth.

(454.) Now the sun is 400 times more remote from the earth than the moon is. And, as gravity increases as the squares of the distances decrease, it must follow that, at *equal* distances, the intensity of solar would exceed that of terrestrial gravity in the above proportion, augmented in the further ratio of the square of 400 to 1; that is, in the proportion of 354936 to 1; and therefore, if we grant that the intensity of the gravitating energy is commensurate with the mass or inertia of the attracting body, we are compelled to admit the mass of the earth to be no more than $\frac{1}{354936}$ of that of the sun.

(455.) The argument is, in fact, nothing more than a recapitulation of what has been adduced in Chap. VII. (art. 380.) But it is here re-introduced, in order to show how the mass of a planet which is attended by one or more satellites can be as it were weighed against the sun, provided we have learned from observation the dimensions of the orbits described by the planet about the sun, and by the satellites about the planet, and also

---

times; then the arcs in question (A and a) are to each other as $\frac{R}{P}$ to $\frac{r}{p}$; and since the versed sines are as the squares of the arcs directly and the radii inversely, these are to each other as $\frac{R}{P^2}$ to $\frac{r}{p^2}$; and in this ratio are the forces acting on the revolving bodies in either case.

the periods in which these orbits are respectively described. It is by this method that the masses of Jupiter, Saturn, and Uranus have been ascertained. (See Synoptic Table.)

(456.) Jupiter, as already stated, is attended by four satellites, Saturn by seven; and Uranus, certainly by two, and perhaps by six. These, with their respective *primaries* (as the central planets are called), form in each case miniature systems, entirely analogous, in the general laws by which their motions are governed, to the great system in which the sun acts the part of the primary, and the planets of its satellites. In each of these systems the laws of Kepler are obeyed, in the sense, that is to say, in which they are obeyed in the planetary system—approximately, and without prejudice to the effects of mutual perturbation, of extraneous interference, if any, and of that small but not imperceptible correction which arises from the elliptic form of the central body. Their orbits are circles or ellipses of very moderate eccentricity, the primary occupying one focus. About this they describe areas very nearly proportional to the times; and the squares of the periodical times of all the satellites belonging to each planet are in proportion to each other as the cubes of their distances. The tables at the end of the volume exhibit a synoptic view of the distances and periods in these several systems, so far as they are at present known; and to all of them it will be observed that the same remark respecting their proximity to their primaries holds good, as in the case of the moon, with a similar reason for such close connection.

(457.) Of these systems, however, the only one which has been studied with great attention is that of Jupiter; partly on account of the conspicuous brilliancy of its four attendants, which are large enough to offer visible and measurable discs in telescopes of great power; but more for the sake of their eclipses, which, as they happen very frequently, and are easily observed, afford signals of considerable use for the determination of terrestrial longitudes (art. 218.). This method, indeed, until thrown into the back ground by the greater facility and exactness now attainable by lunar observations (art. 219.), was the best, or rather the only one which could be relied on for great distances and long intervals.

(458.) The satellites of Jupiter revolve from west to east (following the analogy of the planets and moon), in planes very nearly, although not exactly, coincident with that of the equator of the planet, or parallel to its belts. This latter plane is inclined 3° 5′ 30″ to the orbit of the planet, and is therefore but little different from the plane of the ecliptic. Accordingly, we see their orbits projected very nearly into straight lines, in which they appear to oscillate to and fro, sometimes passing before Jupiter, and casting shadows on his disc, (which are very visible in good telescopes, like small round ink spots,) and sometimes disappearing behind the body, or being eclipsed in its shadow at a distance from it. It is by these eclipses that we are furnished with accurate data for the construction of tables of the satellites' motions, as well as with signals for determining differences of longitude.

(459.) The eclipses of the satellites, in their general conception, are perfectly analogous to those of the moon, but in their detail they differ in several particulars. Owing to the much greater distance of Jupiter from the sun, and its greater magnitude, the cone of its shadow or umbra (art. 355.) is greatly more elongated, and of far greater dimension, than that of the earth. The satellites are, moreover, much less in proportion to their primary, their orbits less inclined to *its* ecliptic, and of (comparatively) smaller dimensions, than is the case with the moon. Owing to these causes, the three interior satellites of Jupiter pass through the shadow, and are totally eclipsed, every revolution; and the fourth, though, from the greater inclination of its orbit, it sometimes escapes eclipse, and *may* occasionally graze as it were the border of the shadow, and suffer partial eclipse, yet this is comparatively rare, and, ordinarily speaking, its eclipses happen, like those of the rest, each revolution.

(460.) These eclipses, moreover, are not seen, as is the case with those of the moon, from the center of their motion, but from a remote station, and one whose situation with respect to the line of shadow is variable. This, of course, makes no difference in the *times* of the eclipses, but a very great one in their visibility, and in their apparent situations with respect to the planet at the moments of their entering and quitting the shadow

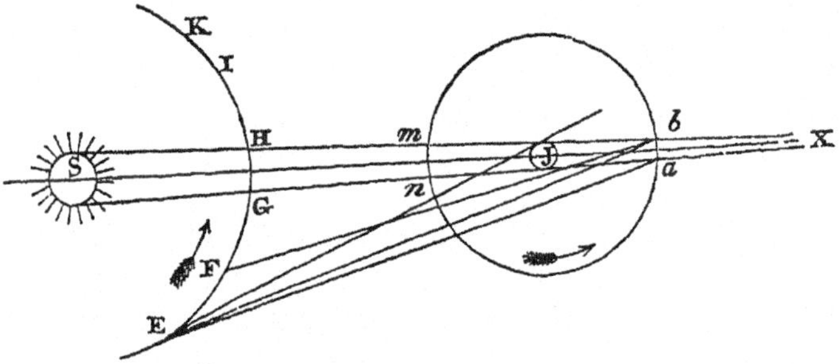

(461.) Suppose S to be the sun, E the earth in its orbit E F G K, J Jupiter, and *a b* the orbit of one of its satellites. The cone of the shadow, then, will have its vertex at X, a point far beyond the orbits of all the satellites; and the penumbra, owing to the great distance of the sun, and the consequent smallness of the angle its disc subtends at Jupiter, will hardly extend, within the limits of the satellites' orbits, to any perceptible distance beyond the shadow,— for which reason it is not represented in the figure. A satellite revolving from west to east (in the direction of the arrows) will be eclipsed when it enters the shadow at *a*, but not suddenly, because, like the moon, it has a considerable diameter seen from the planet; so that the time elapsing from the first perceptible loss of light to its total extinction will be that which it occupies in describing about Jupiter an angle equal to its apparent diameter as seen from the center of the planet, or rather somewhat more, by reason of the penumbra; and the same remark applies to its emergence at *b*. Now, owing to the difference of telescopes and of eyes, it is not possible to assign the *precise* moment of incipient obscuration, or of total extinction at *a*, nor that of the first glimpse of light falling on the satellite at *b*, or the complete recovery of its light. The observation of an eclipse, then, in which only the immersion, or only the emersion, is seen, is incomplete, and inadequate to afford any precise information, theoretical or practical. But, if both the immersion and emersion can be observed *with the same telescope, and by the same person,* the interval of the times will give the duration, and their mean the exact middle of the eclipse, when the satellite is in the line S J X, *i.e.* the true moment of its opposition to the sun. Such observations, and such only, are of use for determining the periods and other particulars of the motions of the satellites, and for affording data of any material use for the calculation of terrestrial longitudes. The intervals of the eclipses, it will be observed, give the *synodic* periods of the satellites' revolutions; from which their sidereal periods must be concluded by the method in art. 353. (note.)

(462.) It is evident, from a mere inspection of our figure, that the eclipses take place to the west of the planet, when the earth is situated to the west of the line S J, *i. e.* before the opposition of Jupiter; and to the east, when in the other half of its orbit, or after the opposition. When the earth approaches the opposition, the visual line becomes more and more nearly coincident with the direction of the shadow, and the apparent place where the eclipses happen will be continually nearer and nearer to the body of the planet. When the earth comes to F, a point determined by drawing *b* F to touch the body of the planet, the *emersions* will cease to be visible, and will thenceforth, to an equal distance on the other side of the opposition, happen *behind* the disc of the planet. When the earth arrives at G (or H) the immersion (or emersion) will happen at the very edge of the visible disc, and when between G and H (a very small space) the satellites will *pass uneclipsed behind the limb* of the planet.

(463.) When the satellite comes to *m*, *its* shadow will be thrown on Jupiter, and will appear to move across it as a black spot till the satellite comes to *n*. But the satellite itself will not appear to enter on the disc till it comes up to the line drawn from E to the eastern edge of the disc, and will not leave it till it attains a similar line drawn to the western edge. It appears then that the shadow will *precede* the satellite in its progress over the disc *before* the opposition, and *vice versâ*. In these transits of the satellites, which, with very powerful telescopes, may be observed with great precision, it frequently happens that the satellite itself is discernible *on* the disc as a bright spot if projected on a dark belt; but occasionally also as a dark spot of smaller dimensions than the shadow. This curious fact (observed by Schroeter and Harding) has led to a conclusion that certain of the satellites have occasionally *on their* own bodies, or in their atmospheres, obscure spots of great extent. We say of great extent; for the satellites of Jupiter, small as they appear to us, are really bodies of considerable size, as the following comparative table will show.*

	Mean apparent diameter.	Diameter in miles.	Mass.†
Jupiter	38″·327	87000	1·0000000
1st satellite	1·105	2508	0·0000173
2d ——	0·911	2068	0·0000232
3d ——	1·488	3377	0·0000885
4th ——	1·273	2890	0·0000427

(464.) An extremely singular relation subsists between the mean angular velocities or *mean motions* (as they are termed) of the three first satellites of Jupiter.

* Struve, Mem. Ast. Soc. iii. 301.

† Laplace, Mec. Cel. liv. viii. § 27.

If the mean angular velocity of the first satellite be added to twice that of the third, the sum will equal three times that of the second. From this relation it follows, that if from the mean longitude of the first added to twice that of the third, be subducted three times that of the second, the remainder will always be the same, or constant, and observation informs us that this constant is 180°, or two right angles; so that, the situations of any two of them being given, that of the third may be found. It has been attempted to account for this remarkable fact, on the theory of gravity by their mutual action. One curious consequence is, that these three satellites cannot be all eclipsed at once; for, in consequence of the last-mentioned relation, when the second and third lie in the *same* direction from the centre, the first must lie on the *opposite*; and therefore, when the first is eclipsed, the other two must lie between the sun and planet, throwing its shadow on the disc, and *vice versâ*. One instance only (so far as we are aware) is on record when Jupiter has been seen *without satellites;* viz. by Molyneux, Nov. 2. (old style) 1681.*

(465.) The discovery of Jupiter's satellites by Galileo, one of the first-fruits of the invention of the telescope, forms one of the most memorable epochs in the history of astronomy. The first astronomical solution of the great problem of "*the longitude*" — the most important for the interests of mankind which has ever been brought under the dominion of strict scientific principles, dates immediately from their discovery. The final and conclusive establishment of the Copernican system of astronomy may also be considered as referable to the discovery and study of this exquisite miniature system, in which the laws of the planetary motions, as ascertained by Kepler, and especially that which connects their periods and distances, were speedily traced, and found to be satisfactorily maintained. And (as if to accumulate historical interest on this point) it is to the observation of their eclipses that we owe the grand dis-

* Molyneux, Optics, p. 271.

covery of the aberration of light, and the consequent determination of the enormous velocity of that wonderful element. This we must explain now at large.

(466.) The earth's orbit being concentric with that of Jupiter and interior to it (see *fig.* art. 460.), their mutual distance is continually varying, the variation extending from the *sum* to the *difference* of the radii of the two orbits, and the difference of the greater and least distances being equal to a diameter of the earth's orbit. Now, it was observed by Roemer, (a Danish astronomer, in 1675,) on comparing together observations of eclipses of the satellites during many successive years, that the eclipses at and about the opposition of Jupiter (or its nearest point to the earth) took place *too soon*—sooner, that is, than, by calculation from an average, he expected them; whereas those which happened when the earth was in the part of its orbit most remote from Jupiter were always *too late*. Connecting the observed error in their computed times with the variation of distance, he concluded, that, to make the calculation on an average period correspond with fact, an allowance in respect of time behoved to be made proportional to the excess or defect of Jupiter's distance from the earth above or below its average amount, and such that a difference of distance of one diameter of the earth's orbit should correspond to $16^{m}\,26^{s}\!\cdot\!6$ of time allowed. Speculating on the probable physical cause, he was naturally led to think of the gradual instead of an instantaneous propagation of light. This explained every particular of the observed phenomenon, but the velocity required (192000 miles per second) was so great as to startle many, and, at all events, to require confirmation. This has been afforded since, and of the most unequivocal kind, by Bradley's discovery of the aberration of light (art. 275.). The velocity of light deduced from this last phænomenon differs by less than one eightieth of its amount from that calculated from the eclipses, and even this difference will no doubt be destroyed by nicer and more rigorously reduced observations.

(467.) The orbits of Jupiter's satellites are but little eccentric, those of the two interior, indeed, have no perceptible eccentricity; their mutual action produces in them perturbations analogous to those of the planets about the sun, and which have been diligently investigated by Laplace and others. By assiduous observation it has been ascertained that they are subject to marked fluctuations in respect of brightness, and that these fluctuations happen periodically, according to their position with respect to the sun. From this it has been concluded, apparently with reason, that they turn on their axes, like our moon, in periods equal to their respective sidereal revolutions about their primary.

(468.) The satellites of Saturn have been much less studied than those of Jupiter. The most distant is by far the largest, and is probably not much inferior to Mars in size. Its orbit is also materially inclined to the plane of the ring, with which those of all the rest nearly coincide. It is the only one of the number whose theory has been at all enquired into, further than suffices to verify Kepler's law of the periodic times, which holds good, *mutatis mutandis*, and under the requisite reservations, in this as in the system of Jupiter. It exhibits, like those of Jupiter, periodic defalcations of light, which prove its revolution on its axis in the time of a sidereal revolution about Saturn. The next in order (proceeding inwards) is tolerably conspicuous; the three next very minute, and requiring pretty powerful telescopes to see them; while the two interior satellites, which just skirt the edge of the ring, and move exactly in its plane, have never been discerned but with the most powerful telescopes which human art has yet constructed, and then only under peculiar circumstances. At the time of the disappearance of the ring (to ordinary telescopes) they have been seen * threading like beads

* By my Father, in 1789, with a reflecting telescope four feet in aperture.

the almost infinitely thin fibre of light to which it is then reduced, and for a short time advancing off it at either end, speedily to return, and hastening to their habitual concealment. Owing to the obliquity of the ring, and of the orbits of the satellites to Saturn's ecliptic, there are no eclipses of the satellites (the interior ones excepted) until near the time when the ring is seen edgewise.

(469.) With the exception of the two interior satellites of Saturn, the attendants of Uranus are the most difficult objects to obtain a sight of, of any in our system. Two undoubtedly exist, and four more have been suspected. These two, however, offer remarkable and, indeed, quite unexpected and unexampled peculiarities. Contrary to the unbroken analogy of the whole planetary system—whether of primaries or secondaries—the planes of their orbits *are nearly perpendicular to the ecliptic,* being inclined no less than 78° 58′ to that plane, and in these orbits their motions *are retrograde;* that is to say, their positions, when projected on the ecliptic, instead of advancing *from west to east* round the center of their primary, as is the case with every other planet and satellite, move in the opposite direction. Their orbits are nearly or quite circular, and they do not appear to have any sensible, or, at least, any rapid motion of nodes, or to have undergone any material change of inclination, in the course, at least, of half a revolution of their primary round the sun.*

* These anomalous peculiarities, which seem to occur at the extreme limits of our system, as if to prepare us for further departure from all its analogies, in other systems which may yet be disclosed to us, have hitherto rested on the sole testimony of their discoverer, who alone had ever obtained a view of them. I am happy to be able, from my own observations from 1828 to the present time, to confirm in the amplest manner my Father's results.—*Author.*

CHAP. X.

OF COMETS.

GREAT NUMBER OF RECORDED COMETS. — THE NUMBER OF UNRECORDED PROBABLY MUCH GREATER. — DESCRIPTION OF A COMET. — COMETS WITHOUT TAILS. — INCREASE AND DECAY OF THEIR TAILS. — THEIR MOTIONS. — SUBJECT TO THE GENERAL LAWS OF PLANETARY MOTION. — ELEMENTS OF THEIR ORBITS. — PERIODIC RETURN OF CERTAIN COMETS. — HALLEY'S. — ENCKE'S. — BIELA'S. — DIMENSIONS OF COMETS. — THEIR RESISTANCE BY THE ETHER, GRADUAL DECAY, AND POSSIBLE DISPERSION IN SPACE.

(470.) The extraordinary aspect of comets, their rapid and seemingly irregular motions, the unexpected manner in which they often burst upon us, and the imposing magnitudes which they occasionally assume, have in all ages rendered them objects of astonishment, not unmixed with superstitious dread to the uninstructed, and an enigma to those most conversant with the wonders of creation and the operations of natural causes. Even now, that we have ceased to regard their movements as irregular, or as governed by other laws than those which retain the planets in their orbits their intimate nature, and the offices they perform in the economy of our system, are as much unknown as ever. No rational or even plausible account has yet been rendered of those immensely voluminous appendages which they bear about with them, and which are known by the name of their tails, (though improperly, since they often precede them in their motions,) any more than of several other singularities which they present.

(471.) The number of comets which have been astronomically observed, or of which notices have been recorded in history, is very great, amounting to several hundreds*; and when we consider that in the earlier ages of astronomy, and indeed in more recent times, before the invention of the telescope, only large and conspicuous ones were noticed; and that, since due attention has been paid to the subject, scarcely a year has passed without the observation of one or two of these bodies, and that sometimes two and even three have appeared at once; it will be easily supposed that their actual number must be at least many thousands. Multitudes, indeed, must escape all observation, by reason of their paths traversing only that part of the heavens which is above the horizon in the daytime. Comets so circumstanced can only become visible by the rare coincidence of a total eclipse of the sun,—a coincidence which happened, as related by Seneca, 60 years before Christ, when a large comet was actually observed very near the sun. Several, however, stand on record as having been bright enough to be seen in the daytime, even at noon and in bright sunshine. Such were the comets of 1402 and 1532, and that which appeared a little before the assassination of Cæsar, and was (*afterwards*) supposed to have predicted his death.

(472.) That feelings of awe and astonishment should be excited by the sudden and unexpected appearance of a great comet, is no way surprising; being, in fact, according to the accounts we have of such events, one of the most brilliant and imposing of all natural phænomena. Comets consist for the most part of a large and splendid but ill defined nebulous mass of light, called the head, which is usually much brighter towards its center, and offers the appearance of a vivid *nucleus*, like a star or planet. From the head, and in a direction *opposite to that in which the sun is situated* from the comet

* See catalogues in the Almagest of Riccioli; Pingré's Cometographia; Delambre's Astron. vol. iii.; Astronomische Abhandlungen, No. 1. (which contains the elements of all the orbits of comets which have been computed to the time of its publication, 1823); also, a catalogue now in progress, by the Rev. T. J. Hussey. Lond. & Ed. Phil. Mag. vol. ii. No. 9. *et seq.* In a list cited by Lalande from the 1st vol. of the Tables de Berlin, 700 comets are enumerated.

appear to diverge two streams of light, which grow broader and more diffused at a distance from the head, and which sometimes close in and unite at a little distance behind it, sometimes continue distinct for a great part of their course; producing an effect like that of the trains left by some bright meteors, or like the diverging fire of a sky-rocket (only without sparks or perceptible motion). This is the tail. This magnificent appendage attains occasionally an immense apparent length. Aristotle relates of the tail of the comet of 371 A. C., that it occupied a third of the hemisphere, or 60°; that of A. D. 1618 is stated to have been attended by a train no less than 104° in length. The comet of 1680, the most celebrated of modern times, and on many accounts the most remarkable of all, with a head not exceeding in brightness a star of the second magnitude, covered with its tail an extent of more than 70° of the heavens, or, as some accounts state, 90°. The figure (*fig.* 2., Plate II.) is a very correct representation of the comet of 1819—by no means one of the most considerable, but the latest which has been conspicuous to the naked eye.

(473.) The tail is, however, by no means an invariable appendage of comets. Many of the brightest have been observed to have short and feeble tails, and not a few have been entirely without them. Those of 1585 and 1763 offered no vestige of a tail; and Cassini describes the comet of 1682 as being as round and as bright as Jupiter. On the other hand, instances are not wanting of comets furnished with many tails or streams of diverging light. That of 1744 had no less than six, spread out like an immense fan, extending to a distance of nearly 30° in length. The tails of comets, too, are often curved, bending, in general, towards the region which the comet has left, as if moving somewhat more slowly, or as if resisted in their course.

(474.) The smaller comets, such as are visible only in telescopes, or with difficulty by the naked eye, and which are by far the most numerous, offer very frequently no appearance of a tail, and appear only as round or somewhat oval vaporous masses, more dense towards the center, where, however, they appear to have no distinct nucleus, or any thing which seems entitled to be considered as a solid body. Stars of the smallest magnitudes remain distinctly visible, though covered by what appears to be the densest portion of their substance; although the same stars would be completely obliterated by a moderate fog, extending only a few yards from the surface of the earth. And since it is an observed fact, that even those larger comets which have presented the appearance of a nucleus have yet exhibited *no phases*, though we cannot doubt that they shine by the reflected solar light, it follows that even these can only be regarded as great masses of thin vapour, susceptible of being penetrated through their whole substance by the sunbeams, and reflecting them alike from their interior parts and from their surfaces. Nor will any one regard this explanation as forced, or feel disposed to resort to a phosphorescent quality in the comet itself, to account for the phænomena in question, when we consider (what will be hereafter shown) the enormous magnitude of the space thus illuminated, and the extremely small *mass* which there is ground to attribute to these bodies. It will then be evident that the most unsubstantial clouds which float in the highest regions of our atmosphere, and seem at sunset to be drenched in light, and to glow throughout their whole depth as if in actual ignition, without any shadow or dark side, must be looked upon as dense and massive bodies compared with the filmy and all but spiritual texture of a comet. Accordingly, whenever powerful telescopes have been turned on these bodies, they have not failed to dispel the illusion which attributes solidity to that more condensed part of the head, which appears to the naked eye as a nucleus; though it is true that in some, a very minute stellar point *has* been seen, indicating the existence of a solid body.

(475.) It is in all probability to the feeble coercion of the elastic power of their gaseous parts, by the gravitation of so small a central mass, that we must attribute this extraordinary developement of the atmospheres of comets. If the earth, retaining its present size, were reduced, by any internal change (as by hollowing out its central parts) to one thousandth part of its actual mass, its coercive power over the atmosphere would be diminished in the same proportion, and in consequence the latter would expand to a thousand times its actual bulk; and indeed much more, owing to the still farther diminution of gravity, by the recess of the upper parts from the center.

(476.) That the luminous part of a comet is something in the nature of a smoke, fog, or cloud, suspended in a transparent atmosphere, is evident from a fact which has been often noticed, viz.—that the portion of the tail where it comes up to, and surrounds the head, is yet separated from it by an interval less luminous, as if sustained and kept off from contact by a transparent stratum, as we often see one layer of clouds laid over another with a considerable clear space between. These, and most of the other facts observed in the history of comets, appear to indicate that the structure of a comet, as seen in section in the direction of its length, must be that of a hollow envelope, of a parabolic form, enclosing near its vertex the nucleus and head, something as represented in the annexed figure. This would account for the ap-

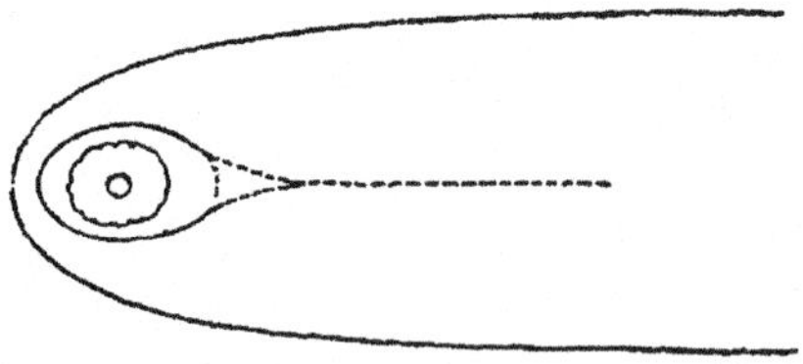

parent division of the tail into two principal lateral branches, the envelope being oblique to the line of sight at its borders, and therefore a greater depth of illuminated matter being there exposed to the eye. In all probability, however, they admit great varieties of structure, and among them may very possibly be bodies of widely different physical constitution.

(477.) We come now to speak of the motions of comets. These are apparently most irregular and capricious. Sometimes they remain in sight for only a few days, at others for many months; some move with extreme slowness, others with extraordinary velocity; while not unfrequently, the two extremes of apparent speed are exhibited by the same comet in different parts of its course. The comet of 1472 described an arc of the heavens of 120° in extent in a single day. Some pursue a direct, some a retrograde, and others a tortuous and very irregular course; nor do they confine themselves, like the planets, within any certain region of the heavens, but traverse indifferently every part. Their variations in apparent size, during the time they continue visible, are no less remarkable than those of their velocity: sometimes they make their first appearance as faint and slow moving objects, with little or no tail; but by degrees accelerate, enlarge, and throw out from them this appendage, which increases in length and brightness till (as always happens in such cases) they approach the sun, and are lost in his beams. After a time they again emerge, on the other side, receding from the sun with a velocity at first rapid, but gradually decaying. It is after thus passing the sun, and not till then, that they shine forth in all their splendour, and that their tails acquire their greatest length and developement; thus indicating plainly the action of the sun's rays as the exciting cause of that extraordinary emanation. As they continue to recede from the sun, their motion diminishes and the tail dies away, or is absorbed into the head, which itself grows continually feebler, and is at length altogether lost sight of, in by far the greater number of cases never to be seen more.

(478.) Without the clue furnished by the theory of gravitation, the enigma of these seemingly irregular and capricious movements might have remained for ever unresolved. But Newton, having demonstrated the possibility of any conic section whatever being described about the sun, by a body revolving under the dominion of that law, immediately perceived the applicability of the general proposition to the case of cometary orbits: and the great comet of 1680, one of the most remarkable on record, both for the immense length of its tail and for the excessive closeness of its approach to the sun (within one sixth of the diameter of that luminary), afforded him an excellent opportunity for the trial of his theory. The success of the attempt was complete. He ascertained that this comet described about the sun as its focus an elliptic orbit of so great an excentricity as to be undistinguishable from a parabola, (which is the extreme, or limiting form of the ellipse when the axis becomes infinite,) and that in this orbit the areas described about the sun were, as in the planetary ellipses, proportional to the times. The representation of the apparent motions of this comet by such an orbit, throughout its whole observed course, was found to be as complete as those of the motions of the planets in their nearly circular paths. From that time it became a received truth, that the motions of comets are regulated by the same general laws as those of the planets, — the difference of the cases consisting only in the extravagant elongation of their ellipses, and in the absence of any limit to the inclinations of their planes to that of the ecliptic, — or any general coincidence in the direction of their motions from west to east, rather than from east to west, like what is observed among the planets.

(479.) It is a problem of pure geometry, from the general laws of elliptic or parabolic motion, to find the situation and dimensions of the ellipse or parabola which shall represent the motion of any given comet. In general, three complete observations of its right ascension and declination, with the times at which they were made, suffice for the solution of this problem, (which is, however, a very difficult one,) and for the determination of the elements of the orbit. These consist, *mutatis mutandis*, of the same data as are required for the computation of the motion of a planet; and, once deter mined, it becomes very easy to compare them with the whole observed course of the comet, by a process exactly similar to that of art. 426., and thus at once to ascertain their correctness, and to put to the severest trial the truth of those general laws on which all such calculations are founded.

(480.) For the most part, it is found that the motions of comets may be sufficiently well represented by parabolic orbits, — that is to say, ellipses whose axes are of infinite length, or, at least, so very long that no appreciable error in the calculation of their motions, during all the time they continue visible, would be incurred by supposing them actually infinite. The parabola is that conic section which is the limit between the ellipse on the one hand, which returns into itself, and the hyperbola on the other, which runs out to infinity. A comet, therefore, which should describe an elliptic path, however long its axis, must *have* visited the sun before, and must again return (unless disturbed) in some determinate period, — but should its orbit be of the hyperbolic character, when once it had passed its perihelion, it could never more return within the sphere of our observation, but must run off to visit other systems, or be lost in the immensity of space. A very few comets have been ascertained to move in hyperbolas, but many more in ellipses. These then, in so far as their orbits can remain unaltered by the attractions of the planets, must be regarded as permanent members of our system.

(481.) The most remarkable of these is the comet of Halley, so called from the celebrated Edmund Halley, who, on calculating its elements from its perihelion passage in 1682, when it appeared in great splendour, with a tail 30° in length, was led to conclude its identity with the great comets of 1531 and 1607, whose elements he had also ascertained. The intervals of these successive apparitions being 75 and 76 years, Halley was encouraged to *predict* its re-appearance about the year 1759. So remarkable a prediction could not fail to attract the attention of all astronomers, and, as the time approached, it became extremely interesting to know whether the attractions of the larger planets might not materially interfere with its orbitual motion. The computation of their influence from the Newtonian law of gravity, a most difficult and intricate piece of calculation, was undertaken and accomplished by Clairaut, who found that the action of Saturn would retard its return by 100 days, and that of Jupiter by no less than 518, making in all 618 days, by which the expected return would happen later than on the supposition of its retaining an unaltered period,—and that, in short, the time of the expected perihelion passage would take place within a month, one way or other, of the middle of April, 1759. —It actually happened on the 12th of March in that year. Its next return to the perihelion has been calculated by Messrs. Damoiseau and Pontecoulant, and fixed by the former on the fourth, and by the latter on the seventh of November, 1835, about a month or six weeks before which time it may be expected to become visible in our hemisphere; and, as it will approach pretty near the earth, will very probably exhibit a brilliant appearance, though, to judge from the successive degradations of its apparent size and the length of its tail in its several returns since its first appearances on record, (in 1305, 1456, &c.) we are not now to expect any of those vast and awful phænomena which threw our remote ancestors of the middle ages into agonies of superstitious terror, and caused public prayers to be put up in the churches against the comet and its malignant agencies.

(482.) More recently, two comets have been especially identified as having performed several revolutions about the sun, and as having been not only observed and recorded in preceding revolutions, without knowledge of this remarkable peculiarity, but have had already several times their returns predicted, and have scrupulously kept to their appointments. The first of these is the comet of Encke, so called from Professor Encke, of Berlin, who first ascertained its periodical return. It revolves in an ellipse of great excentricity, inclined at an angle of about 13° 22′ to the plane of the ecliptic, and in the short period of 1207 days, or about $3\frac{1}{3}$ years. This remarkable discovery was made on the occasion of its fourth recorded appearance, in 1819. From the ellipse then calculated by Encke, its return in 1822 was predicted by him, and observed at Paramatta, in New South Wales, by M. Rümker, being invisible in Europe: since which it has been re-predicted, and re-observed in all the principal observatories, both in the northern and southern hemispheres, in 1825, 1828, and 1832. Its next return will be in 1835.

(483.) On comparing the intervals between the successive perihelion passages of this comet, after allowing in the most careful and exact manner for all the disturbances due to the actions of the planets, a very singular fact has come to light, viz. that the periods are continually diminishing, or, in other words, the mean distance from the sun, or the major axis of the ellipse, dwindling by slow but regular degrees. This is evidently the effect which would be produced by a resistance experienced by the comet from a very rare ethereal medium pervading the regions in which it moves; for such resistance, by diminishing its actual velocity, would diminish also its centrifugal force, and thus give the sun more power over it to draw it nearer. Accordingly (no other mode of accounting for the phænomenon in question appearing), this is the solution proposed by Encke, and generally received. It will, therefore, probably fall ultimately into the sun, should it not first be dissipated altogether, — a thing no way improbable, when the lightness of its materials is considered, and which seems authorised by the observed fact of its having been less and less conspicuous at each re-appearance.

(484.) The other comet of short period which has lately been discovered is that of *Biela*, so called from M. Biela, of Josephstadt, who first arrived at this interesting conclusion. It is identical with comets which appeared in 1789, 1795, &c., and describes its moderately excentric ellipse about the sun in $6\frac{3}{4}$ years; an the last apparition having taken place according to the prediction in 1832, the next will be in 1838. It is a small insignificant comet, without a tail, or any appearance of a solid nucleus whatever. Its orbit, by a remarkable coincidence, very nearly intersects that of the earth; and had the latter, at the time of its passage in 1832, been a month in advance of its actual place, it would have passed through the comet, — a singular rencontre, perhaps not unattended with danger.*

(485.) Comets in passing among and near the planets are materially drawn aside from their courses, and in some cases have their orbits entirely changed. This is remarkably the case with Jupiter, which seems by some strange fatality to be constantly in their way, and to serve as a perpetual stumbling block to them. In the case of the remarkable comet of 1770, which was found by Lexell to revolve in a moderate ellipse in the period of about 5 years, and whose return was predicted by him accordingly, the prediction was disappointed by the comet actually getting entangled among the satellites of Jupiter, and being completely thrown out of its orbit by the attraction of that planet, and forced into a much larger ellipse. By this extraordinary rencontre, *the motions of the satellites suffered not the least perceptible derangement*, — a sufficient proof of the smallness of the comet's mass.

(486.) It remains to say a few words on the actual dimensions of comets. The calculation of the diameters of their heads, and the lengths and breadths of their tails,

* Should calculation establish the fact of a resistance experienced also by this comet, the subject of periodical comets will assume an extraordinary degree of interest. It cannot be doubted that many more will be discovered, and by their resistance questions will come to be decided, such as the following: — What is the law of density of the resisting medium which surrounds the sun? Is it at rest or in motion? If the latter, in what direction does it move? Circularly round the sun, or traversing space? If circularly, in what plane? It is obvious that a circular or vorticose motion of the ether would *accelerate some comets and retard others*, according as their revolution was, relative to such motion, direct or retrograde. Supposing the neighbourhood of the sun to be filled with a material fluid, it is not conceivable that the circulation of the planets in it for ages should not have impressed upon it some degree of rotation in their own direction. And this may preserve them from the extreme effects of accumulated resistance. — *Author.*

offers not the slightest difficulty when once the elements of their orbits are known, for by these we know their real distances from the earth at any time, and the true direction of the tail, which we see only foreshortened. Now calculations instituted on these principles lead to the surprising fact, that the comets are by far the most voluminous bodies in our system. The following are the dimensions of some of those which have been made the subjects of such enquiry.

(487.) The tail of the great comet of 1680, immediately after its perihelion passage, was found by Newton to have been no less than 20000000 of leagues in length, and to have occupied only two days in its emission from the comet's body! a decisive proof this of its being darted forth by some active force, the origin of which, to judge from the direction of the tail, must be sought in the sun itself. Its greatest length amounted to 41000000 leagues, a length much exceeding the whole interval between the sun and earth. The tail of the comet of 1769 extended 16000000 leagues, and that of the great comet of 1811, 36000000. The portion of the head of this last comprised within the transparent atmospheric envelope which separated it from the tail was 180000 leagues in diameter. It is hardly conceivable that matter once projected to such enormous distances should ever be collected again by the feeble attraction of such a body as a comet — a consideration which accounts for the rapid progressive diminution of the tails of such as have been frequently observed.

(488.) A singular circumstance has been remarked respecting the change of dimensions of the comet of Encke in its progress to and retreat from the sun: viz. that the real diameter of the visible nebulosity undergoes a rapid contraction as it approaches, and an equally rapid dilatation as it recedes from the sun. M. Valz, who, among others, had noticed this fact, has accounted for it by supposing a real compression or condensation of volume, owing to the pressure of an ethereal medium growing more dense in the sun's neighbourhood. It is very possible, however, that the change may consist in no real expansion or condensation of volume (further than is due to the convergence or divergence of the different parabolas described by each of its molecules to or from a common vertex), but may rather indicate the alternate conversion of evaporable materials in the upper regions of a transparent atmosphere, into the states of visible cloud and invisible gas, by the mere effects of heat and cold. But it is time to quit a subject so mysterious, and open to such endless speculation.

CHAP. XI.

OF PERTURBATIONS.

SUBJECT PROPOUNDED. — SUPERPOSITION OF SMALL MOTIONS. — PROBLEM OF THREE BODIES. — ESTIMATION OF DISTURBING FORCES. — MOTION OF NODES. — CHANGES OF INCLINATION. — COMPENSATION OPERATED IN A WHOLE REVOLUTION OF THE NODE. — LAGRANGE'S THEOREM OF THE STABILITY OF THE INCLINATIONS. — CHANGE OF OBLIQUITY OF THE ECLIPTIC. — PRECESSION OF THE EQUINOXES. — NUTATION. — THEOREM RESPECTING FORCED VIBRATIONS. — OF THE TIDES. — VARIATION OF ELEMENTS OF THE PLANET'S ORBITS — PERIODIC AND SECULAR. — DISTURBING FORCES CONSIDERED AS TANGENTIAL AND RADIAL. — EFFECTS OF TANGENTIAL FORCE. — 1ST, IN CIRCULAR ORBITS; 2DLY, IN ELLIPTIC. — COMPENSATIONS EFFECTED. — CASE OF NEAR COMMENSURABILITY OF MEAN MOTIONS. — THE GREAT INEQUALITY OF JUPITER AND SATURN EXPLAINED. — THE LONG INEQUALITY OF VENUS AND THE EARTH. — LUNAR VARIATION. — EFFECTS OF THE RADIAL FORCE. — MEAN EFFECT ON THE PERIOD AND DIMENSIONS OF THE DISTURBED ORBIT. — VARIABLE PART OF ITS EFFECT. — LUNAR EVECTION. — SECULAR ACCELERATION OF THE MOON'S MOTION. — INVARIABILITY OF THE AXES AND PERIODS. — THEORY OF THE SECULAR VARIATIONS OF THE EXCENTRICITIES AND PERIHELIA. — MOTION OF THE LUNAR APSIDES. — LAGRANGE'S THEOREM OF THE STABILITY OF THE EXCENTRICITIES. — NUTATION OF THE LUNAR ORBIT. — PERTURBATIONS OF JUPITER'S SATELLITES.

(489.) In the progress of this work, we have more than once called the reader's attention to the existence of inequalities in the lunar and planetary motions not included in the expression of Kepler's laws, but in some sort supplementary to them, and of an order so far subordinate to those leading features of the celestial movements, as to require, for their detection, nicer observations, and longer continued comparison between facts and theories, than suffice for the establishment and verification of the elliptic theory. These inequalities are known, in physical astronomy, by the name of *perturbations*. They arise, in the case of the primary planets, from the mutual gravitations of these planets towards each other, which derange their elliptic motions round the sun; and in that of the secondaries, partly from the mutual gravitation of the secondaries of the same system similarly deranging their elliptic motions round their common primary, and partly from the unequal attraction of the sun on them and on their primary. These perturbations, although small, and, in most instances, insensible in short intervals of time, yet, when accumulated, as some of them may become, in the lapse of ages, alter very greatly the original elliptic relations, so as to render the same elements of the planetary orbits, which at one epoch represented perfectly well their movements, inadequate and unsatisfactory after long intervals of time.

(490.) When Newton first reasoned his way from the broad features of the celestial motions, up to the law of universal gravitation, as affecting all matter, and rendering every particle in the universe subject to the influence of every other, he was not unaware of the modifications which this generalization would induce into the results of a more partial and limited application of the same law to the revolutions of the planets about the sun, and the satellites about their primaries, as their *only* centers of attraction. So far from it, that his extraordinary sagacity enabled him to perceive very distinctly how several of the most important of the lunar inequalities take their origin, in this more general way of conceiving the agency of the attractive power, especially the retrograde motion of the nodes, and the direct revolution of the apsides of her orbit. And if he did not extend his investigations to the mutual perturbations of the planets, it was not for want of perceiving that such perturbations *must* exist, and *might* go the length of producing great derangements from the actual state of the system, but owing to the then undeveloped state of the practical part of astronomy, which had not yet attained the precision requisite to make such an attempt inviting, or indeed feasible. What Newton left undone, however, his successors have accomplished; and, at this day, there is not a single perturbation, great or small, which observation has ever detected, which has not been traced up to its origin in the mutual gravitation of the parts of our system, and been minutely accounted for, in its numerical amount and value, by strict calculation on Newton's principles.

(491.) Calculations of this nature require a very high analysis for their successful performance, such as is far beyond the scope and object of this work to attempt exhibiting. The reader who would master them must prepare himself for the undertaking by an extensive course of preparatory study, and must ascend by steps which we must not here even digress to point out. It will be our object, in this chapter, however, to give some general insight into the nature and manner of operation of the acting forces, and to point out what are the circumstances which, in some cases, give them a high degree of efficiency—a sort of *purchase* on the balance of the system; while, in others, with no less amount of intensity, their effective agency in producing extensive and lasting changes is compensated or rendered abortive; as well as to explain the nature of those admirable results respecting the stability of our system, to which the researches of geometers have conducted them;

and which, under the form of mathematical theorems of great beauty, simplicity, and elegance, involve the history of the past and future state of the planetary orbits during ages, of which, contemplating the subject in this point of view, we neither perceive the beginning nor the end.

(492.) Were there no other bodies in the universe but the sun and one planet, the latter would describe an exact ellipse about the former (or both round their common centers of gravity), and continue to perform its revolutions in one and the same orbit for ever; but the moment we add to our combination a third body, the attraction of this will draw both the former bodies out of their mutual orbits, and, by acting on them unequally, will disturb their relation to each other, and put an end to the rigorous and mathematical exactness of their elliptic motions, either about one another or about a fixed point in space. From this way of propounding the subject, we see that it is not the whole attraction of the newly introduced body which produces perturbation, but *the difference* of its attractions on the two originally present.

(493.) Compared to the sun, all the planets are of extreme minuteness; the mass of Jupiter, the greatest of them all, being not more than one 1300th part that of the sun. Their attractions on each other, therefore, are all very feeble, compared with the presiding central power, and the effects of their disturbing forces are proportionally minute. In the case of the secondaries, the chief agent by which their motions are deranged is the sun itself, whose mass is indeed great, but whose disturbing influence is immensely diminished by their near proximity to their primaries, compared to their distances from the sun, which renders the *difference* of attractions on both extremely small, compared to the whole amount. In this case, the greatest part of the sun's attraction, viz. that which is common to both, is exerted to retain both primary and secondary in their common orbit about itself, and prevent their parting company. The small overplus of force only acts as a disturbing power. The mean value of this overplus, in the case of the moon disturbed by the sun, is calculated by Newton to amount to no higher a fraction than $\frac{1}{638000}$ of gravity at the earth's surface, or $\frac{1}{179}$ of the principal force which retains the moon in its orbit.

(494.) From this extreme minuteness of the intensities of the disturbing, compared to the principal forces, and the consequent smallness of their *momentary* effects, it happens that we can estimate each of these effects separately, as if the others did not take place, without fear of inducing error in our conclusions beyond the limits necessarily incident to a first approximation. It is a principle in mechanics, immediately flowing from the primary relations between forces and the motions they produce, that when a number of very minute forces act at once on a system, their joint effect is the sum or aggregate of their separate effects, at least within such limits, that the original relation of the parts of the system shall not have been materially changed by their action. Such effects supervening on the greater movements due to the action of the primary forces may be compared to the small ripplings caused by a thousand varying breezes on the broad and regular swell of a deep and rolling ocean, which run on as if the surface were a plane, and cross in all directions, without interfering, each as if the other had no existence. It is only when their effects become accumulated in lapse of time, so as to alter the primary relations or data of the system that it becomes necessary to have especial regard to the changes correspondingly introduced into the estimation of their momentary efficiency, by which the *rate* of the subsequent changes is affected, and periods or cycles of immense length take their origin. From this consideration arise some of the most curious theories of physical astronomy.

(495.) Hence it is evident, that in estimating the disturbing influence of several bodies forming a system, in which one has a remarkable preponderance over all the rest, we need not embarrass ourselves with combinations of the disturbing powers one among another, unless where immensely long periods are concerned; such as consist of many thousands of revolutions of the bodies in question about their common centers. So that, in effect, the problem of the investigation of the perturbations of a system, however numerous, constituted as ours is, reduces itself to that of a system of three bodies: a predominant central body, a disturbing, and a disturbed; the two latter of which may exchange denominations, according as the motions of the one or the other are the subject of enquiry.

(496.) The intensity of the disturbing force is continually varying, according to the relative situation of the disturbing and disturbed body with respect to the sun. If the attraction of the disturbing body M, on the central body S, and the disturbed body P, (by which designations, for brevity, we shall hereafter indicate them,) were equal, and acted in parallel lines, whatever might otherwise be its law of variation, there would be no deviation caused in the elliptic motion of P about S, or of each about the other. The case would be strictly that of art. 385.; the attraction of M, so circumstanced, being at every moment exactly analogous in its effects to terrestrial gravity, which acts in parallel lines, and is equally *intense* on all bodies, great and small. But this is not the case of nature. Whatever is stated in the subsequent article to that last cited, of the disturbing effect of the sun and moon, is, *mutatis mutandis*, applicable to every case of perturbation; and it must be now our business to enter, somewhat more in detail, into the general heads of the subject there merely hinted at.

(497.) We shall begin with that part of the disturbing force which tends to draw the disturbed body out of the plane in which its orbit would be performed if undisturbed, and, by so doing, causes it to describe a curve, of which no two adjacent portions lie in one plane, or, as it is called in geometry, a curve of double curvature. Suppose, then, A P N to be the orbit which P would describe about S, if undisturbed, and suppose it to arrive at P, at any instant of time, and to be about to describe in the next instant the undisturbed arc P *p*, which, prolonged in the direction of its tangent P *p* R, will intersect the plane of the orbit M L of the disturbing body, somewhere in the line of nodes S L, suppose in R. This would be the case if M exerted no disturbing power. But suppose it to do so, then, since it draws both S and P towards it, in directions not coincident with the plane of P's orbit, it will cause them *both*, in the next instant of time, to quit that plane, but *unequally* : — first, because it does not draw them both in parallel lines; secondly, because they, being unequally distant from M, are unequally attracted by it, by reason of the general law of gravitation. Now, it is by the difference of the motions thus generated that the relative orbit of P about S is changed; so that, if we continue to refer its motion to S as a fixed center, the disturbing part of the impulse which it receives from M will impel it to deviate from the plane P S N, and describe in the next instant of time, not the arc P *p*, but an arc P *q*, lying either above or below P *p*, according to the preponderance of the forces exerted by M on P and S.

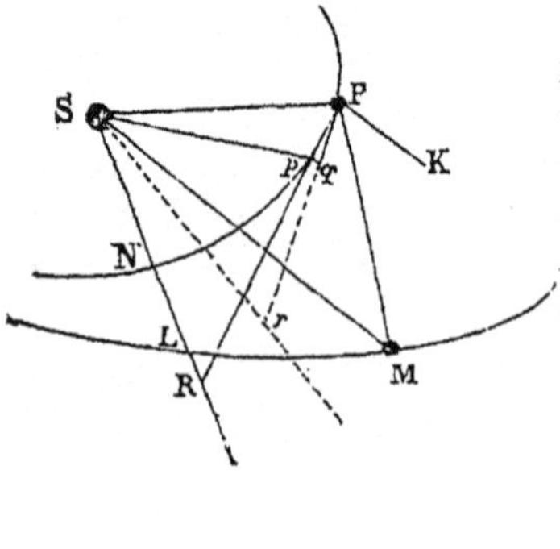

(498.) The disturbing force acts in the plane of the triangle S P M, and may be considered as resolved into two; one of which urges P to or from S, or along the line S P, and, therefore, increases or diminishes, in so far as it is effective, the direct attraction of S or P; the other along a line P K, parallel to S M, and which may be regarded as either *pulling* P in the direction P K, or *pushing* it in a contrary direction; these terms being well understood to have only a relative meaning as referring to a supposed fixity of S, and transfer of the whole effective power to P. The former of these forces, acting always in the plane of P's motion, cannot tend to urge it out of that plane: the latter only is so effective, and that not wholly; another *resolution of forces* being needed to estimate its effective part. But with this we shall not concern ourselves, the object here proposed being only to explain the *manner* in which the motion of the nodes arises, and not to estimate its amount.

(499.) In the situation, or *configuration*, as it is termed, represented in the figure, the force, in the direction P K, is a *pulling* force; and as P K, being parallel to S M, lies *below* the plane of P's orbit (taking that of M's orbit for a ground plane), it is clear that the disturbed arc P *q*, described in the next moment by P, must lie *below* P *p*. When prolonged, therefore, to intersect the plane of M's orbit, it will meet it in a point *r*, *behind* R, and the line S *r*, which will be the line of intersection of the plane S P *q*, (now, for an instant, that of P's disturbed motion,) or its new line of nodes, will fall *behind* S R, the undisturbed line of nodes; that is to say, the line of nodes will have *retrograded* by the angle R S *r*, the motions of P and M being regarded as direct.

(500.) Suppose, now, M to lie to the left instead of the right of the line of nodes, P retaining its situation, then will the disturbing force, in the direction P K, tend to *raise* P out of its orbit, to throw P *q* *above* P *p*, and *r* in advance of R. In this configuration, then, the node will *advance*; but so soon as P passes the node, and comes to the lower side of M's orbit, although the same disposition of the forces will subsist, and P *q* will, in consequence, continue to lie above P *p*, yet, in this case, the little arc P q will have to be *prolonged backwards* to meet our *ground plane*, and, when so prolonged, will lie *below* the similar prolongation of P p, so that, in this case again, the node will retrograde.

(501.) Thus we see that the effect of the disturbing force, in the different states of configuration which the bodies P and M may assume with respect to the node, is to keep the line of nodes in a continual state of fluctuation to and fro; and it will depend on the excess of cases favourable to its advance over those which favour its recess, in an average of all the possible configurations, whether, on the whole, an advance or recess of the node shall take place.

(502.) If the orbit of M be very large compared with that of P, so large that M P may, without material error, be regarded as parallel to M S, which is the case with the moon's orbit disturbed by the sun, it will be very readily seen, on an examination of all the possible varieties of configuration, and having due regard to the direction of the disturbing force, that during every single complete revolution of P, the cases favourable to a retrograde motion of the node preponderate over those of a contrary tendency, the retrogradation taking place over a larger extent of the whole orbit, and being at the same time more rapid, owing to a more intense and favourable action of the force than the recess. Hence it follows that, on *the whole*, during every revolution of the moon about the earth, the nodes of her orbit *recede* on the ecliptic, conformable to experience, with a velocity varying from lunation to lunation. The amount of this retrogradation, when calculated, as it may be, by an exact estimation of all the acting forces, is found to coincide with perfect precision with that immediately derived from observation, so that not a doubt can subsist as to this being the real process by which so remarkable an effect is produced.

(503.) Theoretically speaking, we cannot estimate correctly the recess of the intersection of the moon's orbit with the ecliptic, from a mere consideration of the disturbance of one of these planes. It is a compound phænomenon; both planes are in motion with respect to an imaginary fixed ecliptic, and, to obtain the compound effect, we must also regard the earth as disturbed in its relative orbit about the sun by the moon. But, on account of the excessive distance of the sun, the intensity of the moon's attraction *on it* is quite evanescent, compared with its attraction on the earth; so that the *perturbative* effect in this case, which is the difference of the moon's attraction on the sun and earth, is equal to the whole attraction of the moon on the earth. The effect of this is to produce a monthly displacement of the center on either side of the ecliptic, whose amount is easily calculated by regarding their common center of gravity as lying strictly in the ecliptic. From this it appears, that the displacement in question cannot exceed a small fraction of the earth's radius in its *whole* amount; and, therefore, that its momentary variation, on which the motion of the node of the ecliptic on the moon's orbit depends, must be utterly insensible.

(504.) It is otherwise with the mutual action of the planets. In this case, both the orbits of the disturbed and disturbing planet must be regarded as in motion. Precisely on the above stated principles it may be shown, that the effect of each planet's attraction on the orbit of every other, is to cause a retrogradation of the node of the one orbit on the other in certain configurations, and a recess in others, terminating, like that of the moon, on the average of many revolutions in a regular retrogradation of the node of each orbit on *every* other. But since this is the case with every pair into which the planets can be combined, the motion ultimately arising from their joint action on any one orbit, taking into the account the different situations of all their planes, becomes a singular and complicated phænomenon, whose law cannot be very easily expressed in words, though reducible to strict numerical statement, and being in fact a mere geometrical result of what is above stated.

(505.) The nodes of all the planetary orbits on the *true* ecliptic then are retrograde, although (which is a most material circumstance) they are not all so on a fixed plane, such as we may conceive to exist in the planetary system, and to be a plane of reference unaffected by their mutual disturbances. It is, however, to the ecliptic, that we are under the necessity of referring their movements from our station in the system; and if we would transfer our ideas to a fixed plane, it becomes necessary to take account of the variation of the ecliptic itself, produced by the joint action of all the planets.

(506.) Owing to the smallness of the masses of the planets, and their great distances from each other, the revolutions of their nodes are excessively slow, being in every case less than a single degree per century, and in most cases not amounting to half that quantity. So far as the physical condition of each planet is concerned, it is evident that the position of their nodes can be of little importance. It is otherwise with the mutual inclinations of their orbits, with respect to each other, and to the equator of each. A variation in the position of the ecliptic, for instance, by which its pole should shift its distance from the pole of the equator, would disturb our easons. Should the plane of the earth's orbit, for instance, ever be so changed as to bring the ecliptic to coincide with the equator, we should have perpetual spring over all the world; and, on the other hand, should it coincide with a meridian, the extremes of summer and winter would become intolerable. The enquiry, then, of the variations of inclination of the planetary orbits *inter se*, is one of much higher practical interest than those of their nodes.

(507.) Referring to the figure of art. 498., it is evident that the plane S P q, in which the disturbed body moves during an instant of time from its quitting P, is differently inclined to the orbit of M, or to a fixed plane, from the original or undisturbed plane P S p. The difference of absolute position of these two planes in space is the angle made between the planes P S R and P S r, and is therefore calculable by spherical trigonometry, when the angle R S r or the momentary recess of the node is known, and also the inclination of the planes of the orbits to each other. We perceive, then, that between the momentary change of inclination, and the momentary recess of the node there exists an intimate relation, and that the research of the one is in fact bound up in that of the other. This may be, perhaps, made clearer, by considering the orbit of M to be not merely an imaginary line, but an actual circular or elliptic hoop of some rigid material, without inertia, on which, as on a wire, the body P may slide as a bead. It is evident that the position of this hoop will be determined at any instant, by its inclination to the ground plane to which it is referred, and by the place of its intersection therewith, or node. It will also be determined by the momentary direction of P's motion, which (having no inertia) it must obey; and any change by which P should, in the next instant, alter its orbit, would be equivalent to a shifting, bodily, of the whole hoop, changing at once its inclination and nodes.

(508.) One immediate conclusion from what has been pointed out above, is that where the orbits, as in the case of the planetary system and the moon, are slightly inclined to one another, the momentary variations of the inclination are of an order much inferior in magnitude to those in the place of the node. This is evident on a mere inspection of our figure, the angle R P r being, by reason of the small inclination of the planes S P R and R S r, necessarily much smaller than the angle R S r. In proportion as the planes of the orbits are brought to coincidence, a very trifling angular movement of P p about P S as an axis will make a great variation in the situation of the point r, where its prolongation intersects the ground plane.

(509.) To pass from the momentary changes which take place in the relations of nature to the accumulated effects produced in considerable lapses of time by the continued action of the same causes, under circumstances varied by these very effects, is the business of the integral calculus. Without going into any calculations, however, it will be easy for us to trace, by a few cases, the varying influence of differences of position of the disturbing and disturbed body with respect to each other and to the node, and from these to demonstrate the two leading features in this theory — the periodic nature of the change and re-establishment of the original inclinations, and the small limits within which these changes are confined.

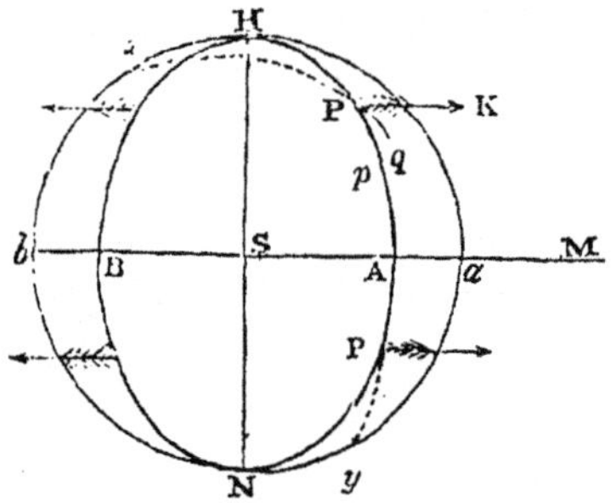

(510.) Case 1. — When the disturbing body M is situated in a direction perpendicular to the line of nodes, or the nodes are in *quadrature* with it: M being the disturbing body, and S N the line of nodes, the disturbing force will act at P, in the direction P K; being a *pulling* force when P is in any part of the semicircle H A N, and a pushing force in the whole of the opposite semicircle. And it is easily seen that this force is greatest at A and B, and evanescent at H and N. Hence, in the whole semicircle H A, P *q* will lie below P *p*, and being produced backwards in the quadrant H A, and forwards in A N, will meet the circle S *b* N *a* in the plane of M's orbit, in points behind the nodes S N, the nodes being retrograde in both cases. But the new inclination of the disturbed orbit is, in the former case, P *x* A, w ich is less than P H *a*; and in the latter, P *y a*, which is greater than P N *a*. In the other semicircle the direction of the disturbing force is changed; but that of the motion, with respect to the plane of M's orbit, being also in each quadrant reversed, the same variations of node and inclination will be caused. In this situation of M, then, the nodes recede during every part of the revolution of P, but the inclination diminishes throughout the quadrant S A, increases again by the same identical degrees in the quadrant A N, decreases throughout the quadrant N *b*, and is finally restored to its pristine value at S. On the average of a revolution of P, supposing M unmoved, the nodes will have retrograded with their utmost speed, but the inclination will remain unaltered

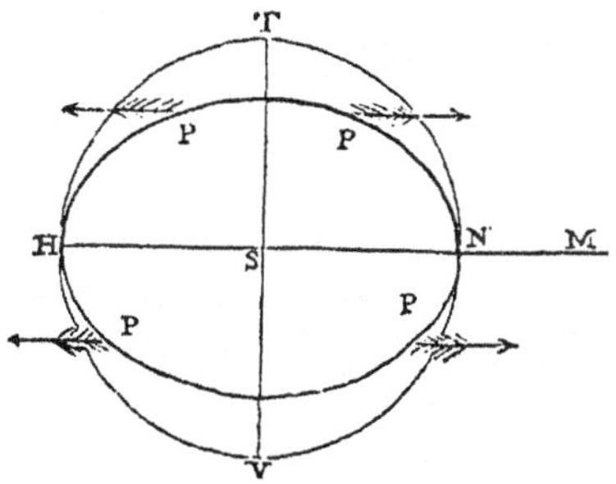

(511.) Case 2. — Suppose the disturbing body now to be fixed in the line of nodes, or the nodes to be in *syzygy*, as in the annexed figure. In this situation the direction of the disturbing force, which is always parallel to S M, lies constantly in the plane of P's orbit, and therefore produces neither variation of inclination nor motion of nodes.

(512.) Case 3. — Let us take now an intermediate situation of M, and indicating by the arrows the directions of the disturbing forces (which are pulling ones throughout all the semi-orbit which lies towards M, and *pushing* in the opposite,) it will readily appear that the reasoning of art. 510. will hold good in all that part of the orbit which lies between T and N, and between V and H, but that the effect will be reversed by the reversal of the direction of the motion with respect to the plane of M's orbit, in the intervals H T and N V. In these portions, however, the disturbing force is feebler than in the others, being evanescent in the *line of quadratures* T V, and increasing to its maximum

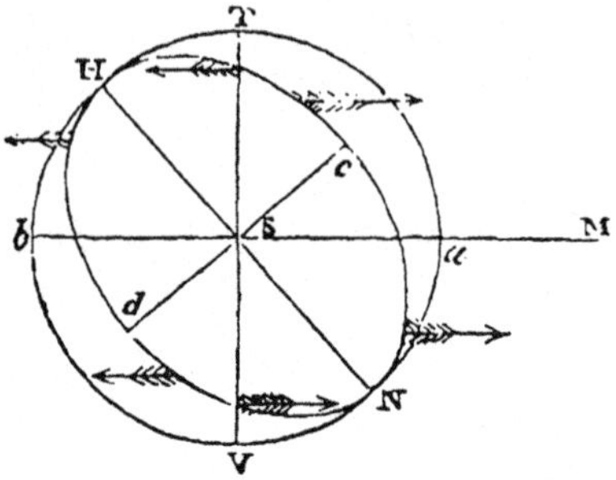

in the *syzygies a b*. The nodes then will recede rapidly in the former intervals, and advance feebly in the latter; but since, as H approaches to *a*, the disturbing force, by acting obliquely to the plane of P's orbit, is again diminished in efficacy, still, on the average of a whole revolution, the nodes recede. On the other hand, the inclination will now diminish during the motion of P from T to *c*, a point 90° distant from the node, while it increases not only during its whole motion over the quadrant *c* N, but also in the rest of its half revolution N V, and so for the other half. There will, therefore be an uncompensated increase of inclination in this position of M, on the average of a whole revolution.

(513.) But this increase is converted into diminution when the line of nodes stands on the other side of S M, or in the quadrants V *b*, T *a*; and still regarding M as fixed, and supposing that the change of circumstances arises not from the motion of M but from that of the node, it is evident that so soon as the line of nodes in its retrograde motion has got past *a*, the circumstances will be all exactly reversed, and the inclination will again be augmented in each revolution by the very same steps taken in reverse order by which it before diminished. On the average, therefore, of A WHOLE REVOLUTION OF THE NODE, the inclination will be restored to its original state. In fact, so far as the mean or average effect on the inclination is concerned, instead of supposing M fixed in one position, we might conceive it at every instant divided into four equal parts, and placed at equal angles on either side of the line of nodes, in which case it is evident that the effect of two of the parts would be to precisely annihilate that of the others in each revolution of P.

(514.) In what is said, we have supposed M at rest; but the same conclusion, as to the mean and final results, holds good if it be supposed in motion: for in the course of a revolution of the nodes, which, owing to the extreme smallness of their motion, in the case of the planets, is of immense length, amounting, in most cases, to several hundred centuries, and in that of the moon is not less than 237 lunations, the disturbing body M is presented *by its own motion*, over and over again, in every variety of situation to the line of nodes. Before the node can have materially changed its position, M has performed a complete revolution, and is restored to its place; so that, in fact (that small difference excepted which arises from the recess of the node in one synodical revolution of M), we may regard it as occupying at every instant every point of its orbit, or rather as having its mass distributed uniformly like a solid ring over its whole circumference. Thus the compensation which we have shown would take place in a whole revolution of the node, does, in fact, take place in every synodic period of M, *that minute difference only excepted* which is due to the cause just mentioned. This difference, then, and *not the whole* disturbing effect of M, is what produces the effective variation of the inclinations, whether of the lunar or planetary orbits; and this difference, which remains uncompensated by the motion of M, is in its turn compensated by the motion of the node during its whole revolution.

(515.) It is clear, therefore, that the total variation of the planetary inclinations must be comprised within very narrow limits indeed. Geometers have accordingly demonstrated, by an accurate analysis of all the circumstances, and an exact estimation of the acting forces, that such is the case; and this is what is meant by asserting the stability of the planetary system as to the mutual inclinations of its orbits. By the researches of Lagrange (of whose analytical conduct it is impossible here to give any idea), the following elegant theorem has been demonstrated:—

"*If the mass of every planet be multiplied by the square root of the major axis of its orbit, and the product by the square of the tangent of its inclination to a fixed plane, the sum of all these products will be constantly the same under the influence of their mutual attraction.*" If the present situation of the plane of the ecliptic be taken for that fixed plane (the ecliptic itself being variable like the other orbits), it is found that this sum is actually very small: it must, therefore, always remain so. This remarkable theorem alone, then, would guarantee the stability of the orbits of the greater planets; but from what has above been shown, of the tendency of each planet to work out a compensation on every other, it is evident that the minor ones are not excluded from this beneficial arrangement.

(516.) Meanwhile, there is no doubt that the plane of the ecliptic does actually vary by the actions of the planets. The amount of this variation is about 48″ per century, and has long been recognized by astronomers, by an increase of the latitudes of all the stars in certain situations, and their diminution in the opposite regions. Its effect is to bring the ecliptic by so much per annum nearer to coincidence with the equator; but from what we have above seen, this diminution of the obliquity of the ecliptic will not go on beyond certain very moderate limits, after which (although in an immense period of ages, being a compound cycle resulting from the joint action of all the planets,) it will again increase, and thus oscillate backward and forward about a mean position, the extent of its deviation to one side and the other being less than 1° 21'.

(517.) One effect of this variation of the plane of the ecliptic,—that which causes its nodes on a fixed plane to change,—is mixed up with the precession of the equinoxes (art. 261.), and undistinguishable from it, except in theory This last-mentioned phænomenon is, however, due to another cause, analogous, it is true, in a general point of view to those above considered, but singularly modified by the circumstances under which it is produced. We shall endeavour to render these modifications intelligible, as far as they can be made so, without the intervention of analytical formulæ.

(518.) The precession of the equinoxes, as we have shown in art. 266., consists in a continual retrogradation of the node of the earth's equator on the ecliptic, and is, therefore, obviously an effect so far analogous to the general phænomenon of the retrogradation of the nodes of the orbits on each other. The immense distance of the planets, however, compared with the size of the earth, and the smallness of their masses compared to that of the sun, puts *their* action out of the question in the enquiry of its cause, and we must, therefore, look to the massive though distant sun, and to our near though minute neighbour, the moon, for its explanation. This will, accordingly, be found in their disturbing action on the redundant matter accumulated on the equator of the earth, by which its figure is rendered spheroidal, combined with the earth's rotation on its axis. It is to the sagacity of Newton that we owe the discovery of this singular mode of action.

(519.) Suppose in our figures (arts. 509, 510, 511.) that instead of one body, P, revolving round S, there were a succession of particles not coherent, but forming a kind of fluid ring, free to change its form by any force applied. Then, while this ring revolved round S in its own plane, under the disturbing influence of the distant body M, (which now represents the moon or the sun, as P does one of the particles of the earth's equator,) two things would happen:—1st, Its figure would be bent out of a plane into an undulated form, those parts of it within the arcs V *c* and T *d* (*fig.* art. 511.) being rendered more inclined to the plane of M's orbit, and those within the arcs *c* T, *d* V, less so than they would otherwise be. 2dly, The nodes of this ring, regarded as a whole, without respect to its change of figure would retreat upon that plane.

(520.) But suppose this ring, instead of consisting of discrete molecules free to move independently, to be rigid and incapable of such flexure, like the *hoop* we have supposed in art. 507., then it is evident that the effort of those parts of it which tend to become more inclined will act through the medium of the ring itself (as a mechanical engine or lever) to counteract the effort of those which have *at the same instant* a contrary tendency. In so far only, then, as there exists an excess on the one or the other side will the inclination change, an average being struck at every moment of the ring's motion; just as was shown to happen in the view we have taken of the inclinations, in every complete revolution of a single disturbed body, under the influence of a fixed disturbing one.

(521.) Meanwhile, however, the nodes of the rigid ring will retrograde, the general or average tendency of the nodes of every molecule being to do so. Here, as in the other case, a struggle will take place by the counteracting efforts of the molecules contrarily disposed, propagated through the solid substance of the ring; and thus, at every instant of time, an average will be struck, which average being identical in its nature with that effected in the complete revolution of a single disturbed body, will, in every case, be in favour of a recess of the node, save only when the disturbing body, be it sun or moon, is situated in the plane of the earth's equator, or in the case of the *fig.* art. 510.

(522.) This reasoning is evidently independent of any consideration of the cause which maintains the rotation of the ring; whether the particles be small satellites retained in circular orbits under the equilibrated action of attractive and centrifugal forces, or whether they be small masses conceived as attached to a set of imaginary spokes as of a wheel, centering in S, and free only to shift their planes by a motion of those spokes perpendicular to the plane of the wheel. This makes no difference in the *general* effect; though the different velocities of rotation, which may be impressed on such a system, may and will have a very great influence both on the absolute and relative magnitudes of the two effects in question—the motion of the nodes and change of inclination. This will be easily understood, if we suppose the ring *without* a rotatory motion, in which extreme case it is obvious, that so long as M remained fixed there would take place no recess of nodes at all, but only a tendency of the ring to tilt its plane round a diameter perpendicular to the position of M, bringing it towards the line S M.

(*523.*) The motion of such a ring, then, as we have been considering, would imitate, so far as the recess of the nodes goes, the precession of the equinoxes, only that its nodes would retrograde far more rapidly than the observed precession, which is excessively slow. But now conceive this ring to be loaded with a spherical mass enormously heavier than itself, placed concentrically within it, and cohering firmly to it, but indifferent, or very nearly so, to any such cause of motion; and suppose, moreover, that instead of one such ring there are a vast multitude heaped together around the equator of such a globe, so as to form an elliptical protuberance, enveloping it like a shell on all sides, but whose mass, taken together, should form but a very minute fraction of the whole spheroid. We have now before us a tolerable representation of the case of nature*; and

* That a perfect sphere would be so inert and indifferent as to a revolution of the nodes of its equator under the influence of a distant attracting body appears from this,—that the direction of the resultant attraction of such a body, or of that single force which, opposed, would neutralize and destroy its whole action, is necessarily in a line passing through the center of the sphere, and, therefore, can have no tendency to turn the sphere one way or other. It may be objected by the reader, that the whole sphere may be conceived as consisting of rings parallel to its equator, of every possible diameter, and that, therefore, its nodes should retrograde even without a protuberant equator. The inference is incorrect, but our limits will not allow us to go intc an exposition of the fallacy. We should, however, caution him, generally, that no dynamical subject is open to more mistakes of this kind, which nothing but the closest attention, in every varied point of view, will detect.

it is evident that the rings, having to drag round with them in their nodal revolution this great inert mass, will have their velocity of retrogradation proportionally diminished. Thus, then, it is easy to conceive how a motion, similar to the precession of the equinoxes, and, like it, characterized by extreme slowness, will arise from the causes in action.

(524.) Now a recess of the node of the earth's equator, upon a given plane, corresponds to a conical motion of its axis round a perpendicular to that plane. But in the case before us, that plane is not the ecliptic, but the moon's orbit for the time being; and it may be asked how we are to reconcile this with what is stated in art. 266. respecting the nature of the motion in question. To this we reply, that the nodes of the lunar orbit, being in a state of continual and rapid retrogradation, while its inclination is preserved nearly invariable, the point in the sphere of the heavens round which the pole of the earth's axis revolves (with that extreme slowness characteristic of the precession) is itself in a state of continual circulation round the pole of the ecliptic, with that much more rapid motion which belongs to

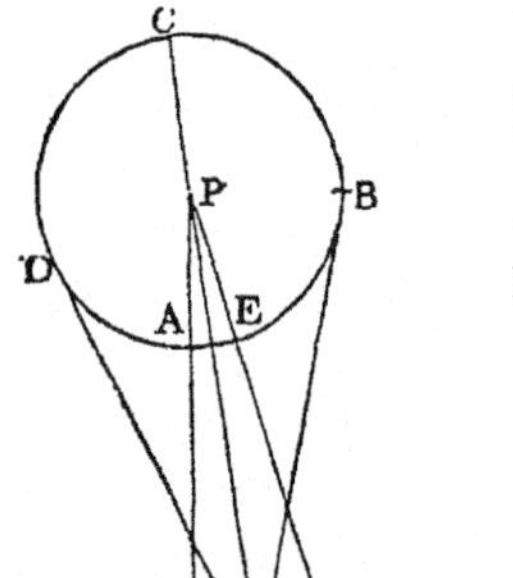

the lunar node. A glance at the annexed figure will explain this better than words. P is the pole of the ecliptic, A the pole of the moon's orbit, moving round the small circle A B C D in 19 years; *a* the pole of the earth's equator, which at each moment of its progress has a *direction* perpendicular to the varying position of the line A *a*, and a *velocity* depending on the varying intensity of the acting causes during the period of the nodes. This velocity, however, being extremely small, when A comes to B, C, D, E, the line A *a* will have taken up the positions B *b*, C *c*, D *d*, E *e*, and the earth's pole *a* will thus, in one tropical revolution of the node, have arrived at *e*, having described not an exactly circular arc, but a single undulation of a wave-shaped or epicycloidal curve, *a b c d e*, with a velocity alternately greater and less than its mean motion, and this will be repeated in every succeeding revolution of the node.

(525.) Now this is precisely the kind of motion which, as we have seen in art. 272., the pole of the earth's equator really has round the pole of the ecliptic, in consequence of the joint effects of precession and nutation, which are thus uranographically represented. If we superadd to the effect of lunar precession that of the solar, which alone would cause the pole to describe a circle uniformly about P, this will only affect the undulations of our waved curve, by extending them in length, but will produce no effect on the depth of the waves, or the excursions of the earth's axis to and from the pole of the ecliptic. Thus we see that the two phænomena of nutation and precession are intimately connected, or rather, both of them essential constituent parts of one and the same phænomenon. It is hardly necessary to state that a rigorous analysis of this great problem, by an exact estimation of all the acting forces and summation of their dynamical effects*, leads to the precise value of the co-efficients of precession and nutation, which observation assigns to them. The solar and lunar portions of the precession of the equinoxes, that is to say, those portions which are uniform, are to each other in the proportion of about 2 to 5.

(526.) In the nutation of the earth's axis we have an example (the first of its kind which has occurred to us), of a periodical movement in one part of the system, giving rise to a motion having the same precise period in another. The motion of the moon's nodes is here, we see, represented, though under a very different form, yet in the same exact periodic time, by the movement of a peculiar oscillatory kind impressed on the solid mass of the earth. We must not let the opportunity pass of generalizing the principle involved

* Vide Prof. Airy's Mathematical Tracts, 2d ed. p. 200, &c.

in this result, as it is one which we shall find again and again exemplified in every part of physical astronomy, nay, in every department of natural science. It may be stated as "the principle of forced oscillations, or of forced vibrations," and thus generally announced: —

*If one part of any system connected either by material ties, or by the mutual attractions of its members, be continually maintained by any cause, whether inherent in the constitution of the system or external to it, in a state of regular periodic motion, that motion will be propagated throughout the whole system, and will give rise, in every member of it, and in every part of each member, to periodic movements executed in equal periods with that to which they owe their origin, though not necessarily synchronous with them in their maxima and minima.**

The system may be favourably or unfavourably constituted for such a transfer of periodic movements, or favourably in some of its parts and unfavourably in others; and, accordingly as it is the one or the other, the *derivative* oscillation (as it may be termed) will be imperceptible in one case, of appreciable magnitude in another, and even more perceptible in its visible effects than the original cause, in a third; of this last kind we have an instance in the moon's acceleration, to be hereafter noticed.

(527.) It so happens that our situation on the earth, and the delicacy which our observations have attained, enable us to make it as it were an instrument to *feel* these forced vibrations, — these derivative motions, communicated from various quarters, especially from our near neighbour, the moon, much in the same way as we detect, by the trembling of a board beneath us, the secret transfer of motion by which the sound of an organ pipe is dispersed through the air, and carried down into the earth. Accordingly, the monthly revo-

* See a demonstration of this theorem for the forced vibrations of systems connected by material ties of imperfect elasticity, in my treatise on Sound, Encyc. Metrop. art. 323. The demonstration is easily extended and generalized to take in other systems. — *Author.*

lution of the moon, and the annual motion of the sun, produce, each of them, small *nutations* in the earth's axis, whose periods are respectively half a month and half a year, each of which, in this view of the subject, is to be regarded as one portion of a period consisting of two equal and similar parts. But the most remarkable instance, by far, of this propagation of periods, and one of high importance to mankind, is that of the tides, which are forced oscillations, excited by the rotation of the earth in an ocean disturbed from its figure by the varying attractions of the sun and moon, each revolving in its own orbit, and propagating its own period into the joint phænomenon.

(528.) The tides are a subject on which many persons find a strange difficulty of conception. That the moon, by her attraction, should heap up the waters of the ocean under her, seems to most persons very natural, — that the same cause should, at the same time, heap them up on the opposite side, seems to many palpably absurd. Yet nothing is more true, nor indeed more evident, when we consider that it is not by her *whole* attraction, but by the differences of her attractions at the two surfaces and at the center that the waters are raised, — that is to say, by forces directed precisely as the arrows in our figure, art. 510., in which we may suppose M the moon, and P a particle of water on the earth's surface. A drop of water existing alone would take a spherical form, by reason of the attraction of its parts; and if the same drop were to fall freely in a vacuum under the influence of an *uniform* gravity, since every part would be equally accelerated, the particles would retain their relative positions, and the spherical form be unchanged. But suppose it to fall under the influence of an attraction acting on each of its particles independently, and increasing in intensity at every step of the descent, then the parts nearer the center of attraction would be attracted more than the central, and the central than the more remote, and the whole would be drawn out in the direction of the motion into an oblong form; the tendency to separation being, however, counteracted by the attraction of the particles on each other, and a form of equilibrium being thus established. Now, in fact, the earth *is* constantly falling to the moon, being continually drawn by it out of its path, the nearer parts more and the remoter less so than the central; and thus, at every instant, the moon's attraction acts to force down the water at the sides, at right angles to her direction, and raise it at the two ends of the diameter pointing towards her. Geometry corroborates this view of the subject, and demonstrates that the form of equilibrium assumed by a layer of water covering a sphere, under the influence of the moon's attraction, would be an oblong ellipsoid, having the semi-axis directed towards the moon longer by about 58 inches than that transverse to it.

(529.) There is never time, however, for this spheroid to be fully formed. Before the waters can take their level, the moon has advanced in her orbit, both diurnal and monthly (for in this theory it will answer the purpose of clearness better if we suppose the earth's diurnal motion transferred to the sun and moon in the contrary direction), the vertex of the spheroid has shifted on the earth's surface, and the ocean has to seek a new bearing. The effect is to produce an immensely broad and excessively flat wave (not a circulating *current*), which follows, or endeavours to follow, the apparent motions of the moon, and must, in fact, if the principle of forced vibrations be true, imitate by equal, though not by *synchronous*, periods, all the periodical inequalities of that motion. When the higher or lower parts of this wave strike our coasts, they experience what we call high and low water.

(530.) The sun also produces precisely such a wave, whose vertex tends to follow the apparent motion of the sun in the heavens, and also to imitate its periodic inequalities. This solar wave co-exists with the lunar — is sometimes superposed on it, sometimes transverse to it, so as to partly neutralize it, according to the monthly synodical configuration of the two luminaries. This alternate mutual reinforcement and destruction of the solar and lunar tides cause what are called the spring and neap tides — the former being their sum, the latter their difference. Although the real amount of either tide is, at present, hardly within the reach of exact calculation, yet their proportion at any one place is probably not very remote from that of the ellipticities which would belong to their respective spheroids, could an equilibrium be attained. Now these ellipticities, for the solar and lunar spheroids, are respectively about two and five feet; so that the average spring tide will be to the neap as 7 to 3, or thereabouts.

(531.) Another effect of the combination of the solar and lunar tides is what is called the *priming* and *lagging* of the tides. If the moon alone existed, and moved in the plane of the equator, the tide-day (*i. e.* the interval between two successive arrivals at the same place of the same vertex of the tide-wave) would be the lunar day (art. 115.), formed by the combination of the moon's sidereal period and that of the earth's diurnal motion. Similarly, did the sun alone exist, and move always on the equator, the tide-day would be the mean solar day. The actual tide-day, then, or the interval of the occurrence of two successive *maxima* of their superposed waves, will vary as the separate waves approach to or recede from coincidence; because, when the vertices of two waves do not coincide, their joint height has its maximum at a point intermediate between them. This variation from uniformity in the lengths of successive tide-days is particularly to be remarked about the time of the new and full moon.

(532.) Quite different in its origin is that deviation of the time of high and low water at any port or harbour, from the culmination of the luminaries, or of the theoretical maximum of their superposed spheroids, which is called the "establishment" of that port. If the water were without inertia, and free from obstruction, either owing to the friction of the bed of the sea, — the narrowness of channels along which the wave has to travel before reaching the port,—their length, &c. &c., the times above distinguished would be identical. But all these causes tend to create a difference, and to make that difference not alike at all ports. The observation of the establishments of harbours is a point of great maritime importance; nor is it of less consequence, theoretically speaking, to a knowledge of the true distribution of the tide waves over the globe.* In making such observations, care must be taken not to confound the time of "slack water," when the current caused by the tide ceases to flow visibly one way or the other, and that of *high* or *low water*, when the level of the surface ceases to rise or fall. These are totally distinct phænomena, and depend on entirely different causes, though it is true they may sometimes coincide in point of time. They are, it is feared, too often mistaken one for the other by practical men; a circumstance which, whenever it occurs, must produce the greatest confusion in any attempt to reduce the system of the tides to distinct and intelligible laws.

(533.) The declination of the sun and moon materially affects the tides at any particular spot. As the vertex of the tide-wave tends to place itself vertically under the luminary which produces it, when this vertical changes its point of incidence on the surface, the tide-wave must tend to shift accordingly, and thus, by monthly and annual periods, must tend to increase and diminish alternately the principal tides. The period of the moon's nodes is thus introduced into this subject; her excursions in declination in one part of that period being 29°, and in another only 17°, on either side the equator.

(534.) Geometry demonstrates that the efficacy of a

* The recent investigations of Mr. Lubbock, and those highly interesting ones in which Mr. Whewell is understood to be engaged, will, it is to be hoped, not only throw theoretical light on the very obscure subject of the tides, but (what is at present quite as much wanted) arouse the attention of observers, and at the same time give it that right direction, by pointing out *what ought to be observed*, without which all observation is lost labour.

luminary in raising tides is inversely proportional to the cube of its distance. The sun and moon, however, by reason of the ellipticity of their orbits, are alternately nearer to and farther from the earth than their mean distances. In consequence of this, the efficacy of the sun will fluctuate between the extremes 19 and 21, taking 20 for its mean value, and that of the moon between 43 and 59. Taking into account this cause of difference, the highest spring tide will be to the lowest neap as $59 + 21$ to $43 - 19$, or as 80 to 24, or 10 to 3. Of all the causes of differences in the height of tides, however, local situation is the most influential. In some places, the tide-wave rushing up a narrow channel, is suddenly raised to an extraordinary height. At Annapolis, for instance, in the Bay of Fundy, it is said to rise 120 feet.* Even at Bristol, the difference of high and low water occasionally amounts to 50 feet.

(535.) The action of the sun and moon, in like manner, produces tides in the atmosphere, which delicate observations have been able to render sensible and mea surable. This effect, however, is extremely minute.

(536.) To return, now, to the plantary perturbations Let us next consider the changes induced by their mutual action on the magnitudes and forms of their orbits, and in their positions therein in different situations with respect to each other. In the first place, however, it will be proper to explain the conventions under which geometers and astronomers have alike agreed to use the language and laws of the elliptic system, and to continue to apply them to disturbed orbits, although those orbits so disturbed are no longer, in mathematical strictness, ellipses, or any known curves. This they do, partly on account of the convenience of conception and calculation which attaches to this system, but much more for this reason,—that it is found, and may be demonstrated from the dynamical relations of the case, that the departure of each planet from its ellipse, as de-

* Robison's Lectures on Mechanical Philosophy.

termined at any epoch, is capable of being truly represented, by supposing the ellipse itself to be slowly variable, to change its magnitude and excentricity, and to shift its position and the plane in which it lies according to certain laws, while the planet all the time continues to move in this ellipse, just as it would do if the ellipse remained invariable and the disturbing forces had no existence. By this way of considering the subject, the whole permanent effect of the disturbing forces is regarded as thrown upon the orbit, while the relations of the planet to that orbit remain unchanged, or only liable to brief and comparatively momentary fluctuation. This course of procedure, indeed, is the most natural, and is in some sort forced upon us by the extreme slowness with which the variations of the elements develope themselves. For instance, the fraction expressing the excentricity of the earth's orbit changes no more than 0·00004 in its amount in a *century;* and the place of its perihelion, as referred to the sphere of the heavens, by only 19′ 39″ in the same time. For several years, therefore, it would be next to impossible to distinguish between an ellipse so varied and one that had not varied at all; and in a single revolution, the difference between the original ellipse and the curve really represented by the varying one, is so excessively minute, that, if accurately drawn on a table, six feet in diameter, the nicest examination with microscopes, continued along the whole outlines of the two curves, would hardly detect any perceptible interval between them. Not to call a motion so minutely conforming itself to an elliptic curve, *elliptic,* would be affectation, even granting the existence of trivial departures alternately on one side or on the other; though, on the other hand, to neglect a variation, which continues to accumulate from age to age, till it forces itself on our notice, would be wilful blindness.

(537.) Geometers, then, have agreed in each single revolution, or for any moderate interval of time, to regard the motion of each planet as elliptic, and performed according to Kepler's laws, with a reserve in favour of certain very small and transient fluctuations, but at the same time to regard all the *elements* of each ellipse as in a continual, though extremely slow, state of change; and, in tracing the effects of perturbation on the system, they take account principally, or entirely, of this change of the elements, as that upon which, after all, any material change in the great features of the system will ultimately depend.

(538.) And here we encounter the distinction between what are termed secular variations, and such as are rapidly periodic, and are compensated in short intervals. In our exposition of the variation of the inclination of a disturbed orbit (art. 514.), for instance, we showed that, in each single revolution of the disturbed body, the plane of its motion underwent fluctuations to and fro in its inclination to that of the disturbing body, which nearly compensated each other; leaving, however, a portion outstanding, which again is nearly compensated by the revolution of the disturbing body, yet still leaving outstanding and uncompensated a minute portion of the change, which requires a whole revolution of the node to compensate and bring it back to an average or mean value. Now, the two first compensations which are operated by the planets going through the succession of configurations with each other, and therefore in comparatively short periods, are called periodic variations; and the deviations thus compensated are called *inequalities depending on configurations;* while the last, which is operated by a period of the node (one of the *elements*), has nothing to do with the configurations of the individual planets, requires an immense period of time for its consummation, and is, therefore, distinguished from the former by the term *secular* variation.

(539.) It is true, that, to afford an exact representation of the motions of a disturbed body, whether planet or satellite, both periodical and secular variations, with their corresponding inequalities, require to be expressed; and, indeed, the former even more than the latter; seeing that the secular inequalities are, in fact, nothing but what remains after the mutual destruction of a much larger amount (as it very often is) of periodical. But these are in their nature transient and temporary: they disappear, and leave no trace. The planet is temporarily drawn from its orbit (its slowly varying orbit), but forthwith returns to it, to deviate presently as much the other way, while the varied orbit accommodates and adjusts itself to the average of these excursions on either side of it; and thus continues to present, for a succession of indefinite ages, a kind of medium picture of all that the planet has been doing in their lapse, in which the expression and character is preserved; but the individual features are merged and lost. These periodic inequalities, however, are, as we have observed, by no means to be neglected, but they are taken account of by a separate process, independent of the secular variations of the elements.

(540.) In order to avoid complication, while endeavouring to give the reader an insight into both kinds of variations, we shall henceforward conceive all the orbits to lie in one plane, and confine our attention to the case of two only, that of the disturbed and disturbing body, a view of the subject which (as we have seen) comprehends the case of the moon disturbed by the sun, since any one of the bodies may be regarded as fixed at pleasure, provided we conceive all its motions transferred

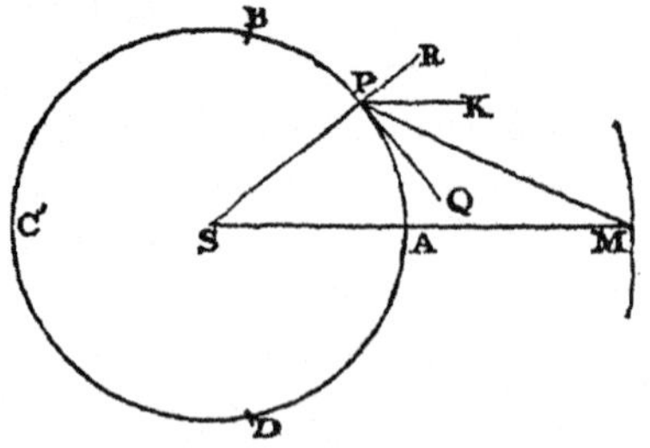

in a contrary direction to each of the others. Suppose, therefore, S to be the central, M the disturbing, and P the disturbed body. Then the attraction of M acts on

P in the direction P M, and on S in the direction S M. And the disturbing part of M's attraction, being the difference only of these forces, will have no fixed direction, but will act on P very differently, according to the configurations of P and M. It will therefore be necessary, in analyzing its effect, to resolve it, according to mechanical principles, into forces acting according to some certain directions; viz., along the radius vector S P, and perpendicular to it. The simplest way to do this, is to resolve the attractions of M on both S and P in these directions, and take, in both cases, their difference, which is the disturbing part of M's effect. In this estimation, it will be found then that two distinct disturbing powers originate; one, which we shall call the *tangential* force, acting in the direction P Q, perpendicular to S P, and therefore in that of a tangent to the orbit of P, supposed nearly a circle—the other, which may be called the *radial* disturbing force, whose direction is always either to or from S.

(541.) It is the former alone (art. 419.) which disturbs the equable description of areas of P about S, and is therefore the chief cause of its angular deviations from the elliptic place. For the equable description of areas depends on no particular law of central force, but only requires that the acting force, whatever it be, should be directed to the center; whatever force does not conform to this condition, must disturb the areas.

(542.) On the other hand, the radial portion of the disturbing force, though, being always directed to or from the center, it does not affect the equable description of areas, yet, as it does not conform in its law of variation to that simple law of gravity by which the elliptic figure of the orbits is produced and maintained, has a tendency to disturb this form; and, causing the disturbed body P, now to approach the center nearer, now to recede farther from it, than the laws of elliptic motion would warrant, and to have its points of nearest approach and farthest recess otherwise situated than they would be in the undisturbed orbit, tends to derange the magnitude, excentricity, and position of the axis of P's ellipse.

(543.) If we consider the variation of the tangential force in the different relative positions of M and P, we shall find that, generally speaking, it vanishes when P is at A or C, see *fig.* to art. 540. *i. e.* in conjunction with M, and also at two points, B and D, where M is equidistant from S and P (or very nearly in the quadratures of P with M); and that, between A and B, or D, tends to urge P towards A, while, in the rest of the orbit, its tendency is to urge it towards C. Consequently, the general effect will be, that in P's progress through a complete *synodical* revolution round its orbit from A, it will first be accelerated from A up to B — thence retarded till it arrives at C — thence again accelerated up to D, and again retarded till its re-arrival at the conjunction A.

(544.) If P's orbit were an exact circle, as well as M's, it is evident that the retardation which takes place during the description of the arc A B would be exactly compensated by the acceleration in the arc D A, these arcs being just equal, and similarly disposed with respect to the disturbing forces; and similarly, that the acceleration through the arc B C would be exactly compensated by the retardation along C D. Consequently, on the average of each revolution of P, a compensation would take place; the period would remain unaltered, and all the errors in longitude would destroy each other.

(545.) This exact compensation, however, depends evidently on the exact symmetry of disposal of the parts of the orbits on either side of the line C S M. If that symmetry be broken, it will no longer take place, and inequalities in P's motion will be produced, which extend beyond the limit of a single revolution, and must await their compensation, if it ever take place at all, in a reversal of the relations of configuration which produced them. Suppose, for example, that, the orbit of P being circular, that of M were elliptic and that, at the moment when P set out from A, M were at its greatest distance from P; suppose, also, that M were so distant as to make only a small part of its whole revolution during a revolution of P. Then it is clear that, during the whole revolution of P, M's disturbing force would be on the increase by the approach of M, and that, in consequence, the disturbance arising in each succeeding quadrant of its motion, would over-compensate that produced in the foregoing; so that, when P had come round again to its conjunction with M, there would be found on the whole to have taken place an over-compensation in favour of an acceleration in the orbitual motion. This kind of action would go on so long as M continued to approach S; but when, in the progress of its elliptic motion, it began again to recede, the reverse effect would take place, and a retardation of P's orbitual motion would happen; and so on alternately, until at length, in the average of a great many revolutions of M, in which the place of P in its ellipse at the moment of conjunction should have been situated in every variety of distance, and of approach and recess, a compensation of a higher and remoter order, among all those successive over and under-compensations, would have taken place, and a mean or average angular motion would emerge, the same as if no disturbance had taken place.

(546.) The case is only a little more complicated, but the reasoning very nearly similar, when the orbit of the disturbed body is supposed elliptic. In an elliptic orbit, the angular velocity is not uniform. The disturbed body then remains in some parts of its revolution longer, in others for a shorter time, under the influence of the accelerating and retarding tangential forces, than is necessary for an exact compensation; independent, then, of any approach or recess of M, there would, on this account alone, take place an over or under compensation, and a surviving, unextinguished perturbation at the end of a synodic period; and, *if the conjunctions always took place on the same point of* P's *ellipse*, this cause would constantly act one way, and an inequality would arise, having no compensation, and which would at length, and permanently, change the *mean* angular motion of P. But this can never be the case in the planetary system. The mean motions (*i. e.* the mean angular velocities) of the planets in their orbits, are *incommensurable* to one another. There are no two planets, for instance, which perform their orbits in times exactly double, or triple, the one of the other, or of which the one performs exactly two revolutions while the other performs exactly three, or five, and so on. If there were, the case in point would arise. Suppose, for example, that the mean motions of the disturbed and disturbing planet were exactly in the proportion of two to five; then would a cycle, consisting of five of the shorter periods, or two of the longer, bring them back exactly to the same configuration. It would cause their conjunction, for instance, to happen once in every such cycle, in the same precise points of their orbits, *while in the intermediate* periods of the cycle the other configurations kept shifting round. Thus, then, would arise the very case we have been contemplating, and a permanent derangement would happen.

(547.) Now, although it is true that the mean motions of no two planets are exactly commensurate, yet cases are not wanting in which there exists an approach to this adjustment. And, in particular, in the case of Jupiter and Saturn,—that cycle we have taken for our example in the above reasoning, viz. a cycle composed of five periods of Jupiter and two of Saturn,—although it does not *exactly* bring about the same configuration, does so pretty nearly. Five periods of Jupiter are 21663 days, and two periods of Saturn 21518 days. The difference is only 145 days, in which Jupiter describes, on an average, 12°, and Saturn about 5°, so that after the lapse of the former interval they will only be 5° from a conjunction in the same parts of their orbits as before. If we calculate the time which will exactly bring about, on the average, three conjunctions of the two planets, we shall find it to be 21760 days, their synodical period being 7253·4 days. In this interval Saturn will have described 8° 6′ in excess of two sidereal revolutions, and Jupiter the same angle in excess of five. Every third conjunction, then, will take place 8° 6′ in advance of the preceding, which is near enough to establish, not, it is true, an identity with, but still a great approach to the case in question. The excess of action, for several such triple conjunctions (7 or 8) in succession, will lie the same way, and at each of them the motion of P will be similarly influenced, so as to accumulate the effect upon its longitude; thus giving rise to an irregularity of considerable magnitude and very long period, which is well known to astronomers by the name of the great inequality of Jupiter and Saturn.

(548.) The arc 8° 6′ is contained $44\frac{4}{9}$ times in the whole circumference of 360°; and accordingly, if we trace round this particular conjunction, we shall find it will return to the same point of the orbit in so many times 21760 days, or in 2648 years. But the conjunction we are now considering, is only one out of three. The other two will happen at points of the orbit about 123° and 246° distant, and *these points also will advance* by the same arc of 8° 6′ in 21760 days. Consequently, the period of 2648 years will bring them *all* round, and in that interval each of them will pass through that point of the two orbits from which we commenced: hence *a conjunction* (one or other of the three) will happen at that point once in one third of this period, or in 883 years; and this is, therefore, the cycle in which the "great inequality" would undergo its full compensation, did the elements of the orbits continue all that time invariable. Their variation, however, is considerable in so long an interval; and, owing to this cause, the period itself is prolonged to about 918 years.

(549.) We have selected this inequality as a proper instance of the action of the tangential disturbing force.

on account of its magnitude, the length of its period, and its high historical interest. It had long been remarked by astronomers, that on comparing together modern with ancient observations of Jupiter and Saturn, the mean motions of these planets did not appear to be uniform. The period of Saturn, for instance, appeared to have been lengthening throughout the whole of the seventeenth century, and that of Jupiter shortening — that is to say, the one planet was constantly lagging behind, and the other getting in advance of its calculated place. On the other hand, in the eighteenth century, a process precisely the reverse seemed to be going on. It is true, the whole retardations and accelerations observed were not very great; but, as their influence went on accumulating, they produced, at length, material differences between the observed and calculated places of both these planets, which, as they could not then be accounted for by any theory, excited a high degree of attention, and were even, at one time, too hastily regarded as almost subversive of the Newtonian doctrine of gravity. For a long while this difference baffled every endeavour to account for it, till at length Laplace pointed out its cause in the near commensurability of the mean motions, as above shown and succeeded in calculating its period and amount.

(550.) The inequality in question amounts, at its maximum, to an alternate retardation and acceleration of about 0° 49′ in the longitude of Saturn, and a corresponding acceleration or retardation of about 0° 21′ in that of Jupiter. That an acceleration in the one planet must necessarily be accompanied by a retardation in the other, and *vice versâ*, is evident, if we consider, that action and reaction being equal, and in contrary directions, whatever momentum Jupiter communicates to Saturn in the direction PM, the same momentum must Saturn communicate to Jupiter in the direction MP. The one, therefore, will be dragged forward, whenever the other is pulled back in its orbit. Geometry demonstrates, that, on the average of each revolution, the proportion in which this reaction will affect the longitudes of the two planets is that of their masses multiplied by the square roots of the major axes of their orbits, inversely, and this result of a very intricate and curious calculation is fully confirmed by observation.

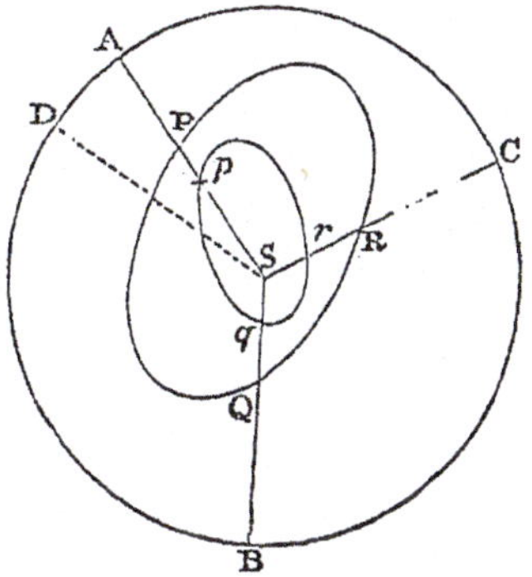

(551.) The inequality in question would be much greater, were it not for the partial compensation which is operated in it in every triple conjunction of the planets. Suppose PQR to be Saturn's orbit, and *pqr* Jupiter's; and suppose a conjunction to take place at P*p*; on the line SA; a second at 123° distance, on the line SB; a third at 246° distance, on SC; and the next at 368°, on SD. This last-mentioned conjunction, taking place nearly in the situation of the first, will produce nearly a repetition of the first effect in retarding or accelerating the planets; but the other two, being in the most remote situations possible from the first, will happen under entirely different circumstances as to the position of the perihelia of the orbits. Now, we have seen that a presentation of the one planet to the other in conjunction, in a variety of situations, tends to produce compensation; and, in fact, the greatest possible amount of compensation which can be produced by only three configurations is when they are thus equally distributed round the center. Three positions of conjunction compensate more than two, four than three, and so on. Hence we see that it is not the whole amount of perturbation, which is thus accumulated in each triple conjunction, but only that small part which is left uncompensated by the intermediate ones. The reader, who possesses already some acquaintance with the subject, will not be at a loss to perceive how this consideration is, in fact, equivalent to that part of the geometrical investigation of this inequality which leads us to seek its expression in terms of the third order, or involving the cubes and products of three dimensions of the excentricities; and how the continual accumulation of small quantities, during long periods, corresponds to what geometers intend when they speak of small terms receiving great accessions of magnitude by integration.

(552.) Similar considerations apply to every case of approximate commensurability which can take place among the mean motions of any two planets. Such, for instance, is that which obtains between the mean motion of the earth and Venus,—13 times the period of Venus being very nearly equal to 8 times that of the earth. This gives rise to an extremely near coincidence of every fifth conjunction, in the same parts of each orbit (within $\frac{1}{240}$th part of a circumference), and therefore to a correspondingly extensive accumulation of the resulting uncompensated perturbation. But, on the other hand, the part of the perturbation thus accumulated is only that which remains outstanding after passing the equalizing ordeal of five conjunctions equally distributed round the circle; or, in the language of geometers, is dependent on powers and products of the excentricities and inclinations of the fifth order. It is, therefore, extremely minute, and the whole resulting inequality, according to the recent elaborate calculations of professor Airy, to whom it owes its detection, amounts to no more than a few seconds at its maximum, while its period is no less than 240 years. This example will serve to show to what minuteness these enquiries have been carried in the planetary theory.

(553.) In the theory of the moon, the tangential force gives rise to many inequalities, the chief of which is that called the variation, which is the direct and principal effect of that part of the disturbance arising from the alternate acceleration and retardation of the areas from the syzigies to the quadratures of the orbit, and *vice versâ*, combined with the elliptic form of the orbit; in consequence of which, the same area described about the focus will, in different parts of the ellipse, correspond to different amounts of angular motion. This inequality, which at its maximum amounts to about 37′, was first distinctly remarked as a periodical correction of the moon's place by Tycho Brahe, and is remarkable in the history of the lunar theory, as the first to be explained by Newton from his theory of gravitation.

(554.) We come now to consider the effects of that part of the disturbing force which acts in the direction of the radius vector, and tends to alter the law of gravity, and therefore to derange, in a more direct and sensible manner than the tangential force, the form of the disturbed orbit from that of an ellipse, or, according to the view we have taken of the subject in art. 536, to produce a change in its magnitude, excentricity, and position in its own plane, or in the place of its perihelion.

(555.) In estimating the disturbing force of M* on P, we have seen that the difference only of M's accelerative attraction on S and P is to be regarded as effective as such, and that the first resolved portion of M's attraction,—that, namely, which acts at P in the direction P S,—not finding in the power which M exerts on P any corresponding part, by which its effects may be nullified, is wholly effective to urge P towards S in addition to its natural gravity. This force is called the *addititious* part of the disturbing force. There is, besides this, another power, acting also in the direction of the radius S P, which is that arising from the difference of actions of M on S and P, estimated first in the direction P K, parallel to S M, and then resolved into two forces; one of which is the tangential force, already considered, in the direction P Q; the other perpendicular to it, or in the direction P R. This part of M's action is termed

* See Fig. Art. (540.)

the *ablatitious* force, because it tends to diminish the gravity of P towards S; and it is the excess of the one of these resolved portions over the other, which, in any assigned position of P and M, constitutes the *radial* part of the disturbing force, and respecting whose effects we are now about to reason.

(556.) The estimation of these forces is a matter of no difficulty when the dimensions of the orbits are given, but they are too complicated in their expressions to find any place here. It will suffice for our purpose to point out their general tendency; and, in the first place, we shall consider their mean or average effect. In order to estimate, what, in any one position of P, will be the mean action of M in all the situations it can hold with respect to P, we have nothing to do but to suppose M broken up, and distributed in the form of a thin ring round the circumference of its orbit. If we would take account of the elliptic motion of M, we might conceive the thickness of this ring in its different parts to be proportional to the *time* which M occupies in every part of its orbit, or in the inverse proportion of its angular motion. But into this nicety we shall not go, but content ourselves, in the first instance, with supposing M's orbit circular and its motion uniform. Then it is clear that the mean disturbing effect on P will be the difference of attractions of that ring on the two points P and S, of which the latter occupies its center, the former is excentric. Now the attraction of a ring on its center is manifestly equal in all directions, and therefore, estimated in any one direction, is zero. On the other hand, on a point P out of its center, if *within* the ring, the resulting attraction will always be *outwards*, towards the nearest point of the ring, or directly from the center.*

* As this is a proposition which the equilibrium of Saturn's ring renders not merely speculative or illustrative, it will be well to demonstrate it; which may be done very simply, and without the aid of any calculus. Conceive a spherical shell, and a point within it: every line passing through the point, and terminating both ways in the shell, will, of course, be equally inclined to its surface at either end, being a chord of a spherical surface, and, therefore, symmetrically related to all its parts. Now, conceive a small double cone, or pyramid, having its apex at the point, and formed by the conical motion of such a line round the point. Then will the two portions

But if P lie without the ring, the resulting force will act always *inwards*, urging P towards its center. Hence it appears that the mean effect of the radial force will be different in its direction, according as the orbit of the disturbing body is exterior or interior to that of the disturbed. In the former case it will diminish, in the latter will increase, the central gravity.

(557.) Regarding, still, only the mean effect, as produced in a great number of revolutions of both bodies, it is evident that an increase of central force must be accompanied with a diminution of periodic time, and a contraction of dimension of the orbit of a body revolving with a stated velocity, and *vice versâ*. This, then, is the first and most obvious effect of the radial part of the disturbing force. It alters permanently, and by a certain mean and invariable amount, the dimensions of all the orbits and the periodic times of all the bodies composing the planetary system, from what they would be, did each planet circulate about the sun uninfluenced by the attraction of the rest; the angular motion of the interior bodies of the system being thus rendered less, and those of the exterior greater, than on that supposition. The latter effect, indeed, might be at once concluded from this obvious consideration, — that all the planets revolving interiorly to any orbit may be considered as adding to the general aggregate of the

---

of the spherical shell, which form the bases of both the cones, or pyramids, be similar and equally inclined to their axes. Therefore their areas will be to each other as the squares of their distances from the common apex. Therefore their attractions on it will be equal, because the attraction is as the attracting matter directly, and the square of its distance inversely. Now, these attractions act in opposite directions, and, therefore, counteract each other. Therefore, the point is in equilibrium between them; and as the same is true of every such pair of areas into which the spherical shell can be broken up, therefore the point will be in equilibrium, *however situated within* such a spherical shell. Now take a ring, and treat it similarly, breaking its circumference up into pairs of elements, the bases of *triangles* formed by lines passing through the attracted point. Here the attracting elements, being *lines*, not *surfaces*, are in the *simple* ratio of the distances, not the *duplicate*, as they should be to maintain the equilibrium. Therefore it will not be maintained, but the *nearest* elements will have the superiority, and the point will, on the whole, be urged towards the nearest part of the ring. The same is true of every *linear* ring, and is, therefore, true of any assemblage of concentric ones forming a flat annulus, like the ring of Saturn.

attracting matter within, which is not the less efficient for being distributed over space, and maintained in a state of circulation.

(558.) This effect, however, is one which we have no means of measuring, or even of detecting, otherwise than by calculation. For our knowledge of the periods of the planets, and the dimensions of their orbits, is drawn from observations made on them in their actual state, and, therefore, under the influence of this *constant part* of the perturbative action. Their observed mean motions are, therefore, affected by the whole amount of its influence; and we have no means of distinguishing this from the direct effect of the sun's attraction, with which it is blended. Our knowledge, however, of the masses of the planets assures us that it is extremely small; and this, in fact, is all which it is at all important to us to know, in the theory of their motions.

(559.) The action of the sun upon the moon, in like manner, tends, by its mean influence during many successive revolutions of both bodies, to dilate permanently the moon's orbit, and increase her periodic time. But this general average is not established, either in the case of the moon or planets, without a series of subordinate fluctuations due to the elliptic forms of their orbits, which we have purposely neglected to take account of in the above reasoning, and which obviously tend, in the average of a great multitude of revolutions, to neutralize each other. In the lunar theory, however, many of these subordinate fluctuations are very sensible to observation, and of great importance to a correct knowledge of her motions. For example: — The sun's orbit (referred to the earth as fixed) is elliptic, and requires thirteen lunations for its description, during which the distance of the sun undergoes an alternate increase and diminution, each extending over at least six complete lunations. Now, as the sun approaches the earth, its disturbing forces of every kind are increased in a high ratio, and *vice versâ*. Therefore the dilatation it produces on the lunar orbit, and the diminution of the moon's periodic time, will be kept in a continual state of fluctuation, increasing as the sun approaches its perigee, and diminishing as it recedes. And this is consonant to fact,—the observed difference between a lunation in January (when the sun is nearest the earth) and in July (when it is farthest) being no less than 35 minutes.

(560.) Another very remarkable and important effect of this cause, in one of its subordinate fluctuations, (extending, however, over an immense period of time,) is what is called the *secular acceleration of the moon's mean motion*. It had been observed by Dr. Halley, on comparing together the records of the most ancient lunar eclipses of the Chaldean astronomers with those of modern times, that the period of the moon's revolution at present is sensibly shorter than at that remote epoch; and this result was confirmed by a further comparison of both sets of observations with those of the Arabian astronomers of the eighth and ninth centuries. It appeared from these comparisons, that the rate at which the moon's mean motion increases is about 11 seconds per century,—a quantity small in itself, but becoming considerable by its accumulation during a succession of ages. This remarkable fact, like the great equation of Jupiter and Saturn, had been long the subject of toilsome investigation to geometers. Indeed, so difficult did it appear to render any exact account of, that while some were on the point of again declaring the theory of gravity inadequate to its explanation, others were for rejecting altogether the evidence on which it rested, although quite as satisfactory as that on which most historical events are credited. It was in this dilemma that Laplace once more stepped in to rescue physical astronomy from its reproach, by pointing out the real cause of the phænomenon in question, which, when so explained, is one of the most curious and instructive in the whole range of our subject,—one which leads our speculations further into the past and future, and points to longer vistas in the dim perspective of changes which our system has undergone and is yet to undergo, than any other which observation assisted by theory has developed.

(561.) If the solar ellipse were invariable, the alternate dilatation and contraction of the moon's orbit, explained in art. 559., would, in the course of a great many revolutions of the sun, at length effect an exact compensation in the distance and periodic time of the moon, by bringing every possible step in the sun's change of distance to correspond to every possible elongation of the moon from the sun in her orbit. But this is not, in fact, the case. The solar ellipse is kept (as we have already hinted in art. 536., and as we shall very soon explain more fully) in a continual but excessively slow state of change, by the action of the planets on the earth. Its axis, it is true, remains unaltered, but its excentricity is, and has been since the earliest ages, diminishing; and this diminution will continue (there is little reason to doubt) till the excentricity is annihilated altogether, and the earth's orbit becomes a perfect circle; after which it will again open out into an ellipse, the excentricity will again increase, attain a certain moderate amount, and then again decrease. The time required for these evolutions, though calculable, has not been calculated, further than to satisfy us that it is not to be reckoned by hundreds or by thousands of years. It is a period, in short, in which the whole history of astronomy and of the human race occupies but as it were a point, during which all its changes are to be regarded as uniform. Now, it is by this variation in the excentricity of the earth's orbit that the secular acceleration of the moon is caused. The compensation above spoken of (which, if the solar ellipse remained unaltered, would be effected in a few years or a few centuries at furthest in the mode already stated) will now, we see, be only imperfectly effected, owing to this slow shifting of one of the essential data. The steps of restoration are no longer identical with, nor equal to, those of change. The same reasoning, in short, applies, with that by which we explained the long inequalities produced by the tangential force. The struggle up hill is not maintained on equal terms with the downward tendency. The ground is all the while slowly sliding beneath the feet of the antagonists. During the whole time that the earth's excentricity is diminishing, a preponderance is given to the action over the re-action; and it is not till that diminution shall cease, that the tables will be turned, and the process of ultimate restoration will commence. Meanwhile, a minute, outstanding, and uncompensated effect is left at each recurrence, or near recurrence, of the same configurations of the sun, the moon, and the solar and lunar perigee. These accumulate, influence the moon's periodic time and mean motion, and thus becoming repeated in every lunation, at length affect her longitude to an extent not to be overlooked.

(562.) The phænomenon of which we have now given an account is another and very striking example of the propagation of a periodic change from one part of a system to another. The planets have no direct, appreciable action on the lunar motions as referred to the earth. Their masses are too small, and their distances too great, for their difference of action on the moon and earth, ever to become sensible. Yet their effect on the earth's orbit is thus, we see, propagated through the sun to that of the moon; and what is very remarkable, the transmitted effect thus indirectly produced on the angle described by the moon round the earth is more sensible to observation than that directly produced by them on the angle described by the earth round the sun.

(563.) The dilatation and contraction of the lunar and planetary orbits, then, which arise from the action of the radial force, and which tend to affect their mean motions, are distinguishable into two kinds; — the one permanent, depending on the distribution of the attracting matter in the system, and on the order which each planet holds in it; the other periodic, and which operates in length of time its own compensation. Geometers have demonstrated (it is to Lagrange that we owe this most important discovery) that, besides these, there exists no third class of effects, whether arising from the radial or tangential disturbing forces, or from their combination, such as can go on for ever increasing in one direction without self-compensation; and, in particular, that the major axes of the planetary ellipses are not liable even to those slow secular changes by which the inclinations, nodes, and all the other elements of the system, are affected, and which, it is true, are periodic, but in a different sense from those long inequalities which depend on the mutual configurations of the planets *inter se.* Now, the periodic time of a planet in its orbit about the sun depends only on the masses of the sun and planet, and on the major axis of the orbit it describes, without regard to its degree of excentricity, or to any other element. The mean sidereal periods of the planets, therefore, such as result from an average of a sufficient number of revolutions to allow of the compensation of the last-mentioned inequalities, are unalterable by lapse of time. The length of the sidereal year, for example, if concluded at this present time from observations embracing a thousand revolutions of the earth round the sun, (such, in short, as we now possess it,) is the same with that which (if we can stretch our imagination so far) must result from a similar comparison of observations made a million of years hence.

(564.) This theorem is justly regarded as the most important, as a single result, of any which have hitherto rewarded the researches of mathematicians. We shall, therefore, endeavour to make clear to our readers, at least the principle on which its demonstration rests; and although the complete application of that principle cannot be satisfactorily made without entering into details of calculation incompatible with our objects, we shall have no difficulty in leading them up to that point where those details must be entered on, and in giving such an insight into their general nature as will render it evident what must be their result when gone through.

(565.) It is a property of elliptic motion performed under the influence of gravity, and in conformity with Kepler's laws, that if the velocity with which a planet moves at any point of its orbit be given, and also the distance of that point from the sun, the major axis of the orbit is thereby also given. It is no matter in what *direction* the planet may be moving at that moment. This will influence the excentricity and the position of its ellipse, but not its length. This property of elliptic motion has been demonstrated by Newton, and is one of the most obvious and elementary conclusions from his theory. Let us now consider a planet describing an indefinitely small arc of its orbit about the sun, under the joint influence of its attraction, and the disturbing power of another planet. This arc will have some certain curvature and direction, and, therefore, may be considered as an arc of a certain ellipse described about the sun as a focus, for this plain reason,— that whatever be the curvature and direction of the arc in question, an ellipse may always be assigned, whose focus shall be in the sun, and which shall coincide with it throughout the whole interval (supposed indefinitely small) between its extreme points. This is a matter of pure geometry. It does not follow, however, that the ellipse thus instantaneously determined will have the same elements as that similarly determined from the arc described in either the previous or the subsequent instant. If the disturbing force did not exist, this would be the case; but, by its action, a variation of the elements from instant to instant is produced, and the ellipse so determined is in a continual state of change. Now, when the planet has reached the end of the small arc under consideration, the question whether it will in the next instant describe an arc of an ellipse having the same or a varied axis will depend, not on the new direction impressed upon it by the acting forces,—for the axis, as we have seen, is independent of that direction, —not on its change of distance from the sun, while describing the former arc,—for the elements of that arc are accommodated to it, so that one and the same axis must belong to its beginning and its end. The question, in short, whether in the next arc it shall take up a new major axis, or go on with the old one, will depend solely on this, — whether the *velocity* has undergone, by the action of the disturbing force, a change incompatible with the continuance of the same axis. We say by the action of the disturbing force, because the central force residing in the focus can impress on it no such change of velocity as to be incompatible with the permanence of *any* ellipse in which it may at *any* instant be freely moving about that focus.

(566.) Thus we see that the momentary variation of the major axis depends on nothing but the momentary deviation from the law of elliptic velocity produced by the disturbing force, without the least regard to the direction in which that extraneous velocity is impressed, or the distance from the sun at which the planet may be situated in consequence of the variation of the other elements of its orbit. And as this is the case at every instant of its motion, it will follow that, after the lapse of any time however great, the amount of change which the axis may have undergone will be determined by the total deviation from the original elliptic velocity produced by the disturbing force; without any regard to alterations which the action of that force may have produced in the other elements, except in so far as the velocity may be thereby modified. This is the point at which the exact estimation of the effect must be intrusted to the calculations of the geometer. We shall be at no loss, however, to perceive that these calculations can only terminate in demonstrating the periodic nature and ultimate compensation of all the variations of the axis which can thus arise, when we consider that the circulation of two planets about the sun, in the same direction and in incommensurable periods, cannot fail to ensure their presentation to each other in every state of approach and recess, and under every variety as to their mutual distance and the consequent intensity of their mutual action. Whatever velocity, then, may be generated in one by the disturbing action of the other, in one situation, will infallibly be destroyed by it in another, by the mere effect of change of configuration.

(567.) It appears, then, that the variations in the major axes of the planetary orbits depend entirely on cycles of configuration, like the great inequality of Jupiter and Saturn, or the long inequality of the Earth and Venus above explained, which, indeed, may be regarded as due to such periodic variations of their axes. In fact, the mode in which we have seen those inequalities arise, from the accumulation of imperfectly compensated actions of the tangential force, brings them directly under the above reasoning: since the efficacy of this force falls almost wholly upon the *velocity* of the disturbed planet, whose motion is always nearly coincident with or opposite to its direction.

(568.) Let us now consider the effect of perturbation in altering the excentricity and the situation of the axis of the disturbed orbit in its own plane. Such a change of position (as we have observed in art. 318.) actually takes place, although very slowly, in the axis of the earth's orbit, and much more rapidly in that of the moon's (art. 360.); and these movements we are now to account for.

(569.) The motion of the apsides of the lunar and planetary orbits may be illustrated by a very pretty mechanical experiment, which is otherwise instructive in giving an idea of the mode in which orbitual motion is carried on under the action of central forces variable according to the situation of the revolving body. Let a leaden weight be suspended by a brass or iron wire to a hook in the under side of a firm beam, so as to allow of its free motion on all sides of the vertical, and so that when in a state of rest it shall just clear the floor of the room, or a table placed ten or twelve feet beneath the hook. The point of support should be well secured from wagging to and fro by the oscillation of the weight, which should be sufficient to keep the wire as tightly stretched as it will bear, with the certainty of not breaking. Now, let a very small motion be communicated to the weight, not by merely withdrawing it from the vertical and letting it fall, but by giving it a slight impulse sideways. It will be seen to describe a regular ellipse about the point of rest as its center. If the weight be heavy, and carry attached to it a pencil, whose point lies exactly in the direction of the string, the ellipse may be transferred to paper lightly stretched and gently pressed against it. In these circumstances, the situation of the major and minor axes of the ellipse will remain for a long time very nearly the same, though the resistance of the air and the stiffness of the wire will gradually diminish its dimensions and excentricity. But if the impulse communicated to the weight be considerable, so as to carry it out to a great angle (15° or 20° from the vertical), this permanence of situation of the ellipse will no longer subsist. Its axis will be seen to shift its position at every revolution of the weight, advancing in the same direction with the weight's motion, by an uniform and regular progression, which at length will entirely reverse its situation, bringing the direction of the longest excursions to coincide with that in which the shortest were previously made; and so on, round the whole circle; and, in a word, imitating to the eye, very completely, the motion of the apsides of the moon's orbit.

(570.) Now, if we enquire into the cause of this progression of the apsides, it will not be difficult of detection. When a weight is suspended by a wire, and drawn aside from the vertical, it is urged to the lowest point (or rather in a direction at every instant perpendicular to the wire) by a force which varies as the sine of the deviation of the wire from the perpendicular. Now, the sines of very small arcs are nearly in the proportion of the arcs themselves; and the more nearly, as the arcs are smaller. If, therefore, the deviations from the vertical are so small that we may neglect the curvature of the spherical surface in which the weight moves, and regard the curve described as coincident with its projection on a horizontal plane, it will be then moving under the same circumstances as if it were a revolving body attracted to a center by a force varying directly as the distance; and, in this case, the curve described would be an ellipse, having its center of attraction not in the focus, but in the center *, and the apsides of this ellipse would remain fixed. But if the excursions of the weight from the vertical be considerable, the force urging it towards the center will deviate in its law from the simple ratio of the distances; being as the *sine*, while the distances are as the *arc*. Now the sine, though it continues to increase as the arc increases, yet does not increase so fast. So soon as the arc has any sensible extent, the sine begins to fall somewhat short of the magnitude which an exact numerical proportionality would require; and therefore the force urging the weight towards its center or point of rest, at great distances falls, in like proportion, somewhat short of that which would keep the body in its precise elliptic orbit. It will no longer, therefore, have, at those greater distances, the same command over the weight, *in proportion to its speed*, which would enable it to deflect it from its rectilinear tangential course into an ellipse. The true path which it describes will be *less curved in the remoter parts* than is consistent with the elliptic figure, as in the annexed cut; and, therefore, it will not so soon have its

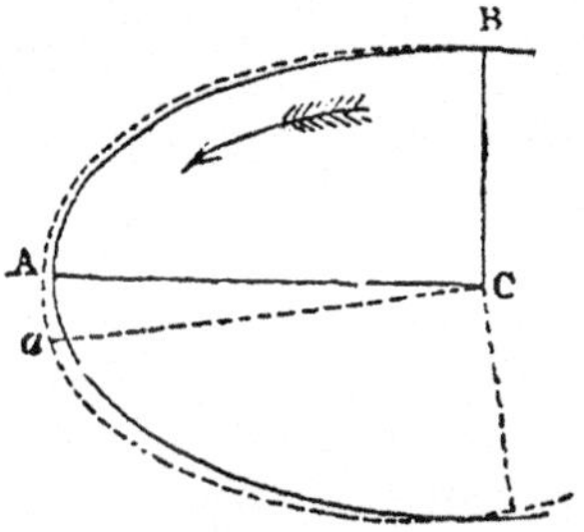

motion brought to be again at right angles to the radius. It will require a longer continued action of the central

* Newton, Princip. i. 47

force to do this ; and before it is accomplished, more than a quadrant of its revolution must be passed over in angular motion round the center. But this is only stating at length, and in a more circuitous manner, that fact which is more briefly and summarily expressed by saying that *the apsides of its orbit are progressive.*

(571.) Now, this is what takes place, *mutatis mutandis,* with the lunar and planetary motions. The action of the sun on the moon, for example, as we have seen, besides the tangential force, whose effects we are not now considering, produces a force in the direction of the radius vector, whose law is not that of the earth's direct gravity. When compounded, therefore, with the earth's attraction, it will deflect the moon into an orbit deviating from the elliptic figure, being either too much curved, or too little, in its recess from the perigee, to bring it to an apogee at exactly 180° from the perigee; —too much, if the compound force thus produced decrease at a slower rate than the inverse square of the distance (*i. e.* be too strong in the remoter distances), too little, if the joint force decrease faster than gravity; or more rapidly than the inverse square, and be therefore too weak at the greater distance. In the former case, the curvature, being excessive, will bring the moon to its apogee sooner than would be the case in an elliptic orbit; in the latter, the curvature is insufficient,

Fig. 1.

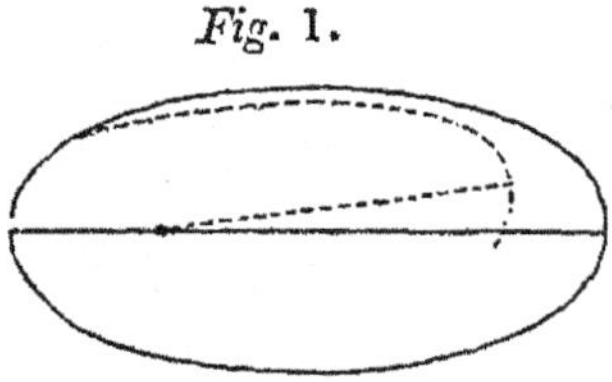

Fig. 2.

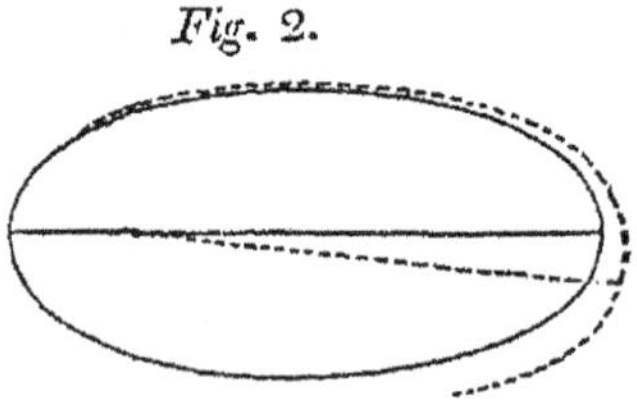

and will therefore bring it later to an apogee. In the former case, then, the line of apsides will retrograde ; in the latter, advance. (See *fig.* 1. and *fig.* 2.)

(572.) Both these cases obtain in different configurations of the sun and moon. In the syzigies, the effect of the sun's attraction is to weaken the gravity of the earth by a force, whose law of variation, instead of the inverse square, follows the direct proportional relation of the distance; while, in the quadratures, the reverse takes place,—the whole effect of the radial disturbing force here conspiring with the earth's gravity, but the portion added being still, as in the former case, in the direct ratio of the distance. Therefore the motion of the moon, in and near the first of these situations, will be performed in an ellipse, whose apsides are in a state of advance; and in and near the latter, in a state of recess. But, as we have already seen (art. 556.), the average effect arising from the mutual counteraction of these temporary values of the disturbing force gives the preponderance to the ablatitious or enfeebling power. On the average, then, of a whole revolution, the lunar apogee will advance.

(573.) The above reasoning renders a satisfactory enough general account of the advance of the lunar apogee; but it is not without considerable difficulty that it can be applied to determine numerically the rapidity of such advance: nor, when so applied, does it account for the whole amount of the movement in question, as assigned by observation—not more, indeed, than about one half of it; the remaining part is produced by the tangential force. It is evident, that an increase of velocity in the moon will have the same effect in diminishing the curvature of its orbit as the decrease of central force, and *vice versâ*. Now, the direct effect of the tangential force is to cause a fluctuation of the moon's velocity above and below its *elliptic value*, and therefore an alternate progress and recess of the apogee. This would compensate itself in each synodic revolution, *were the apogee invariable.* But this is not the case; the apogee is kept *rapidly advancing* by the action of the radial force, as above explained. An uncompensated portion of the action of the tangential force, therefore, remains outstanding (according to the reasoning already so often employed in this chapter), and this portion is so distributed over the orbit as to conspire with the former cause, and, in fact, nearly to double its effect. This is what is meant by geometers, when they say that this part of the motion of the apogee is due to the square of the disturbing force. The effect of the tangential force in disturbing the apogee would compensate itself, were it not for the motion which the apogee has already had impressed upon it by the radial force; and we have here, therefore, disturbance re-acting on disturbance.

(574.) The curious and complicated effect of perturbation, described in the last article, has given more trouble to geometers than any other part of the lunar theory. Newton himself had succeeded in tracing that part of the motion of the apogee which is due to the direct action of the radial force; but finding the amount only half what observation assigns, he appears to have abandoned the subject in despair. Nor, when resumed by his successors, did the enquiry, for a very long period, assume a more promising aspect. On the contrary, Newton's result appeared to be even minutely verified, and the elaborate investigations which were lavished upon the subject without success began to excite strong doubts whether this feature of the lunar motions could be explained at all by the Newtonian law of gravitation. The doubt was removed, however, almost in the instant of its origin, by the same geometer, Clairaut, who first gave it currency, and who gloriously repaired the error of his momentary hesitation, by demonstrating the exact coincidence between theory and observation, when the effect of the tangential force is properly taken into the account. The lunar apogee circulates, as already stated (art. 360.), in about nine years.

(575.) The same cause which gives rise to the displacement of the line of apsides of the disturbed orbit produces a corresponding change in its excentricity. This is evident on a glance at our figures 1. and 2. of art. 571. Thus, in fig. 1., since the disturbed body, proceeding from its lower to its upper apsis, is acted on by a force greater than would retain it in an elliptic orbit, and too much curved, its whole course (as far as it is so affected) will lie *within* the ellipse, as shown by the dotted line; and when it arrives at the upper apsis, its distance will be less than in the undisturbed ellipse; that is to say, the excentricity of its orbit, as estimated by the comparative distances of the two apsides from the focus, will be diminished, or the orbit rendered more nearly circular. The contrary effect will take place in the case of fig. 2. There exists, therefore, between the momentary shifting of the perihelion of the disturbed orbit, and the momentary variation of its excentricity, a relation much of the same kind with that which connects the change of inclination with the motion of the nodes; and, in fact, the strict geometrical theories of the two cases present a close analogy, and lead to final results of the very same nature. What the variation of excentricity is to the motion of the perihelion, the change of inclination is to the motion of the node. In either case, the period of the one is also the period of the other; and while the perihelia describe considerable angles by an oscillatory motion to and fro, or circulate in immense periods of time round the entire circle, the excentricities increase and decrease by comparatively small changes, and are at length restored to their original magnitudes. In the lunar orbit, as the rapid rotation of the nodes prevents the change of inclination from accumulating to any material amount, so the still more rapid revolution of its apogee effects a speedy compensation in the fluctuations of its excentricity, and never suffers them to go to any material extent; while the same causes, by presenting *in quick succession* the lunar orbit in every possible situation to all the disturbing forces, whether of the sun, the planets, or the protuberant matter at the earth's equator, prevent any secular accumulation of small changes, by which, in the lapse of ages, its ellipticity might be materially increased or diminished. Accordingly, observation shows the *mean* excentricity of the moon's orbit to be the same now as in the earliest ages of astronomy.

(576.) The movements of the perihelia, and variations of excentricity of the planetary orbits, are interlaced and complicated together in the same manner and nearly by the same laws as the variations of their nodes and inclinations. Each acts upon every other, and every such mutual action generates its own peculiar period of compensation; and every such period, in pursuance of the principle of art. 526., is thence propagated throughout the system. Thus arise cycles upon cycles, of whose compound duration some notion may be formed, when we consider what is the length of one such period in the case of the two principal planets—Jupiter and Saturn. Neglecting the action of the rest, the effect of their mutual attraction would be to produce a variation in the excentricity of Saturn's orbit, from 0·08409, its *maximum*, to 0·01345, its *minimum* value; while that of Jupiter would vary between the narrower limits, 0·06036 and 0·02606: the greatest excentricity of Jupiter corresponding to the least of Saturn, and *vice versâ*. The period in which these changes are gone through, would be 70414 years. After this example, it will be easily conceived that many millions of years will require to elapse before a complete fulfilment of the joint cycle which shall restore the whole system to its original state as far as the excentricities of its orbits are concerned.

(577.) The place of the perihelion of a planet's orbit is of little consequence to its well-being; but its excentricity is most important, as upon this (the axes of the orbits being permanent) depends the mean temperature of its surface, and the extreme variations to which its seasons may be liable. For it may be easily shown that the *mean annual amount* of light and heat received by a planet from the sun is, *cæteris paribus*, as the minor axis of the ellipse described by it.* Any variation, therefore, in the ex-

* "On the Astronomical Causes which may influence Geological Phænomena."—*Geol. Trans.* 1832.

centrity by changing the minor axis, will alter the *mean* temperature of the surface. How such a change will also influence the extremes of temperature appears from art. 315. Now, it may naturally be enquired whether, in the vast cycle above spoken of, in which, at some period or other, conspiring changes may accumulate on the orbit of one planet from several quarters, it may not happen that the excentricity of any one planet — as the earth — may become exorbitantly great, so as to subvert those relations which render it habitable to man, or to give rise to great changes, at least, in the physical comfort of his state. To this the researches of geometers have enabled us to answer in the negative. A relation has been demonstrated by Lagrange between the masses, axes of the orbits, and excentricities of each planet, similar to what we have already stated with respect to their inclinations, viz. *that if the mass of each planet be multiplied by the square root of the axis of its orbit, and the product by the square of its excentricity, the sum of all such products throughout the system is invariable;* and as, in point of fact, this sum is extremely small, so it will always remain. Now, since the axes of the orbits are liable to no secular changes, this is equivalent to saying that no one orbit shall increase its excentricity, unless at the expense of a common fund, the whole amount of which is, and must for ever remain, extremely minute.*

(578.) We have hinted, in our last art. but one, at perturbations produced in the lunar orbit by the protuberant matter of the earth's equator. The attraction of a sphere is the same as if all its matter were condensed into a point in its center; but that is not the case with a spheroid. The attraction of such a mass is neither exactly directed to its center, nor does it exactly

* There is nothing in this relation, however, taken *per se*, to secure the smaller planets — Mercury, Mars, Juno, Ceres, &c. — from a catastrophe, could they accumulate on themselves, or any one of them, the whole amount of this *excentricity fund.* But that can never be: Jupiter and Saturn will always retain the lion's share of it. A similar remark applies to the *inclination fund* of art. 515. These *funds*, be it observed, can never get into debt. Every term of them is essentially positive.

follow the law of the inverse squares of the distances. Hence will arise a series of perturbations, extremely small in amount, but still perceptible, in the lunar motions; by which the node and the apogee will be affected. A more remarkable consequence of this cause, however, is a small nutation of the lunar orbit, exactly analogous to that which the moon causes in the plane of the earth's equator, by its action on the same elliptic protuberance. And, in general, it may be observed, that in the systems of planets which have satellites, the elliptic figure of the primary has a tendency to bring the orbits of the satellites to coincide with its equator, — a tendency which, though small in the case of the earth, yet in that of Jupiter, whose ellipticity is very considerable, and of Saturn especially, where the ellipticity of the body is reinforced by the attraction of the rings, becomes predominant over every external and internal cause of disturbance, and produces and maintains an almost exact coincidence of the planes in question. Such, at least, is the case with the nearer satellites. The more distant are comparatively less affected by this cause, the difference of attractions between a sphere and spheroid diminishing with great rapidity as the distance increases. Thus, while the orbits of all the six interior satellites of Saturn lie almost exactly in the plane of the ring and equator of the planet, that of the external satellite, whose distance from Saturn is between sixty and seventy diameters of the planet, is inclined to that plane considerably. On the other hand, this considerable distance, while it permits the satellite to retain its actual inclination, prevents (by parity of reasoning) the ring and equator of the planet from being perceptibly disturbed by its attraction, or being subjected to any appreciable movements analogous to our nutation and precession. If such exist, they must be much slower than those of the earth; the mass of this satellite (though the largest of its system) being, as far as can be judged by its apparent size, a much smaller fraction of that of Saturn than the moon is of the earth; while the solar precession, by reason of the immense distance of the sun, must be quite inappreciable.

(579.) It is by means of the perturbations of the planets, as ascertained by observation, and compared with theory, that we arrive at a knowledge of the masses of those planets, which, having no satellites, offer no other hold upon them for this purpose. Every planet produces an amount of perturbation in the motions of every other, proportioned to its mass, and to the degree of advantage or *purchase* which its situation in the system gives it over their movements. The latter is a subject of exact calculation; the former is unknown, otherwise than by observation of its effects. In the determination, however, of the masses of the planets by this means, theory lends the greatest assistance to observation, by pointing out the combinations most favourable for eliciting this knowledge from the confused mass of superposed inequalities which affect every observed place of a planet; by pointing out the laws of each inequality in its periodical rise and decay; and by showing how every particular inequality depends for its magnitude on the mass producing it. It is thus that the mass of Jupiter itself (employed by Laplace in his investigations, and interwoven with all the planetary tables) has of late been ascertained, by observations of the derangements produced by it in the motions of the ultra-zodiacal planets, to have been insufficiently determined, or rather considerably mistaken, by relying too much on observations of its satellites, made long ago by Pound and others, with inadequate instrumental means. The same conclusion has been arrived at, and nearly the same mass obtained, by means of the perturbations produced by Jupiter on Encke's comet. The error was one of great importance; the mass of Jupiter being by far the most influential element in the planetary system, after that of the sun. It is satisfactory, then, to have ascertained, — as by his observations Professor Airy is understood to have recently done, — the cause of the error; to have traced it up to its source, in insufficient micrometric measurements of the greatest elongations of the satellites; and to have found it disappear when measures taken with more care, and with infinitely superior instruments, are substituted for those before employed.

(580.) In the same way that the perturbations of the planets lead us to a knowledge of their masses, as compared with that of the sun, so the perturbations of the satellites of Jupiter have led, and those of Saturn's attendants will, no doubt, hereafter lead, to a knowledge of the proportion *their* masses bear to their respective primaries. The system of Jupiter's satellites has been elaborately treated by Laplace; and it is from his theory, compared with innumerable observations of their eclipses, that the masses assigned to them in art. 463. have been fixed. Few results of theory are more surprising, than to see these minute atoms weighed in the same balance which we have applied to the ponderous mass of the sun, which exceeds the least of them in the enormous proportion of 65000000 to 1.

CHAP. XII.

OF SIDEREAL ASTRONOMY.

OF THE STARS GENERALLY. — THEIR DISTRIBUTION INTO CLASSES ACCORDING TO THEIR APPARENT MAGNITUDES. — THEIR DISTRIBUTION OVER THE HEAVENS. — OF THE MILKY WAY. — ANNUAL PARALLAX. — REAL DISTANCES, PROBABLE DIMENSIONS, AND NATURE OF THE STARS. — VARIABLE STARS. — TEMPORARY STARS. — OF DOUBLE STARS. — THEIR REVOLUTION ABOUT EACH OTHER IN ELLIPTIC ORBITS. — EXTENSION OF THE LAW OF GRAVITY TO SUCH SYSTEMS. — OF COLOURED STARS. — PROPER MOTION OF THE SUN AND STARS. — SYSTEMATIC ABERRATION AND PARALLAX. — OF COMPOUND SIDEREAL SYSTEMS. — CLUSTERS OF STARS. — OF NEBULÆ. — NEBULOUS STARS. — ANNULAR AND PLANETARY NEBULÆ. — ZODIACAL LIGHT.

(581.) BESIDES the bodies we have described in the foregoing chapters, the heavens present us with an innumerable multitude of other objects, which are called generally by the name of stars. Though comprehending individuals differing from each other, not merely in brightness, but in many other essential points, they all agree in one attribute,—a high degree of permanence as to apparent relative situation. This has procured them the title of "fixed stars;" an expression which is to be understood in a comparative and not an absolute sense, it being certain that many, and probable that all are in a state of motion, although too slow to be perceptible unless by means of very delicate observations, continued during a long series of years.

(582.) Astronomers are in the habit of distinguishing the stars into classes, according to their apparent brightness. These are termed magnitudes. The brightest stars are said to be of the first magnitude; those which fall so far short of the first degree of brightness as to make a marked distinction are classed in the second, and so on down to the sixth or seventh, which comprise the smallest stars visible to the naked eye, in the clearest and darkest night. Beyond these, however, telescopes continue the range of visibility, and magnitudes from the 8th down to the 16th are familiar to those who are in the practice of using powerful instruments; nor does there seem the least reason to assign a limit to this progression; every increase in the dimensions and power of instruments, which successive improvements in optical science have attained, having brought into view multitudes innumerable of objects invisible before; so that, for any thing experience has hitherto taught us, the number of the stars may be really infinite, in the only sense in which we can assign a meaning to the word.

(583.) This classification into magnitudes, however, it must be observed, is entirely arbitrary. Of a multitude of bright objects, differing probably, intrinsically, both in size and in splendour, and arranged at unequal distances from us, one must of necessity appear the brightest, one next below it, and so on. An order of succession (relative, of course, to our local situation among them) *must* exist, and it is a matter of absolute indifference, where, in that infinite progression downwards, from the one brightest to the invisible, we choose to draw our lines of demarcation. All this is a matter of pure convention. Usage, however, has established such a convention; and though it is impossible to determine exactly, or *à priori*, where one magnitude ends and the next begins, and although different observers have differed in their magnitudes, yet, on the whole, astronomers have restricted their first magnitude to about 15 or 20 principal stars; their second to 50 or 60 next inferior; their third to about 200 yet smaller, and so on; the numbers increasing very rapidly as we descend in the scale of brightness, the whole number of stars already registered down to the seventh magnitude, inclusive, amounting to 15000 or 20000.

(584.) As we do not see the actual disc of a star, but judge only of its brightness by the total impression made upon the eye, the apparent "magnitude" of any star will, it is evident, depend, 1st, on the star's distance from us; 2d, on the absolute magnitude of its illuminated surface; 3d, on the intrinsic brightness of that surface. Now, as we know nothing, or next to nothing, of any of these data, and have every reason for believing that each of them may differ in different individuals, in the proportion of many millions to one, it is clear that we are not to expect much satisfaction in any conclusions we may draw from numerical statements of the number of individuals arranged in our artificial classes. In fact, astronomers have not yet agreed upon any principle by which the magnitudes may be photometrically arranged, though a leaning towards a geometrical progression, of which each term is the half of the preceding, may be discerned.* Nevertheless, it were much to be wished, that, setting aside all such arbitrary subdivisions, a numerical estimate should be formed,

* Struve, Dorpat Catal. of Double Stars, p. xxxv

grounded on precise photometrical experiments, of the apparent brightness of each star. This would afford a definite character in natural history, and serve as a term of comparison to ascertain the changes which may take place in them; changes which we know to happen in several, and may therefore fairly presume to be possible in all. Meanwhile, as a first approximation, the following proportions of light, concluded from Sir William Herschel's* experimental comparisons of a few selected stars, may be borne in mind:—

Light of a star of the average	1st magnitude	= 100
	2d	= 25
	3d	= 12?
	4th	= 6
	5th	= 2
	6th	= 1

By my own experiments, I have found that the light of Sirius (the brightest of all the fixed stars) is about 324 times that of an average star of the 6th magnitude.†

(585.) If the comparison of the apparent magnitudes of the stars with their numbers leads to no definite conclusion, it is otherwise when we view them in connection with their local distribution over the heavens. If indeed we confine ourselves to the three or four brightest classes, we shall find them distributed with tolerable impartiality over the sphere; but if we take in the whole amount visible to the naked eye, we shall perceive a great and rapid increase of number as we approach the borders of the milky way. And when we come to telescopic magnitudes, we find them crowded beyond imagination, along the extent of that circle, and of the branch which it sends off from it; so (art. 253.) that in fact its whole light is composed of nothing but stars, whose average magnitude may be stated at about the tenth or eleventh.

* Phil Tr. 1817.

† Trans. Astron. Soc. iii. 183.

(586.) These phænomena agree with the supposition that the stars of our firmament, instead of being scattered in all directions indifferently through space, form a stratum, of which the thickness is small, in comparison with its length and breadth; and in which the earth occupies a place somewhere about the middle of its thickness, and near the point where it subdivides into two principal laminæ, inclined at a small angle to each other. For it is certain that, to an eye so situated, the apparent density of the stars, supposing them pretty equally scattered through the space they occupy, would be least in a direction of the visual ray (as S A), perpendicular to the lamina, and greatest in that of its breadth, as S B, S C, S D; increasing rapidly in passing from one to the other direction, just as we see a slight haze in the atmosphere thickening into a decided fog bank near the horizon, by the rapid increase of the mere length of the visual ray. Accordingly, such is the view of the construction of the starry firmament taken by Sir William Herschel, whose powerful telescopes have effected a complete analysis of this wonderful zone, and demon-

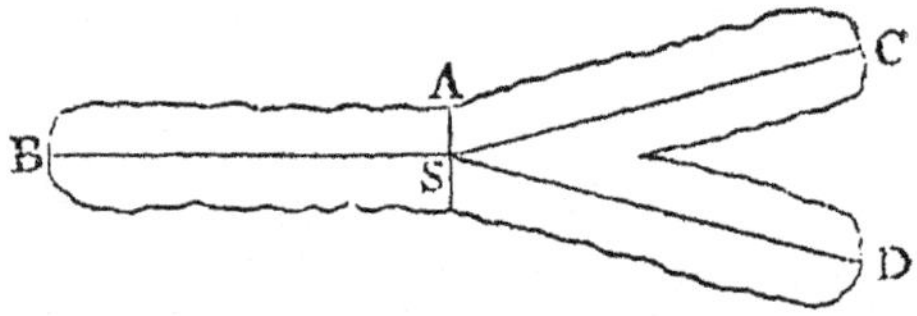

strated the fact of its entirely consisting of stars. So crowded are they in some parts of it, that by counting the stars in a single field of his telescope, he was led to conclude that 50000 had passed under his review in a zone two degrees in breadth, during a single hour's observation. The immense distances at which the remoter regions must be situated will sufficiently account for the vast predominance of small magnitudes which are observed in it.

(587.) When we speak of the comparative remoteness of certain regions of the starry heavens beyond others, and of our own situation in them, the question immediately arises, What is the distance of the nearest fixed star? What is the scale on which our visible firmament is constructed? And what proportion do its dimensions bear to those of our own immediate system? To this, however, astronomy has hitherto proved unable to supply an answer. All we know on this subject is negative. We have attained, by delicate observations and refined combinations of theoretical reasoning to a correct estimate, first, of the dimensions of the earth; then, taking that as a base, to a knowledge of those of its orbit about the sun; and again, by taking our stand, as it were, on the opposite borders of the circumference of this orbit, we have extended our measurements to the extreme verge of our own system, and by the aid of what we know of the excursions of comets, have felt our way, as it were, a step or two beyond the orbit of the remotest known planet. But between that remotest orb and the nearest star there is a gulf fixed, to whose extent no observations yet made have enabled us to assign any distinct approximation, or to name any distance, however immense, which it may not, for any thing we can tell, surpass.

(588.) The diameter of the earth has served us as the base of a triangle, in the *trigonometrical survey* of our system (art. 226.), by which to calculate the distance of the sun: but the extreme minuteness of the sun's parallax (art. 304.) renders the calculation from this "ill-conditioned" triangle (art. 227.) so delicate, that nothing but the fortunate combination of favourable circumstances, afforded by the transits of Venus (art. 409.), could render its results even tolerably worthy of reliance. But the earth's diameter is too small a base for direct triangulation to the verge even of our own system (art. 449.), and we are, therefore, obliged to substitute the *annual parallax* for the diurnal, or, which comes to the same thing, to ground our calculation on the relative velocities of the earth and planets in their orbits (art. 414.), when we would push our triangulation to that extent. It might be naturally enough expected, that by this enlargement of our base to the vast diameter of the earth's orbit, the next step in our survey (art. 227.) would be made at a great advantage;—that our change of station, from side to side of it, would produce a perceptible and measurable amount of annual parallax in the stars, and that by its means we should come to a knowledge of their distance. But, after exhausting every refinement of observation, astronomers have been unable to come to any positive and coincident conclusion upon this head; and it seems, therefore, demonstrated, that the amount of such parallax, even for the nearest fixed star which has hitherto been examined with the requisite attention, remains still mixed up with, and concealed among, the errors incidental to all astronomical determinations. Now, such is the nicety to which these have been carried, that did the quantity in question amount to a single second (*i. e.* did the radius of the earth's orbit subtend at the nearest fixed star that minute angle) it could not possibly have escaped detection and universal recognition.

(589.) Radius is to the sine of 1″, in round numbers, as 200000 to 1. In this proportion, then, *at least*, must the distance of the fixed stars from the sun exceed that of the sun from the earth. The latter distance, as we have already seen, exceeds the earth's radius in the proportion of 24000 to 1; and, lastly, to descend to ordinary standards, the earth's radius is 4000 of our miles. The distance of the stars, then, *cannot be so small* as 4800000000 radii of the earth, or 19200000000000 miles! How much larger it may be, we know not.

(590.) In such numbers, the imagination is lost. The only mode we have of conceiving such intervals at all is by the time which it would require for light to traverse them. Now light, as we know, travels at the rate of 192000 miles per second. It would, therefore, occupy 100000000 seconds, or upwards of three years, in such a journey, at the very lowest estimate. What, then, are we to allow for the distance of those innumerable stars of the smaller magnitudes which the telescope discloses to us! If we admit the light of a star of each magnitude to be half that of the magnitude next above it, it will follow that a star of the first magnitude will require to be removed to 362 times its distance to appear no larger than one of the sixteenth. It follows, therefore, that among the countless multitude of such stars, visible in telescopes, there must be many whose light has taken at least a thousand years to reach us; and that when we observe their places, and note their changes, we are, in fact, reading only their history of a thousand years' date, thus wonderfully recorded. We cannot escape this conclusion, but by adopting as an alternative an intrinsic inferiority of light in *all* the smaller stars of the milky way. We shall be better able to estimate the probability of this alternative, when we have made acquaintance with other sidereal systems, whose existence the telescope discloses to us, and whose analogy will satisfy us that the view of the subject we have taken above is in perfect harmony with the general tenour of astronomical facts.

(591.) Quitting, however, the region of speculation, and confining ourselves within limits which we are sure are less than the truth, let us employ the negative knowledge we have obtained respecting the distances of the stars to form some conformable estimate of their real magnitudes. Of this, telescopes afford us no direct information. The discs which good telescopes show us of the stars are not real, but *spurious*—a mere optical illusion.* Their light, therefore, must be our only guide. Now Dr. Wollaston, by direct photometrical experiments, open, as it would seem, to no objections†, has ascertained the light of Sirius, as received by us, to be to that of the sun as 1 to 20000000000. The sun, therefore, in order that it should appear to us no brighter than Sirius, would require to be removed to 141400 times its actual distance. We have seen, however, that the distance of Sirius cannot be so small as 200000 times

* See Cab. Cyc. Optics.

† Phil. Trans. 1829, p. 24.

that of the sun. Hence it follows, that, upon the lowest possible computation, the light really thrown out by Sirius cannot be so little as double that emitted by the sun; or that Sirius must, in point of intrinsic splendour, be at least equal to two suns, and is in all probability vastly greater.*

(592.) Now, for what purpose are we to suppose such magnificent bodies scattered through the abyss of space? Surely not to illuminate *our* nights, which an additional moon of the thousandth part of the size of our own would do much better, nor to sparkle as a pageant void of meaning and reality, and bewilder us among vain conjectures. Useful, it is true, they are to man as points of exact and permanent reference; but he must have studied astronomy to little purpose, who can suppose man to be the only object of his Creator's care, or who does not see in the vast and wonderful apparatus around us provision for other races of animated beings. The planets, as we have seen, derive their light from the sun; but that cannot be the case with the stars. These doubtless, then, are themselves suns, and may, perhaps, each in its sphere, be the presiding center round which other planets, or bodies of which we can form no conception from any analogy offered by our own system, may be circulating.

(593.) Analogies, however, more than conjectural, are not wanting to indicate a correspondence between the dynamical laws which prevail in the remote regions of the stars and those which govern the motions of our own system. Wherever we can trace the law of periodicity — the regular recurrence of the same phænomena in the same times — we are strongly impressed with the idea of rotatory or orbitual motion. Among the stars are several which, though no way distinguishable from others by any apparent change of place, nor by any difference of appearance in telescopes, yet un-

* Dr. Wollaston, assuming, as we think he is perfectly justified in doing, a much lower limit of *possible* parallax in Sirius than we have adopted in the text, has concluded the intrinsic light of Sirius to be nearly that of fourteen suns.

dergo a regular periodical increase and diminution of lustre, involving, in one or two cases, a complete extinction and revival. These are called *periodical stars.* One of the most remarkable is the star *Omicron,* in the constellation *Cetus,* first noticed by Fabricius in 1596. It appears about twelve times in eleven years, — or, more exactly, in a period of 334 days; remains at its greatest brightness about a fortnight, being then, on some occasions, equal to a large star of the second magnitude; decreases during about three months, till it becomes completely invisible, in which state it remains during about five months, when it again becomes visible, and continues increasing during the remaining three months of its period. Such is the general course of its phases. It does not always, however, return to the same degree of brightness, nor increase and diminish by the same gradations. Hevelius, indeed, relates (Lalande, art. 794.) that during the four years between October, 1672, and December, 1676, it did not appear at all.

(594.) Another very remarkable periodical star is that called Algol, or β Persei. It is usually visible as a star of the second magnitude, and such it continues for the space of $2^d\ 14^h$, when it suddenly begins to diminish in splendour, and in about $3\frac{1}{2}$ hours is reduced to the fourth magnitude. It then begins again to increase, and in $3\frac{1}{2}$ hours more is restored to its usual brightness, going through all its changes in $2^d\ 20^h\ 48^m$, or thereabouts. This remarkable law of variation certainly appears strongly to suggest the revolution round it of some opaque body, which, when interposed between us and Algol, cuts off a large portion of its light; and this is accordingly the view taken of the matter by Goodricke, to whom we owe the discovery of this remarkable fact*, in the year 1782; since which time

* The same discovery appears to have been made nearly about the same time by Palitzch, a farmer of Prolitz, near Dresden, — a peasant by station, an astronomer by nature, — who, from his familiar acquaintance with the aspect of the heavens, had been led to notice among so many thousand stars this one as distinguished from the rest by its variation, and had ascertained its period. The same Palitzch was also the first to rediscover the predicted comet of Halley in 1759, which he saw

the same phænomena have continued to be observed, though with much less diligence, than their high interest would appear to merit. Taken any how, it is an indication of a high degree of *activity*, in regions where, but for such evidences, we might conclude all lifeless. Our own sun requires nine times this period to perform a revolution on its own axis. On the other hand, the periodic time of an opaque revolving body, sufficiently large, which should produce a similar temporary obscuration of the sun, seen from a fixed star, would be less than fourteen hours.

(595.) The following list exhibits specimens of periodical stars of every variety of period, so far as they can be considered to be at present ascertained :—

Star's Name.	Period. D. H. M.	Variation of Magnitude.	Discoverers.
β Persei	2 20 48	2 to 4	Goodricke, 1782. Palitzch, 1783.
δ Cephei	5 8 37	3.4 — 5	Goodricke, 1784.
β Lyræ	6 9 0	3 — 4.5	Goodricke, 1784.
η Antinoi	7 4 15	3.4 — 4.5	Pigott, 1784.
α Herculis	60 6 0	3 — 4	Herschel, 1796.
* Serpentis RA. $15^h\ 41^m$ PD. 74° 15′	180 ——	7? — 0	Harding, 1826.
o Ceti	334 ——	2 — 0	Fabricius, 1596.
χ Cygni	396 21 0	6 — 11	Kirch. 1687.
367 B.* Hydræ	494 ——	4 — 10	Maraldi, 1704.
34 Fl. Cygni	18 years	6 — 0	Janson, 1600.
420 M. Leonis	Many years	7 — 0	Koch, 1782.
κ Sagittarii	Ditto	3 — 6	Halley, 1676.
ψ Leonis	Ditto	6 — 0	Montanari, 1667.

The variations of these stars, however, appear to be affected, perhaps in duration of period, but certainly in extent of change, by physical causes at present unknown.

---

nearly a month before any of the astronomers, who, armed with their telescopes, were anxiously watching its return. These anecdotes carry us back to the era of the Chaldean shepherds.

* These letters B. Fl. and M. refer to the Catalogues of Bode, Flamsteed, and Mayer.

The non-appearance of *o* Ceti, during four years, has already been noticed; and to this instance we may add that of *χ* Cygni, which is stated by Cassini to have been scarcely visible throughout the years 1699, 1700, and 1701, at those times when it ought to have been most conspicuous.

(596.) These irregularities prepare us for other phænomena of stellar variation, which have hitherto been reduced to no law of periodicity, and must be looked upon, in relation to our ignorance and inexperience, as altogether casual; or, if periodic, of periods too long to have occurred more than once within the limits of recorded observation. The phænomena we allude to are those of temporary stars, which have appeared, from time to time, in different parts of the heavens, blazing forth with extraordinary lustre; and after remaining awhile apparently immovable, have died away, and left no trace. Such is the star which, suddenly appearing in the year 125 B. C., is said to have attracted the attention of Hipparchus, and led him to draw up a catalogue of stars, the earliest on record. Such, too, was the star which blazed forth, A. D. 389, near *α* Aquilæ, remaining for three weeks as bright as Venus, and disappearing entirely. In the years 945, 1264, and 1572, brilliant stars appeared in the region of the heavens between Cepheus and Cassiopeia; and, from the imperfect account we have of the places of the two earlier, as compared with that of the last, which was well determined, as well as from the tolerably near coincidence of the intervals of their appearance, we may suspect them to be one and the same star, with a period of about 300, or, as Goodricke supposes, of 150 years. The appearance of the star of 1572 was so sudden, that Tycho Brahe, a celebrated Danish astronomer, returning one evening (the 11th of November) from his laboratory to his dwelling-house, was surprised to find a group of country people gazing at a star, which he was sure did not exist half an hour before. This was the star in question. It was then as bright as

Sirius, and continued to increase till it surpassed Jupiter when brightest, and was visible at mid-day. It began to diminish in December of the same year, and in March, 1574, had entirely disappeared. So, also, on the 10th of October, 1604, a star of this kind, and not less brilliant, burst forth in the constellation of Serpentarius, which continued visible till October, 1605.

(597.) Similar phænomena, though of a less splendid character, have taken place more recently, as in the case of the star of the third magnitude discovered in 1670, by Anthelm, in the head of the Swan; which, after becoming completely invisible, re-appeared, and, after undergoing one or two singular fluctuations of light, during two years, at last died away entirely, and has not since been seen. On a careful re-examination of the heavens, too, and a comparison of catalogues, many stars are now found to be missing; and although there is no doubt that these losses have often arisen from mistaken entries, yet in many instances it is equally certain that there is no mistake in the observation or entry, and that the star has really been observed, and as really has disappeared from the heavens.* This is a branch of practical astronomy which has been too little followed up, and it is precisely that in which amateurs of the science, provided with only good eyes, or moderate instruments, might employ their time to excellent advantage.† It holds out a sure promise of rich discovery, and is one in which astronomers in established observatories are almost of necessity precluded from taking a part by the nature of the observations required. Catalogues of the comparative brightness of the stars in each constellation have been

* The star 42 Virginis is inserted in the Catalogue of the Astronomical Society from Zach's Zodiacal Catalogue. I missed it on the 9th May, 1828, and have since repeatedly had its place in the field of view of my 20 feet reflector, without perceiving it, unless it be one of two equal stars of the 9th magnitude, very nearly in the place it must have occupied.—*Author.*

† "Ces variations des étoiles sont bien dignes de l'attention des observateurs curieux . . . Un jour viendra, peut-être, où les sciences auront assez d'amateurs pour qu'on puisse suffire à ces détails." —*Lalande,* art 824.—Surely that day is now arrived.

constructed by Sir Wm. Herschel, with the express object of facilitating these researches, and the reader will find them, and a full account of his method of comparison, in the Phil. Trans. 1796, and subsequent years.

(598.) We come now to a class of phænomena of quite a different character, and which give us a real and positive insight into the nature of at least some among the stars, and enable us unhesitatingly to declare them subject to the same dynamical laws, and obedient to the same power of gravitation, which governs our own system. Many of the stars, when examined with telescopes, are found to be double, *i. e.* to consist of two (in some cases three) individuals placed near together. This might be attributed to accidental proximity, did it occur only in a few instances; but the frequency of this companionship, the extreme closeness, and, in many cases, the near equality of the stars so conjoined, would alone lead to a strong suspicion of a more near and intimate relation than mere casual juxtaposition. The bright star Castor, for example, when much magnified, is found to consist of two stars of between the third and fourth magnitude, within 5″ of each other. Stars of this magnitude, however, are not so common in the heavens as to render it at all likely that, if scattered at random, any two would fall so near. But this is only one out of numerous such instances. Sir Wm. Herschel has enumerated upwards of 500 double stars, in which the individuals are within half a minute of each other; and to this list Professor Struve of Dorpat, prosecuting the enquiry by the aid of instruments more conveniently mounted for the purpose, has recently added nearly five times that number. Other observers have still further extended the catalogue, already so large, without exhausting the fertility of the heavens. Among these are great numbers in which the interval between the centers of the individuals is less than a single second, of which ϵ Arietis, Atlas Pleiadum, γ Coronæ, η Coronæ, η and ζ Herculis, and τ and λ Ophiuchi, may be cited as instances. They are divided into classes according to their distances — the closest forming the first class.

(599.) When these combinations were first noticed, it was considered that advantage might be taken of them, to ascertain whether or not the annual motion of the earth in its orbit might not produce a relative apparent displacement of the individuals constituting a double star. Supposing them to lie at a great distance one behind the other, and to appear only by casual juxtaposition nearly in the same line, it is evident that any motion of the earth must subtend different angles at the two stars so juxtaposed, and must therefore produce different parallactic displacements of them on the surface of the heavens, regarded as infinitely distant. Every star, in consequence of the earth's annual motion, should appear to describe in the heavens a small ellipse, (distinct from that which it would appear to describe in consequence of the aberration of light, and not to be confounded with it,) being a section, by the concave surface of the heavens, of an oblique elliptic cone, having its vertex in the star, and the earth's orbit for its

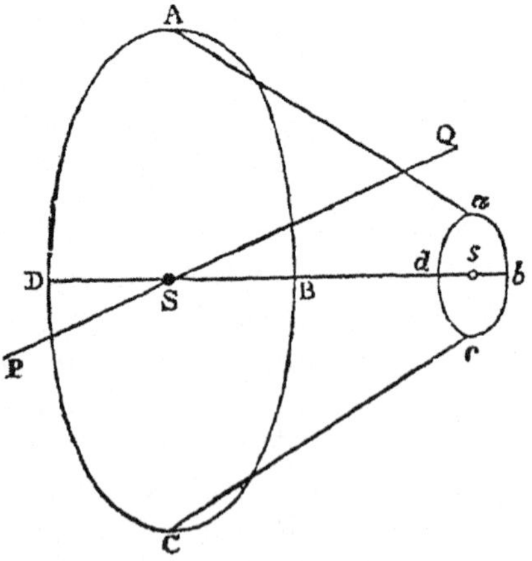

base; and this section will be of less dimensions, the more distant is the star. If, then, we regard two stars, apparently situated close beside each other, but in reality at very different distances, their parallactic ellipses will be similar, but of different dimensions. Suppose, for instance, S and *s* to be the positions of two stars of such an apparently or *optically* double star as seen from the sun, and let A B C D, *a b c d*, be their parallactic ellipses; then, since they will be at all times similarly situated in these ellipses, when the one star is seen at A, the other will be seen at *a*. When the earth has made a quarter of a revolution in its orbit, their apparent places will be B *b*; when another quarter, C *c* and when another, D *d*. If, then, we measure carefully with micrometers adapted for the purpose, their apparent situation with respect to each other, at different times of the year, we should perceive a periodical change, both in the *direction* of the line joining them, and in the *distance* between their centers. For the lines A *a* and C *c* cannot be parallel, nor the lines B *b* and D *d* equal, unless the ellipses be of equal dimensions, *i. e.* unless the two stars have the same parallax, or are equidistant from the earth.

(600.) Now, micrometers, properly mounted, enable us to measure very exactly both the distance between two objects which can be seen together in the same field of a telescope, and the position of the line joining them with respect to the horizon, or the meridian, or any other determinate direction in the heavens. The meridian is chosen as the most convenient; and the situation of the line of junction between the two stars of a double star is referred to its direction, by placing in the focus of the eye-piece of a telescope, equatorially mounted, two cross wires making a right angle, and adjusting their position so that one of the two stars shall just run along it by its diurnal motion, while the telescope remains at rest; noting their situation; and then turning the whole system of wires round in its own plane by a proper mechanical movement, till the other wire becomes exactly parallel to their line of junction, and reading off on a divided circle the angle the wires have moved through. Such an apparatus is called a position micrometer; and by its aid we determine the *angle of position* of a double star, or the angle which their line of junction makes with the meridian; which angle is usually reckoned round the whole circle, from 0 to 360, beginning at the north, and proceeding in the direction north, following (or east) south, preceding (or west).

(601). The advantages which this mode of operation offers for the estimation of parallax are many and great. In the first place, the result to be obtained, being dependent only on the relative apparent displacement of the two stars, is unaffected by almost every cause which would induce error in the separate determination of the place of either by right ascension and declination. Refraction, that greatest of all obstacles to accuracy in astronomical determinations, acts equally on both stars; and is therefore eliminated from the result. We have no longer any thing to fear from errors of graduation in circles from levels or plumb-lines—from uncertainty attending the uranographical reductions of aberration, precession, &c.—all which bear alike on both objects. In a word, if we suppose the stars to have no proper motions of their own by which a *real* change of relative situation may arise, no other cause but their difference of parallax can possibly affect the observation.

(602.) Such were the considerations which first induced Sir William Herschel to collect a list of double stars, and to subject them all to careful measurements of their angles of position and mutual distances. He had hardly entered, however, on these measurements, before he was diverted from the original object of the enquiry (which, in fact, promising as it is, still remains open and untouched, though the only method which seems to offer a chance of success in the research of parallax,) by phænomena of a very unexpected character, which at once engrossed his whole attention. Instead of finding, as he expected, that annual fluctuation to and fro of one star of a double star with respect to the other,—that alternate annual increase and decrease of their distance and angle of position, which the parallax of the earth's annual motion would produce,—he observed, in many instances, a regular progressive change; in some cases bearing chiefly on their distance,—in others on their position, and advancing steadily in one direction, so as clearly to indicate either a real motion of the stars themselves, or a general rectilinear motion of the sun and whole solar system, producing a parallax of a higher order than would arise from the earth's orbitual motion, and which might be called systematic parallax.

(603.) Supposing the two stars in motion independently of each other, and also the sun, it is clear that for the interval of a few years, these motions must be regarded as rectilinear and uniform. Hence, a very slight acquaintance with geometry will suffice to show that the *apparent motion* of one star of a double star, referred to the other as a center, and mapped down, as it were, on a plane in which that other shall be taken for a fixed or zero point, can be no other than a right line. This, at least, must be the case if the stars be independent of each other; but it will be otherwise if they have a physical connection, such as, for instance, real proximity and mutual gravitation would establish. In that case, they would describe orbits round each other, and round their common center of gravity; and therefore the apparent path of either, referred to the other as fixed, instead of being a portion of a straight line, would be bent into a curve concave towards that other. The observed motions, however, were so slow, that many years' observation was required to ascertain this point; and it was not, therefore, until the year 1803, twenty-five years from the commencement of the enquiry, that any thing like a positive conclusion could be come to respecting the rectilinear or orbitual character of the observed changes of position.

(604.) In that, and the subsequent year, it was distinctly announced by Sir William Herschel, in two papers, which will be found in the Transactions of the Royal Society for those years, that there exist sidereal systems, composed of two stars revolving about each other in regular orbits, and constituting what may be termed *binary stars*, to distinguish them from double stars generally so called, in which these physically connected stars are confounded, perhaps, with others only *optically* double, or casually juxtaposed in the heavens at different distances from the eye; whereas the individuals of a binary star are, of course, equidistant from the eye, or, at least, cannot differ more in distance than the semidiameter of the orbit they describe about each other, which is quite insignificant compared with the immense distance between them and the earth. Between fifty and sixty instances of changes, to a greater or less amount, in the angles of position of double stars, are adduced in the memoirs above mentioned; many of which are too decided, and too regularly progressive, to allow of their nature being misconceived. In particular, among the more conspicuous stars,—Castor, γ Virginis, ξ Ursæ, 70 Ophiuchi, σ and η Coronæ, ξ Bootis, η Cassiopeiæ, γ Leonis, ζ Herculis, δ Cygni, μ Bootis, ε 4 and ε 5 Lyræ, λ Ophiuchi, μ Draconis, and ζ Aquarii, are enumerated as among the most remarkable instances of the observed motion; and to some of them even periodic times of revolution are assigned, approximative only, of course, and rather to be regarded as rough guesses than as results of any exact calculation, for which the data were at the time quite inadequate. For instance, the revolution of Castor is set down at 334 years, that of γ Virginis at 708, and that of γ Leonis at 1200 years.

(605.) Subsequent observation has fully confirmed these results, not only in their general tenor, but for the most part in individual detail. Of all the stars above named, there is not one which is not found to be fully entitled to be regarded as binary; and, in fact, this list comprises nearly all the most considerable objects of that description which have yet been detected, though (as attention has been closely drawn to the subject, and observations have multiplied) it has, of late, begun to extend itself rapidly. The number of double stars which are certainly known to possess this peculiar character is between thirty and forty at the time we write, and more are emerging into notice with every fresh mass of observations which come before the public. They require excellent telescopes for their observation, being for the most part so close as to necessitate the use of very high magnifiers, (such as would be considered extremely powerful microscopes if employed to examine objects within our reach,) to perceive an interval between the individuals which compose them.

(606.) It may easily be supposed, that phænomena of this kind would not pass without attempts to connect them with dynamical theories. From their first discovery, they were naturally referred to the agency of some power, like that of gravitation, connecting the stars thus demonstrated to be in a state of circulation about each other; and the extension of the Newtonian law of gravitation to these remote systems was a step so obvious, and so well warranted by our experience of its all-sufficient agency in our own, as to have been expressly or tacitly made by every one who has given the subject any share of his attention. We owe, however, the first distinct system of calculation, by which the elliptic elements of the orbit of a binary star could be deduced from observations of its angle of position and distance at different epochs, to M. Savary, who showed *, that the motions of one of the most remarkable among them (ξ Ursæ) were explicable, within the limits allowable for error of observation, on the supposition of an elliptic orbit described in the short period of 58¼ years. A different process of computation has conducted Professor Encke† to an elliptic orbit for 70 Ophiuchi, described in a period of seventy-four years; and the author of these pages has himself attempted to contribute his mite to these interesting investigations. The following may be stated as the chief results which have been hitherto obtained in this branch of astronomy:—

* Connoiss. des Temps, 1830. † Berlin Ephem. 1832.

Names of Stars.	Period of Revolution.	Major Semi-axis of Ellipse.	Excentricity.
	Years.		
γ Leonis -	1200	———	———
γ Virginis -	628·9000	12″·090	0·83350
61 Cygni -	452 ——	15·430	———
σ Coronæ -	286·6000	8·679	0·61125
Castor - -	252·6600	8·086	0·75820
70 Ophiuchi -	80·3400	4·392	0·46670
ξ Ursæ - -	58·2625	3·857	0·4164
ζ Cancri -	55?	———	———
oronæ -	43·40	———	———

(607.) Of these, perhaps, the most remarkable is γ Virginis, not only on account of the length of its period, but by reason also of the great diminution of apparent distance, and rapid increase of angular motion about each other, of the individuals composing it. It is a bright star of the fourth magnitude, and its component stars are almost exactly equal. It has been known to consist of two stars since the beginning of the eighteenth century, their distance being then between six and seven seconds; so that any tolerably good telescope would resolve it. Since that time they have been constantly approaching, and are at present hardly more than a single second asunder; so that no telescope, that is not of very superior quality, is competent to show them otherwise than as a single star somewhat lengthened in one direction. It fortunately happens, that Bradley, in 1718, noticed, and recorded in the margin of one of his observation books, the apparent direction of their line of junction, as being parallel to that of two remarkable stars, α and δ of the same constellation, as seen by the naked eye; and this note, which has been recently rescued from oblivion by the diligence of Professor Rigaud, has proved of signal service in the investigation of their orbit. They are entered also as distinct stars in Mayer's catalogue; and this affords also another means of recovering their relative situation at the date of his observations, which were made about the year 1756. Without particularizing individual measurements, which will be found in their proper repositories *, it will suffice to remark, that their whole series (which since the beginning of the present century has been very numerous and carefully made, and which embraces an angular motion of 100°, and a diminution of distance to one sixth of its former amount) is represented with a degree of exactness *fully equal to that of observation itself*, by an ellipse of the dimensions and period stated in the foregoing little table, and of which the further requisite particulars are as follows:—

Perihelion passage.	August 18. 1834.
Inclination of orbit to the visual ray - - -	22° 58′
Angle of position of the perihelion projected on the heavens - - - - - -	36° 24′
Angle of position of the line of nodes, or intersection of the plane of the orbit with the surface of the heavens - -	97° 23′

(608.) If the great length of the periods of some of these bodies be remarkable, the shortness of those of others is hardly less so. η Coronæ has already made a complete revolution since its first discovery by Sir William Herschel, and is far advanced in its second period; and ξ Ursæ, ζ Cancri, and 70 Ophiuchi, have all accomplished by far the greater parts of their respective ellipses since the same epoch. If any doubt, therefore, could remain as to the reality of their orbitual motions, or any idea of explaining them by mere parallactic changes, these facts must suffice for their complete dissipation. We have the same evidence, indeed, of their rotations about each other, that we have of those of Uranus and Saturn about the sun; and the correspondence between their calculated and observed places in such very elongated ellipses, must be admitted to carry with it proof of the prevalence of the Newtonian law of

* See them collected in Mem. R. Ast. Soc. vol. v. p. 35.

gravity in their systems, of the very same nature and cogency as that of the calculated and observed places of comets round the central body of our own.

(609.) But it is not with the revolutions of bodies of a planetary or cometary nature round a solar center that we are now concerned; it is with that of sun around sun—each, perhaps, accompanied with its train of planets and *their* satellites, closely shrouded from our view by the splendour of their respective suns, and crowded into a space bearing hardly a greater proportion to the enormous interval which separates *them*, than the distances of the satellites of our planets from their primaries bear to their distances from the sun itself. A less distinctly characterized subordination would be incompatible with the stability of their systems, and with the planetary nature of their orbits. Unless closely nestled under the protecting wing of their immediate superior, the sweep of their other sun in its perihelion passage round their own might carry them off, or whirl them into orbits utterly incompatible with the conditions necessary for the existence of their inhabitants. It must be confessed, that we have here a strangely wide and novel field for speculative excursions, and one which it is not easy to avoid luxuriating in.

(610.) Many of the double stars exhibit the curious and beautiful phænomenon of contrasted or complementary colours.* In such instances, the larger star is usually of a ruddy or orange hue, while the smaller one appears blue or green, probably in virtue of that general law of optics, which provides, that when the retina is under the influence of excitement by any bright, coloured light; feebler lights, which seen alone would produce no sensation but of whiteness, shall for the time appear coloured with the tint complementary to that of the brighter. Thus, a yellow colour predominating in the light of the

* "——— other suns, perhaps,
With their attendant moons thou wilt descry,
Communicating male and female light,
(Which two great sexes animate the world,)
Stored in each orb, perhaps, with some that live."
Paradise Lost, viii. 148.

brighter star, that of the less bright one in the same field of view will appear blue; while, if the tint of the brighter star verge to crimson, that of the other will exhibit a tendency to green — or even appear as a vivid green, under favourable circumstances. The former contrast is beautifully exhibited by ι Cancri — the latter by γ Andromedæ; both fine double stars. If, however, the coloured star be much the less bright of the two, it will not materially affect the other. Thus, for instance, η Cassiopeiæ exhibits the beautiful combination of a large white star, and a small one of a rich ruddy purple. It is by no means, however, intended to say, that in all such cases one of the colours is a mere effect of contrast, and it may be easier suggested in words, than conceived in imagination, what variety of illumination *two suns* — a red and a green, or a yellow and a blue one — must afford a planet circulating about either; and what charming contrasts and "grateful vicissitudes," — a red and a green day, for instance, alternating with a white one and with darkness, — might arise from the presence or absence of one or other, or both, above the horizon. Insulated stars of a red colour, almost as deep as that of blood, occur in many parts of the heavens, but no green or blue star (of any decided hue) has, we believe, ever been noticed unassociated with a companion brighter than itself.

(611.) Another very interesting subject of enquiry, in the physical history of the stars, is their proper motion. *A priori*, it might be expected that apparent motions of some kind or other should be detected among so great a multitude of individuals scattered through space, and with nothing to keep them fixed. Their mutual attractions even, however inconceivably enfeebled by distance, and counteracted by opposing attractions from opposite quarters, must, in the lapse of countless ages, produce *some* movements — some change of internal arrangement — resulting from the difference of the opposing actions. And it is a fact, that such apparent motions do exist, not only among single, but in many of the double stars; which, besides revolving round each other, or round their common center of gravity, are transferred, without parting company, by a progressive motion common to both, towards some determinate region. For example, the two stars of 61 Cygni, which are nearly equal, have remained constantly at the same, or very nearly the same, distance, of 15″, for at least fifty years past. Meanwhile they have shifted their local situation in the heavens, in this interval of time, through no less than 4′ 23″, the annual proper motion of each star being 5″·3; by which quantity (exceeding a third of their interval) this system is every year carried bodily along in some unknown path, by a motion which, for many centuries, must be regarded as uniform and rectilinear. Among stars not double, and no way differing from the rest in any other obvious particular, μ Cassiopeiæ is to be remarked as having the greatest proper motion of any yet ascertained, amounting to 3″·74 of annual displacement. And a great many others have been observed to be thus constantly carried away from their places by smaller, but not less unequivocal motions.

(612.) Motions which require whole centuries to accumulate before they produce changes of arrangement, such as the naked eye can detect, though quite sufficient to destroy that idea of mathematical fixity which precludes speculation, are yet too trifling, as far as practical applications go, to induce a change of language, and lead us to speak of the stars in common parlance as otherwise than fixed. Too little is yet known of their amount and directions, to allow of any attempt at referring them to definite laws. It may, however, be stated generally, that their apparent directions are various, and seem to have no marked common tendency to one point more than to another of the heavens. It was, indeed, supposed by Sir William Herschel, that such a common tendency could be made out; and that, allowing for individual deviations, a general recess could be perceived in the principal stars, *from* that point occupied by the star ζ Herculis, *towards* a point diametrically opposite. This generally tendency was referred by him to a motion of the sun and solar system in the opposite direction. No one, who reflects with due attention on the subject, will be inclined to deny the high probability, nay certainty, that the sun *has* a proper motion in *some* direction; and the inevitable consequence of such a motion, unparticipated by the rest, must be a slow *average* apparent tendency of all the stars to the vanishing point of lines parallel to that direction, and to the region which he is leaving. This is the necessary effect of perspective; and it is certain that it must be detected by such observations, if we knew accurately the apparent proper motions of all the stars, and if we were sure that they were independent, *i. e.* that the whole firmament, or at least all that part which we see in our own neighbourhood, were not drifting along together, by a general *set* as it were, in one direction, the result of unknown processes and slow internal changes going on in the sidereal stratum to which our system belongs, as we see motes sailing in a current of air, and keeping nearly the same relative situation with respect to one another. But it seems to be the general opinion of astronomers, at present, that their science is not yet matured enough to afford data for any secure conclusions of this kind one way or other. Meanwhile, a very ingenious idea has been suggested by the present Astronomer Royal (Mr. Pond), viz. that a solar motion, if it exist, and have a velocity at all comparable to that of light, must necessarily produce a *solar aberration;* in consequence of which we do not see the stars disposed as they really are, but too much crowded in the region the sun is leaving, too open in that he is approaching. (See art. 280.) Now this, so long as the solar velocity continues the same, must be a constant effect which observation cannot detect; but *should it vary* in the course of ages, by a quantity at all commensurate to the velocity of the earth in its orbit, the fact would be detected by a general apparent *rush* of all the stars to the one or other quarter of the heavens, according as the sun's motion were accelerated or retarded; which observation would not fail to indicate, even if it should amount to no more than a very few seconds. This consideration, refined and remote as it is, may serve to give some idea of the delicacy and intricacy of any enquiry into the matter of proper motion; since the last mentioned effect would necessarily be mixed up with the systematic parallax, and could only be separated from it by considering that the nearer stars would be affected more than the distant ones by the one cause, but both near and distant alike by the other.

(613.) When we cast our eyes over the concave of the heavens in a clear night, we do not fail to observe that there are here and there groups of stars which seem to be compressed together in a more condensed manner than in the neighbouring parts, forming bright patches and clusters, which attract attention, as if they were there brought together by some general cause other than casual distribution. There is a group, called the Pleiades, in which six or seven stars may be noticed, if the eye be directed full upon it; and many more if *the eye* be turned carelessly aside, while *the attention* is kept directed* upon the group. Telescopes show fifty or sixty large stars thus crowded together in a very moderate space, comparatively insulated from the rest of the heavens. The constellation called Coma Berenices is another such group, more diffused, and consisting of much larger stars.

(614.) In the constellation Cancer, there is a somewhat similar, but less definite, luminous spot, called Præsepe, or the bee-hive, which a very moderate tele-

* It is a very remarkable fact, that the center of the visual area is by far less sensible to feeble impressions of light, than the exterior portions of the retina. Few persons are aware of the extent to which this comparative insensibility extends, previous to trial. To appreciate it, let the reader look alternately full at a star of the fifth magnitude, and beside it; or choose two, equally bright, and about 3° or 4° apart, and look full at one of them, the probability is, he will see *only the other:* such, at least, is my own case. The fact accounts for the multitude of stars with which we are impressed by a general view of the heavens; their paucity when we come to count them. — *Author.*

scope,—an ordinary night-glass, for instance,—resolves entirely into stars. In the sword-handle of Perseus, also, is another such spot, crowded with stars, which requires rather a better telescope to resolve into individuals separated from each other. These are called clusters of stars; and, whatever be their nature, it is certain that other laws of aggregation subsist in these spots, than those which have determined the scattering of stars over the general surface of the sky. This conclusion is still more strongly pressed upon us, when we come to bring very powerful telescopes to bear on these and similar spots. There are a great number of objects which have been mistaken for comets, and, in fact, have very much the appearance of comets without tails: small round, or oval nebulous specks, which telescopes of moderate power only show as such. Messier has given, in the *Connois. des Temps* for 1784, a list of the places of 103 objects of this sort; which all those who search for comets ought to be familiar with, to avoid being misled by their similarity of appearance. That they are not, however, comets, their fixity sufficiently proves; and when we come to examine them with instruments of great power,—such as reflectors of eighteen inches, two feet, or more in aperture,—any such idea is completely destroyed. They are then, for the most part, perceived to consist entirely of stars crowded together so as to occupy almost a definite outline, and to run up to a blaze of light in the center, where their condensation is usually the greatest. (See *fig*. 1. pl. ii., which represents (somewhat rudely) the thirteenth nebula of Messier's list (described by him as *nebuleuse sans etoiles*), as seen in the 20 feet reflector at Slough).* Many of them, indeed, are of an exactly round figure, and convey the complete idea of a globular space filled full of stars, insulated in the heavens, and constituting in itself a family or society apart from the rest, and subject

* This beautiful object was first noticed by Halley in 1714. It is visible to the naked eye, between the stars η and ζ Herculis In a night-glass it appears exactly like a small round comet.

only to its own internal laws. It would be a vain task to attempt to count the stars in one of these *globular clusters.* They are not to be reckoned by hundreds; and on a rough calculation, grounded on the apparent intervals between them at the borders (where they are seen not projected on each other), and the angular diameter of the whole group, it would appear that many clusters of this description must contain, at least, ten or twenty thousand stars, compacted and wedged together in a round space, whose angular diameter does not exceed eight or ten minutes; that is to say, in an area not more than a tenth part of that covered by the moon.

(615.) Perhaps it may be thought to savour of the gigantesque to look upon the individuals of such a group as suns like our own, and their mutual distances as equal to those which separate our sun from the nearest fixed star: yet, when we consider that their *united* lustre affects the eye with a less impression of light than a star of the fifth or sixth magnitude, (for the largest of these clusters is barely visible to the naked eye,) the idea we are thus compelled to form of their *distance* from us may render even such an estimate of their dimensions familiar to our imagination; at all events, we can hardly look upon a group thus insulated, thus *in seipso totus, teres, atque rotundus,* as not forming a system of a peculiar and definite character. Their round figure clearly indicates the existence of some general bond of union in the nature of an attractive force; and, in many of them, there is an evident acceleration in the rate of condensation as we approach the center, which is not referable to a merely uniform distribution of equidistant stars through a globular space, but marks an intrinsic *density* in their state of aggregation, greater at the center than at the surface of the mass. It is difficult to form any conception of the dynamical state of such a system. On the one hand, without a rotatory motion and a centrifugal force, it is hardly possible not to regard them as in a state of progressive collapse. On the other, granting such a motion and such a force, we find it no less difficult to reconcile the apparent sphericity of their form with a rotation of the whole system round any single axis, without which internal collisions would appear to be inevitable.* The following are the places, for 1830, of a few of the principal of these remarkable objects, as specimens of their class:—

R. A.		N. P. D.		R. A.		N. P. D.	
H.	M.	°	′	H.	M.	°	′
13	5	70	55	17	29	93	8
13	34	60	45	21	22	78	34
15	10	87	16	21	25	91	34
16	36	53	13				

(616.) It is to Sir William Herschel that we owe the most complete analysis of the great variety of those objects which are generally classed under the common head of Nebulæ, but which have been separated by him into—1st, Clusters of stars, in which the stars are clearly distinguishable; and these, again, into globular and irregular clusters. 2d, Resolvable nebulæ, or such as excite a suspicion that they consist of stars, and which any increase of the optical power of the telescope may be expected to resolve into distinct stars; 3d, Nebulæ, properly so called, in which there is no appearance whatever of stars; which, again, have been subdivided into subordinate classes, according to their brightness and size; 4th, Planetary nebulæ; 5th, Stellar nebulæ; and, 6th, Nebulous stars. The great power of his telescopes has disclosed to us the existence of an immense number of these objects, and shown them to be distributed over the heavens, not by any means uniformly, but, generally speaking, with a marked preference to a broad zone crossing the milky way nearly at right angles, and whose general direction is not very remote from that of the hour circle of 0^h and 12^h. In some parts of this zone, indeed,—especially where it

* See a note on this subject at the end of the work, p. 415.

crosses the constellations Virgo, Coma Berenices, and the Great Bear,—they are assembled in great numbers; being, however, for the most part *telescopic*, and beyond the reach of any but the most powerful instruments.

(617.) Clusters of stars are either globular, such as we have already described, or of irregular figure. These latter are, generally speaking, less rich in stars, and especially less condensed towards the center. They are also less definite in point of outline; so that it is often not easy to say where they terminate, or whether they are to be regarded otherwise than as merely richer parts of the heavens than those around them. In some of them the stars are nearly all of a size, in others extremely different; and it is no uncommon thing to find a very red star much brighter than the rest, occupying a conspicuous situation in them. Sir William Herschel regards these as globular clusters in a less advanced state of condensation, conceiving all such groups as approaching, by their mutual attraction, to the globular figure, and assembling themselves together from all the surrounding region, under laws of which we have, it is true, no other proof than the observance of a gradation by which their characters shade into one another, so that is impossible to say where one species ends and the other begins.

(618.) Resolvable nebulæ can, of course, only be considered as clusters either too remote, or consisting of stars intrinsically too faint to affect us by their individual light, unless where two or three happen to be close enough to make a joint impression, and give the idea of a point brighter than the rest. They are almost universally round or oval—their loose appendages, and irregularities of form, being as it were extinguished by the distance, and only the general figure of the more condensed parts being discernible. It is under the appearance of objects of this character that all the greater globular clusters exhibit themselves in telescopes of insufficient optical power to show them well; and the conclusion is obvious, that those which the most powerful can barely render *resolvable*, would be completely *resolved* by a further increase of instrumental force.

(619.) Of nebulæ, properly so called, the variety is again very great. By far the most remarkable are those represented in *figs.* 2. and 3. Plate III., the latter of which represents the nebulæ surrounding the quadruple (or rather sextuple) star θ, in the constellation Orion; the former, that about η, in the southern constellation Robur Caroli: the one discovered by Huygens, in 1656, and figured as seen in the twenty feet reflector at Slough; the other by Lacaille, from a figure by Mr. Dunlop, Phil. Trans. 1827. The nebulous character of these objects, at least of the former, is very different from what might be supposed to arise from the congregation of an immense collection of small stars. It is formed of little flocky masses, like wisps of cloud; and such wisps seem to adhere to many small stars at its outskirts, and especially to one considerable star (represented, in the figure, below the nebula), which it envelopes with a nebulous atmosphere of considerable extent and singular figure. Several astronomers, on comparing this nebula with the figures of it handed down to us by its discoverer, Huygens, have concluded that its form has undergone a perceptible change. But when it is considered how difficult it is to represent such an object duly, and how entirely its appearance will differ, even in the same telescope, according to the clearness of the air, or other temporary causes, we shall readily admit that we have no evidence of change that can be relied on.

(620.) Plate II. *fig.* 3. represents a nebula of a quite different character. The original of this figure is in the constellation Andromeda near the star ν. It is visible to the naked eye, and is continually mistaken for a comet, by those unacquainted with the heavens. Simon Marius, who noticed it in 1612, describes its appearance as that of a candle shining through horn, and the resemblance is not inapt. Its form is a pretty long oval, increasing by insensible gradations of brightness, at first very gradually, but at last more rapidly, up to a central point, which, though very much brighter than the rest, is yet evidently not stellar, but only nebula in a high state of condensation. It has in it a few small stars; but they are obviously casual, and the nebula itself offers not the slightest appearance to give ground for a suspicion of its consisting of stars. It is very large, being nearly half a degree long, and 15 or 20 minutes broad.

(621.) This may be considered as a type, on a large scale, of a very numerous class of nebulæ, of a round or oval figure, increasing more or less in density towards the central point: they differ extremely, however, in this respect. In some, the condensation is slight and gradual; in others great and sudden: so sudden, indeed, that they present the appearance of a dull and blotted star, or of a star with a slight burr round it, in which case they are called stellar nebulæ; while others, again, offer the singularly beautiful and striking phænomenon of a sharp and brilliant star surrounded by a perfectly circular disc, or atmosphere, of faint light in some cases, dying away on all sides by insensible gradations; in others, almost suddenly terminated. These are *nebulous stars*. A very fine example of such a star is 55 Andromedæ R. A. 1^h 43^m, N. P. D. 50° 7′. ε Orionis and ι of the same constellation are also nebulous; but the nebula is not to be seen without a very powerful telescope. In the extent of deviation, too, from the spherical form, which oval nebulæ affect, a great diversity is observed: some are only slightly elliptic; others much extended in length; and in some, the extension so great, as to give the nebula the character of a long narrow, spindle-shaped ray, tapering away at both ends to points. One of the most remarkable specimens of this kind is in R. A. 12^h 28^m; N. P. D. 63° 4′.

(622.) Annular nebulæ also exist, but are among the rarest objects in the heavens. The most conspicuous of this class is to be found exactly half way between the stars β and γ Lyræ, and may be seen with a telescope of moderate power. It is small, and particularly well defined, so as in fact to have much more the appearance of a flat oval solid ring than of a nebula. The axes of the ellipse are to each other in the proportion of about 4 to 5, and the opening occupies about half its diameter: its light is not quite uniform, but has something of a curdled appearance, particularly at the exterior edge; the central opening is not entirely dark, but is filled up with a faint hazy light, uniformly spread over it, like a fine gauze stretched over a hoop.

(623.) Planetary nebulæ are very extraordinary objects. They have, as their name imports, exactly the appearance of planets: round or slightly oval discs, in some instances quite sharply terminated, in others a little hazy at the borders, and of a light exactly equable or only a very little mottled, which, in some of them, approaches in vividness to that of actual planets. Whatever be their nature, they must be of enormous magnitude. One of them is to be found in the parallel of ν Aquarii, and about 5^m preceding that star. Its apparent diameter is about $20''$. Another, in the constellation Andromeda, presents a visible disc of $12''$, perfectly defined and round. Granting these objects to be equally distant from us with the stars, their real dimensions must be such as would fill, on the lowest computation, the whole orbit of Uranus. It is no less evident that, if they be solid bodies of a solar nature, the intrinsic splendour of their surfaces must be almost infinitely inferior to that of the sun's. A circular portion of the sun's disc, subtending an angle of $20''$, would give a light equal to 100 *full moons;* while the objects in question are hardly, if at all, discernible with the naked eye. The uniformity of their discs, and their want of apparent central condensation, would certainly augur their light to be merely superficial, and in the nature of a hollow spherical shell; but whether filled with solid or gaseous matter, or altogether empty, it would be a waste of time to conjecture.

(624.) Among the nebulæ which possess an evident symmetry of form, and seem clearly entitled to be regarded as systems of a definite nature, however mysterious their structure and destination, the most remarkable are the 51st and 27th of Messier's catalogue. The former consists of a large and bright globular nebula surrounded by a double ring, at a considerable distance from the globe, or rather a single ring divided through about two fifths of its circumference into two laminæ, and having one portion, as it were, turned up out of the plane of the rest. The latter consists of two bright and highly condensed round or slightly oval nebulæ, united by a short neck of nearly the same density. A faint nebulous atmosphere completes the figure, enveloping them both, and filling up the outline of a circumscribed ellipse, whose shorter axis is the axis of symmetry of the system about which it may be supposed to revolve, or the line passing through the centers of both the nebulous masses. These objects have never been properly described, the instruments with which they were originally discovered having been quite inadequate to showing the peculiarities above mentioned, which seem to place them in a class apart from all others. The one offers obvious analogies either with the structure of Saturn or with that of our own sidereal firmament and milky way. The other has little or no resemblance to any other known object.

(625.) The nebulæ furnish, in every point of view, an inexhaustible field of speculation and conjecture. That by far the larger share of them consist of stars there can be little doubt; and in the interminable range of system upon system, and firmament upon firmament, which we thus catch a glimpse of, the imagination is bewildered and lost. On the other hand, if it be true, as, to say the least, it seems extremely probable, that a phosphorescent or self-luminous matter also exists, disseminated through extensive regions of space, in the manner of a cloud or fog—now assuming capricious shapes, like actual clouds drifted by the wind, and now concentrating itself like a cometic atmosphere around particular stars;—what, we naturally ask, is the nature and destination of this nebulous matter? Is it absorbed by the stars in whose neighbourhood it is found, to furnish, by its condensation, their supply of light and heat? or is it progressively concentrating itself by the effect of its own gravity into masses, and so laying the foundation of new sidereal systems or of insulated stars? It is easier to propound such questions than to offer any probable reply to them. Meanwhile, appeal to fact, by the method of constant and diligent observation, is open to us; and, as the double stars have yielded to this style of questioning, and disclosed a series of relations of the most intelligible and interesting description, we may reasonably hope that the assiduous study of the nebulæ will, ere long, lead to some clearer understanding of their intimate nature.

(626.) We shall conclude this chapter by the mention of a phænomenon, which seems to indicate the existence of some slight degree of nebulosity about the sun itself, and even to place it in the list of nebulous stars. It is called the zodiacal light, and may be seen any very clear evening soon after sunset, about the months of April and May, or at the opposite season before sunrise, as a cone or lenticular-shaped light, extending from the horizon obliquely upwards, and following, generally, the course of the ecliptic, or rather that of the sun's equator. The apparent angular distance of its vertex from the sun varies, according to circumstances, from 40° to 90°, and the breadth of its base perpendicular to its axis from 8° to 30°. It is extremely faint and ill defined, at least in this climate, though better seen in tropical regions, but cannot be mistaken for any atmospheric meteor or aurora borealis. It is manifestly in the nature of a thin lenticularly-formed atmosphere, surrounding the sun, and extending at least beyond the orbit of Mercury and even of Venus, and may be conjectured to be no other than the denser part of that medium, which, as we have reason to believe resists the motion of comets; loaded, perhaps, with the actual materials of the tails of millions of those bodies, of which they have been stripped in their successive perihelion passages (art. 487.), and which may be slowly subsiding into the sun.

CHAP. XIII.

OF THE CALENDAR

(627.) Time, like distance, may be measured by comparison with standards of any length, and all that is requisite for ascertaining correctly the length of any interval, is to be able to apply the standard to the interval throughout its whole extent, without overlapping on the one hand, or leaving unmeasured vacancies on the other; to determine, without the possible error of a unit, the number of integer standards which the interval admits of being interposed between its beginning and end; and to estimate precisely the fraction, over and above an integer, which remains when all the possible integers are subtracted.

(628.) But though all standard units of time are equally possible, theoretically speaking, all are not, practically, equally convenient. The tropical year and the solar day are natural units, which the wants of man and the business of society force upon us, and compel us to adopt as our greater and lesser standards for the measurement of time, for all the purposes of civil life; and that, in spite of inconveniencies which, did any choice exist, would speedily lead to the abandonment of one or other. The principal of these are their *incommensurability*, and the want of perfect uniformity in one at least of them.

(629.) The mean lengths of the sidereal day and

year, when estimated on an average sufficiently large to compensate the fluctuations arising from nutation in the one, and from inequalities of configuration in the other, are the two most invariable quantities which nature presents us with; the former, by reason of the uniform diurnal rotation of the earth—the latter, on account of the invariability of the axes of the planetary orbits. Hence it follows that the mean solar day is also invariable. It is otherwise with the tropical year. The motion of the equinoctial points varies not only from the retrogradation of the equator on the ecliptic, but also partly from that of the ecliptic on the orbits of all the other planets. It is therefore variable, and this produces a variation in the *tropical* year, which is dependent on the place of the equinox (arts. 517. 328.) The *tropical* year is actually above $4{\cdot}21^s$ shorter than it was in the time of Hipparchus. This absence of the most essential requisite for a standard, viz. invariability, renders it necessary, since we cannot help employing the tropical year in our reckoning of time, to adopt an arbitrary or artificial value for it, so near the truth, as not to admit of the accumulation of its error for several centuries producing any practical mischief, and thus satisfying the ordinary wants of civil life; while, for scientific purposes, the tropical year, so adopted, is considered only as the representative of a certain number of integer days and a fraction—the day being, in effect, the only standard employed. The case is nearly analogous to the reckoning of value by guineas and shillings, an artificial relation of the two coins being fixed by law, near to, but scarcely ever exactly coincident with, the natural one, determined by the relative market price of gold and silver, of which either the one or the other,—whichever is really the most invariable, or the most in use with other nations,—may be assumed as the true theoretical standard of value.

(630.) The other inconvenience of the standards in question is their incommensurability. In our measure, of space, all our subdivisions are into aliquot parts: a yard is three feet, a mile eight furlongs, &c. But a year is no *exact* number of days, nor an integer number with any exact fraction, as one third or one fourth, over and above; but the surplus is an *incommensurable* fraction, composed of hours, minutes, seconds, &c., which produces the same kind of inconvenience in the reckoning of time that it would do, in that of money, if we had gold coins of the value of twenty-one shillings, with odd pence and farthings, and a fraction of a farthing over. For this, however, there is no remedy but to keep a strict register of the surplus fractions; and, when they amount to a whole day, cast them over into the integer account.

(631.) To do this in the simplest and most convenient manner is the object of a well-adjusted calendar. In the Gregorian calendar, which we follow, it is accomplished, with remarkable simplicity and neatness, by carrying a little farther than is done above the principle of an assumed or artificial year, and adopting *two* such years, both consisting of an exact integer number of days, viz. one of 365 and the other of 366, and laying down a simple and easily remembered rule for the order in which these years shall succeed each other in the civil reckoning of time, so that during the lapse of at least some thousands of years the sum of the integer artificial, or Gregorian, years elapsed shall not differ from the same number of real tropical years by a whole day. By this contrivance, the equinoxes and solstices will always fall on days similarly situated, and bearing the same name, in each Gregorian year; and the seasons will for ever correspond to the same months, instead of running the round of the whole year, as they must do upon any other system of reckoning, and used, in fact, to do before this was adopted.

(632.) The Gregorian rule is as follows:—The years are denominated from the birth of Christ, according to one chronological determination of that event. Every year whose number is not divisible by 4 without remainder, consists of 365 days; every year which *is* so divisible, but is not divisible by 100, of 366; every year divisible by 100, but not by 400, again of 365; and every year divisible by 400, again of 366. For example, the year 1833, not being divisible by 4, consists of 365 days; 1836 of 366; 1800 and 1900 of 365 each; but 2000 of 366. In order to see how near this rule will bring us to the truth, let us see what number of days 10000 Gregorian years will contain, beginning with the year 1. Now, in 10000, the numbers not divisible by 4 will be $\frac{3}{4}$ of 10000, or 7500; those divisible by 100, but not by 400, will in like manner be $\frac{3}{4}$ of 100, or 75; so that, in the 10000 years in question, 7575 consist of 366, and the remaining 2425 of 365, producing in all 3652425 days, which would give for an average of each year, one with another, $365^{d}\cdot2425$. The actual value of the tropical year (art. 327.) reduced into a decimal fraction, is 365·24224, so the error of the Gregorian rule on 10000 of the present tropical years is 2·6, or $2^{d}\ 14^{h}\ 24^{m}$; that is to say, less than a day in 3000 years; which is more than sufficient for all human purposes, those of the astronomer excepted, who is in no danger of being led into error from this cause. Even this error might be avoided by extending the wording of the Gregorian rule one step farther than its contrivers probably thought it worth while to go, and declaring that years divisible by 4000 should consist of 365 days. This would take off two integer days from the above-calculated number, and 2·5 from a larger average; making the sum of days in 100000 Gregorian years, 36524225, which differs only by a single day from 100000 real tropical years, such as they exist at present.

(633.) As any distance along a high road might, though in a rather inconvenient and roundabout way, be expressed without introducing error by setting up a series of milestones, at intervals of unequal lengths, so that every fourth mile, for instance, should be a yard longer than the rest, or according to any other fixed rule; taking care only to mark the stones, so as to leave room for no mistake, and to advertise all travellers of the difference of lengths and their order of succession; so may any interval of time be expressed correctly by stating in what Gregorian years it begins and ends, and *whereabouts in each.* For this statement coupled with the declaratory rule, enables us to say how many integer years are to be reckoned at 365, and how many at 366 days. The latter years are called bissextiles, or leap-years, and the surplus days thus thrown into the reckoning are called *intercalary or leap-days.*

(634.) If the Gregorian rule, as above stated, had always been adhered to, nothing would be easier than to reckon the number of days elapsed between the present time and any historical recorded event. But this is not the case; and the history of the calendar, with reference to chronology, or to the calculation of ancient observations, may be compared to that of a clock, going regularly when left to itself, but sometimes forgotten to be wound up; and when wound, sometimes set forward, sometimes backward, and that often to serve particular purposes and private interests. Such, at least, appears to have been the case with the Roman calendar, in which our own originates, from the time of Numa to that of Julius Cæsar, when the lunar year of 13 months, or 355 days, was augmented at pleasure, to correspond to the solar, by which the seasons are determined, by the arbitrary intercalations of the priests, and the usurpations of the decemvirs and other magistrates, till the confusion became inextricable. To Julius Cæsar, assisted by Sosigenes, an eminent Alexandrian astronomer and mathematician, we owe the neat contrivance of the two years of 365 and 366 days, and the insertion of one bissextile after three common years. This important change took place in the 45th year before Christ, which was the first regular year, commencing on the 1st of January, being the day of the new moon immediately following the winter solstice of the year before. We may judge of the state into which the reckoning of time had fallen, by the fact, that, to introduce the new system, it was necessary to enact that the previous year (46 B. C.) should consist of 455 days, a circumstance which obtained it the epithet of "the year of confusion."

(635.) The Julian rule made every fourth year, without exception, a bissextile. This is, in fact, an over-correction; it supposes the length of the tropical year to be $365\frac{1}{4}^d$, which is too great, and thereby induces an error of 7 days in 900 years, as will easily appear on trial. Accordingly, so early as the year 1414, it began to be perceived that the equinoxes were gradually creeping away from the 21st of March and September, where they ought to have always fallen had the Julian year been exact, and happening (as it appeared) too early. The necessity of a fresh and effectual reform in the calendar was from that time continually urged, and at length admitted. The change (which took place under the popedom of Gregory XIII.) consisted in the omission of ten nominal days after the 4th of October, 1582 (so that the next day was called the 15th, and not the 5th), and the promulgation of the rule already explained for future regulation. The change was adopted immediately in all catholic countries; but more slowly in protestant. In England, "the change of style," as it was called, took place after the 2d of September, 1752, eleven nominal days being then struck out; so that, the last day of old style being the 2d, the first of New Style (the next day) was called the 14th, instead of the 3d. The same legislative enactment which established the Gregorian year in England in 1752, shortened the preceding year, 1751, by a full quarter. Previous to that time, the year was held to begin with the 25th March, and the year A. D. 1751 did so accordingly; but that year was not suffered to run out, but was supplanted on the 1st January by the year 1752, which it was enacted should commence on that day, as well as every subsequent year. Russia is now the only country in Europe in which the Old Style is still adhered to, and (another secular year having elapsed) the difference between the European and Russian dates amounts, at present, to 12 days.

(636.) It is fortunate for astronomy that the confusion of dates, and the irreconcilable contradictions which historical statements too often exhibit, when confronted with the best knowledge we possess of the ancient reckonings of time, affect recorded observations but little. An astronomical observation, of any striking and well-marked phænomenon, carries with it, in most cases, abundant means of recovering its exact date, when any tolerable approximation is afforded to it by chronological records; and, so far from being abjectly dependent on the obscure and often contradictory dates which the comparison of ancient authorities indicates, is often itself the surest and most convincing evidence on which a chronological epoch can be brought to rest. Remarkable eclipses, for instance, now that the lunar theory is thoroughly understood, can be calculated back for several thousands of years, without the possibility of mistaking the day of their occurrence. And whenever any such eclipse is so interwoven with the account given by an ancient author of some historical event, as to indicate precisely the interval of time between the eclipse and the event, and at the same time completely to identify the eclipse, that date is recovered and fixed for ever.*

(637.) The days thus parcelled out into years, the next step to a perfect knowledge of time is to secure the identification of each day, by imposing on it a name universally known and employed. Since, however, the days of a whole year are too numerous to admit of loading the memory with distinct names for each, all nations have felt the necessity of breaking them down into parcels of a more moderate extent; giving names to each of these parcels, and particularizing the days in each by numbers, or by some especial indication. The

* See the remarkable calculations of Mr. Baily relative to the celebrated solar eclipse which put an end to the battle between the kings of Media and Lydia, B. C. 610. Sept. 30. Phil. Trans. ci. 220.

lunar month has been resorted to in many instances; and some nations have, in fact, preferred a lunar to a solar chronology altogether, as the Turks and Jews continue to do to this day, making the year consist of 13 lunar months, or 355 days.* Our own division into twelve unequal months is entirely arbitrary, and often productive of confusion, owing to the equivoque between the lunar and calendar month. The intercalary day naturally attaches itself to February as the shortest.

* The Metonic cycle, or the fact, discovered by Meton, a Greek mathematician, that 19 solar years contain just 235 lunations (which in fact they do to a very great degree of approximation), was duly appreciated by the Greeks, as ensuring the correspondence of the solar and lunar years, and honours were decreed to its discoverer.

NOTE

On the Constitution of a Globular Cluster, referred to in page 401.

If we suppose a globular space filled with equal stars, uniformly dispersed through it, and very numerous, each of them attracting every other with a force inversely as the square of the distance, the resultant force by which any one of them (those at the surface alone excepted) will be urged, in virtue of their joint attractions, will be directed towards the common center of the sphere, and will be directly as the distance therefrom. This follows from what Newton has proved of the *internal* attraction of a homogeneous sphere. Now, under such a law of force, each particular star would describe a perfect ellipse about the common center of gravity as its center, and *that*, in whatever plane and whatever direction it might revolve. The condition, therefore, of a rotation of the cluster, as a mass, about a single axis would be unnecessary. Each ellipse, whatever might be the proportion of its axes, or the inclination of its plane to the others, would be invariable *in every particular*, and all would be described in one common period, so that at the end of every such period, or *annus magnus* of the system, every star of the cluster (except the superficial ones) would be exactly re-established in its original position, thence to set out afresh, and run the same unvarying round for an indefinite succession of ages. Supposing their motions, therefore, to be so adjusted at any one moment as that the orbits should not intersect each other, and so that the magnitude of each star, and the sphere of its more intense attraction, should bear but a small proportion to the distance separating the individuals, such a system, it is obvious, might subsist, and realise, in great measure, that abstract and ideal harmony, which Newton, in the 89th Proposition of the First Book of the *Principia*, has shown to characterise a law of force directly as the distance. See also *Quarterly Review*, No. 94. p. 540. — *Author.*

Synoptic Table of the Elements of the Solar System.

N. B. — The data for Vesta, Juno, Ceres, and Pallas are for January 1. 1820. The rest for January 1. 1801.

Planet's name.	Mean distance from Sun, or Semi-axis.	Mean Sidereal Period in Mean Solar Days.	Excentricity in Parts of the Semi-axis.
Mercury	0·3870981	87·9692580	0·2055149
Venus	0·7233316	224·7007869	0·0068607
Earth	1·0000000	365·2563612	0·0167836
Mars	1·5236923	686·9796458	0·0933070
Vesta	2·3678700	1325·7431000	0·0891300
Juno	2·6690090	1592·6608000	0·2578480
Cerés	2·7672450	1681·3931000	0·0784390
Pallas	2·7728860	1686·5388000	0·2416480
Jupiter	5·2027760	4332·5848212	0·0481621
Saturn	9·5387861	10759·2198174	0·0561505
Uranus	19·1823900	30686·8208296	0·0466794

Planet's Name.	Inclination to the Ecliptic.	Longitude of ascending Node.	Longitude of Perihelion.
Mercury	7° 0′ 9″·1	45° 57′ 30″·9	74° 21′ 46″·9
Venus	3 23 28 ·5	74 54 12 ·9	128 43 53 ·1
Earth	———	———	99 30 5 ·0
Mars	1 51 6 ·2	48 0 3 ·5	332 23 56 ·6
Vesta	7 8 9 ·0	103 13 18 ·2	249 33 24 ·4
Juno	13 4 9 ·7	171 7 40 ·4	53 33 46 ·0
Ceres	10 37 26 ·2	80 41 24 ·0	147 7 31 ·5
Pallas	34 34 55 ·0	172 39 26 ·8	121 7 4 ·3
Jupiter	1 18 51 ·3	98 26 18 ·9	11 8 34 ·6
Saturn	2 29 35 ·7	111 56 37 ·4	89 9 29 ·8
Uranus	0 46 28 ·4	72 59 35 ·3	167 31 16 ·1

Planet's Name.	Mean Longitude at the Epoch.	Mass in Billionths of the Sun's.	Equatorial Diameter, the Sun's being 111·454.
Mercury	166° 0′ 48″·6	493628	0·398
Venus	11 33 3 ·0	2463836	0·975
Earth	100 39 10 ·2	2817409	1·000
Mars	64 22 55 ·5	392735	0·517
Vesta	278 30 0 ·4	———	———
Juno	200 16 19 ·1	———	———
Ceres	123 16 11 ·9	———	———
Pallas	108 24 57 ·9	———	———
Jupiter	112 15 23 ·0	953570222	10·860
Saturn	135 20 6 ·5	284738000	9·987
Uranus	177 48 23 ·0	55809812	4·332

SYNOPTIC TABLE OF THE ELEMENTS OF THE ORBITS OF THE SATELLITES, SO FAR AS THEY ARE KNOWN.

N. B.—The distances are expressed in equatorial radii of the primaries. The epoch is Jan. 1. 1801. The periods, &c. are expressed in mean solar days.

I. THE MOON.

Mean distance from earth - - - -	$29^{r}\cdot98217500$
Mean sidereal revolution - - - -	$27^{d}\cdot321661418$
Mean synodical ditto - - - -	$29^{d}\cdot530588715$
Excentricity of orbit - - - -	0·054844200
Mean revolution of nodes - - -	$6793^{d}\cdot391080$
Mean revolution of apogee - -	$3232^{d}\cdot575343$
Mean longitude of node at epoch -	$13^{\circ}\ 53'\ 17''\cdot7$
Mean longitude of perigee at do. -	266 10 7 ·5
Mean inclination of orbit - - -	5 8 47 ·9
Mean longitude of moon at epoch -	118 17 8 ·3
Mass, that of earth being 1, - -	0·0125172
Diameter in miles - - - - -	2160

II. SATELLITES OF JUPITER.

Sat.	Mean Distance.	Sidereal Revolution.	Inclination of Orbit to that of Jupiter.	Mass: that of Jupiter being 1000000000.
1	6·04853	$1^{d}\ 18^{h}\ 28^{m}$	$3^{\circ}\ 5'\ 30''$	17328
2	9·62347	3 13 14	Variable	23235
3	15·35024	7 3 43	Variable	88497
4	26·99835	16 16 32	2 58 48	42659

The excentricities of the 1st and 2d satellite are insensible, that of the 3d and 4th small, but variable in consequence of their mutual perturbations.

III. SATELLITES OF SATURN.

Sat.	Mean Distance.	Sidereal Revolution.	Excentricities and Inclinations.
1	3·351	0d 22h 38m	The orbits of the six interior satellites are nearly circular, and very nearly in the plane of of the ring. That of the seventh is considerably inclined to the rest, and approaches nearer to coincidence with the ecliptic.
2	4·300	1 8 53	
3	5·284	1 21 18	
4	6·819	2 17 45	
5	9·524	4 12 25	
6	22·081	15 22 41	
7	64·359	79 7 55	

IV. SATELLITES OF URANUS.

Sat.	Mean Distance.	Sidereal Period.	Inclination to Ecliptic.
1?	13·120	5d 21h 25m 0s	Their orbits are inclined about 78° 58′ to the ecliptic, and their motion is retrograde. The periods of the 2d and 4th require a trifling correction. The orbits appear to be nearly circles.
2	17·022	8 16 56 5	
3?	19·845	10 23 4 0	
4	22·752	13 11 8 59	
5?	45·507	38 1 48 0	
6?	91·008	107 16 40 0	

INDEX.

T.

U.

V.

Y.

Z.

THE END.

London:
Printed by A. Spottiswoode,
New-Street-Square.

Plate I.

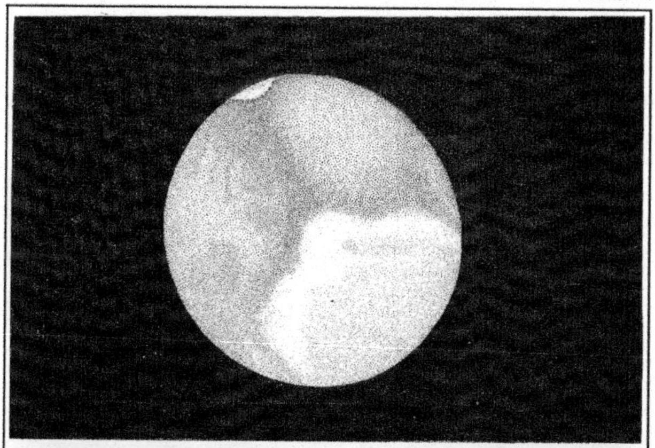

Fig. 1.

Fig. 2.

Fig. 3.

Pub^d by Longman & Co June 1833.

H. Adlard, sc.

Plate II

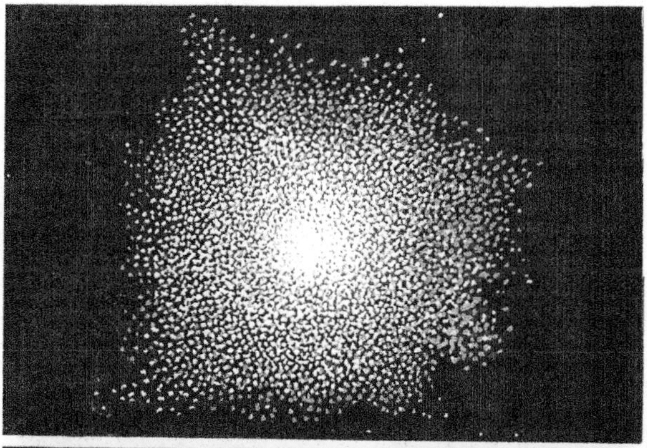

Fig. 1

Fig. 2.

Fig. 3.

Pubd by Longman & Co June 1833.

H. Adlard, sc.

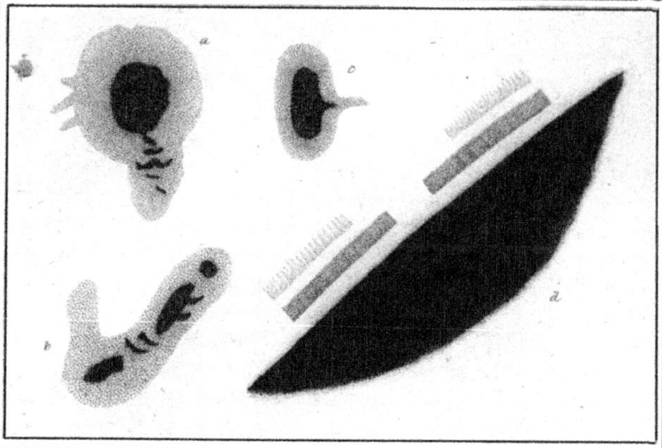

Fig. 1.

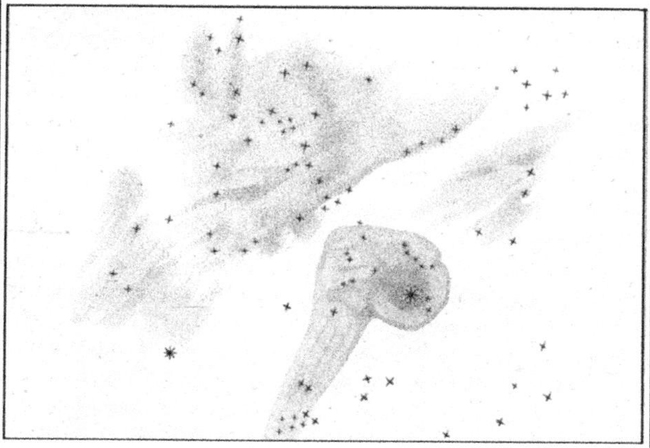

Fig. 2.

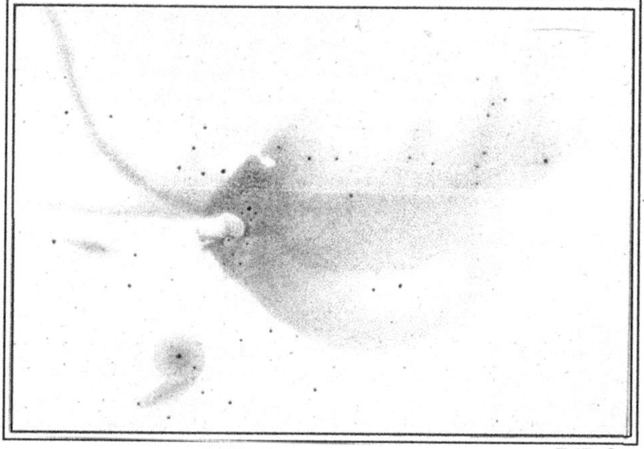

Fig. 3.

Pub^d. by Longman & Co. June 1833.

H. Adlard, sc.

THE

CABINET

OF

NATURAL PHILOSOPHY.

CONDUCTED BY THE

REV. DIONYSIUS LARDNER, LL.D. F.R.S. L.&E.

M.R.I.A. F.R.A.S. F.L.S. F.Z.S. Hon. F.C.P.S. &c. &c.

ASSISTED BY

EMINENT SCIENTIFIC MEN.

ASTRONOMY.

BY

SIR JOHN F. W. HERSCHEL, KN^{T}. GUELP.

F.R.S.L. & E. M.R.I.A. F.R.A.S. F.G.S. M.C.U.P.S.

CORRESPONDENT OF THE ROYAL ACADEMY OF SCIENCES OF PARIS, AND OTHER FOREIGN SCIENTIFIC INSTITUTIONS.

A NEW EDITION.

LONDON:

PRINTED FOR

LONGMAN, ORME, BROWN, GREEN, & LONGMANS,

PATERNOSTER-ROW;

AND JOHN TAYLOR,

UPPER GOWER STREET.

1838.

London:
Printed by A. Spottiswoode,
New-Street-Square.